DATE DUE

NOV 1 7 1983	NOV 1 2000	
FEB 2 8 1984		
OCT 0 3 1984		
JAN 2 1985		
FEB 6 1985		
MAR 2 8 1985		
MAY 6 1985		
OCT 1 5 1985		
JAN 2 1 1987		
FEB 0 9 1987		
JAN 4 1988		
JAN 2 0 1988		
MAR 2 0 '90		
MAR 2 9 '90		
MAR 1 8 '91		
OCT 1 6 '91		
SEP 2 7 1995		
GAYLORD		PRINTED IN U.S.A.

MECHANICS

By

J. P. DEN HARTOG

PROFESSOR OF MECHANICAL ENGINEERING
MASSACHUSETTS INSTITUTE OF TECHNOLOGY

DOVER PUBLICATIONS, INC.
NEW YORK

This Dover edition, first published in 1961, is an
unabridged and corrected republication of the first
edition, published by McGraw-Hill Book Company,
Inc., in 1948.

Standard Book Number: 486-60754-2
Library of Congress Catalog Card Number: 61-1958

Manufactured in the United States of America
Dover Publications, Inc.
180 Varick Street
New York, N. Y. 10014

PREFACE

No self-respecting publishing house will permit a book to appear in print without a title page or without a preface. Most readers agree on the desirability of a title page, but with almost equal unanimity they find a preface superfluous and uninteresting. Good prefaces are rarely written, but it has been done, notably by George Bernard Shaw, Oliver Heaviside, and Henri Bouasse. The incomparable G.B.S. is delightfully entertaining and at the same time full of wisdom; the other two are entertaining but have axes to grind. Oliver Heaviside assails the "Cambridge mathematicians" of whom he reluctantly admits that "even *they* are human"; Henri Bouasse charges with Gallic wit into the members of the French Academy of Sciences, who have more starch in their shirts than is pleasing to that author. However, the writing of those witty prefaces did them little good because they lacked that simple knowledge of the fundamental laws of mechanics that even Sancho Panza possessed. It is recorded that Sancho, when he saw his famous master charge into the windmills, muttered in his beard something about relative motion and Newton's third law (as carefully explained on page 175 and on page 297 of this textbook). Sancho was right: the windmills hit his master just as hard as he hit them, and of course the same thing happened to Heaviside and Bouasse. The gentlemen with the starched shirts never elected Bouasse to be a member of their famous academy, and as to Heaviside and the mathematicians, the "Heaviside Calculus" of a decade ago has become the "Theory of Laplace's Transforms," and the "Heaviside Layer" has acquired the scientific name of "Ionosphere." Thus an author who aspires to wear well-laundered shirts should be very careful about what he writes in the preface to his book and should stick to innocuous and dry facts, such as, for instance, that the reader has now before him a textbook designed for a two-semester course for sophomores or juniors in a regular four-year engineering curriculum. The author further believes that it will do no harm to suggest that the book also can be adapted to a simpler, single-semester course by the omission of Chaps. IV, V, VII, VIII, XI, XVI, and XVII.

He has done his best to write carefully and has placed no deliberate

v

errors in the book, but he has lived long enough to be quite familiar with his own imperfections. He therefore asks his readers to be indulgent and assures them that letters from them calling attention to errors or containing suggestions for improvements of the book will be gratefully received and very much appreciated.

J. P. DEN HARTOG

CAMBRIDGE, MASS.
May, 1948

TABLE OF CONTENTS

CHAPTER VII—SPACE FORCES

CHAPTER VIII—THE METHOD OF WORK

CHAPTER IX—KINEMATICS OF A POINT

CHAPTER X—DYNAMICS OF A PARTICLE

CHAPTER XI—KINEMATICS OF PLANE MOTION

CHAPTER XII—MOMENTS OF INERTIA

CHAPTER XIII—DYNAMICS OF PLANE MOTION

CHAPTER XIV—WORK AND ENERGY

MECHANICS

CHAPTER I

DISCRETE COPLANAR FORCES

1. Introduction. Mechanics is usually subdivided into three parts: statics, kinematics, and dynamics. Statics deals with the distribution of forces in bodies at rest; kinematics describes the motions of bodies and mechanisms without inquiring into the forces or other causes of those motions; finally, dynamics studies the motions as they are caused by the forces acting.

A problem in *statics*, for example, is the question of the compressive force in the boom of the crane of Fig. 17 (page 24) caused by a load of a given magnitude. Other, more complicated problems deal with the forces in the various bars or members of a truss (Fig. 61, page 54) or in the various cables of a suspension bridge (Fig. 67, page 63). Statics, therefore, is of primary importance to the civil engineer and architect, but it also finds many applications in mechanical engineering, for example, in the determination of the tensions in the ropes of pulleys (Fig. 15, page 21), the force relations in screw jacks and levers of various kinds, and in many other pieces of simple apparatus that enter into the construction of a complicated machine.

Statics is the oldest of the engineering sciences. Its first theories are due to Archimedes (250 B.C.), who found the laws of equilibrium of levers and the law of buoyancy. The science of statics as it is known today, however, started about A.D. 1600 with the formulation of the parallelogram of forces by Simon Stevin.

Kinematics deals with motion without reference to its cause and is, therefore, practically a branch of geometry. It is of importance to the mechanical engineer in answering questions such as the relation between the piston speed and the crankshaft speed in an engine, or, in general, the relation between the speed of any two elements in complicated "kinematical" machines used for the high-speed automatic manufacture of razor blades, shoes, or zippers. Another example appears in the design of a quick-return mechanism (Fig. 146, page 168) such as is used in a shaper, where the cutting tool does useful work in one direction only and where it is of practical importance to waste as little time as possible in the return stroke. The design of gears and cams is almost entirely a problem in kinematics.

1

Historically, one of the first applications of the science was "James Watt's parallelogram," whereby the rotating motion of the flywheel of the first steam engine was linked to the rectilinear motion of the piston by means of a mechanism of bars (Fig. 177, page 203). Watt found it necessary to do this because the machine tools of his day were so crude that he could not adopt the now familiar crosshead-guide construction, which is "kinematically" much simpler.

From this it is seen that kinematics is primarily a subject for the mechanical engineer. The civil engineer encounters it as well but to a lesser extent, for instance in connection with the design of drawbridges, sometimes also called "bascule" bridges, where the bridge deck is turned up about a hinge and is held close to static equilibrium in all positions by a large counterweight (Problem 44, page 353). It is sometimes of practical importance to make the motion of this counterweight much smaller than the motion of the tip of the bridge deck, and the design of a mechanism to accomplish this is a typical problem in kinematics.

Finally, *dynamics* considers the motions (or rather, their accelerations) as they are influenced by forces. The subject started with Galileo and Newton, three centuries ago, with applications principally to astronomy. Engineers hardly used the new science before 1880 because the machines in use up to that time ran so slowly that their forces could be calculated with sufficient accuracy by the principles of statics. Two practical dynamical devices used before 1880 are Watt's flyball engine governor and the escape mechanism of clocks. These could be and were put to satisfactory operation without much benefit of dynamical theory. Shortly after 1880 the steam turbine, the internal-combustion engine, and the electric motor caused such increases in speed that more and more questions appeared for which only dynamics could provide an answer, until at the present time the large majority of technical problems confronting the mechanical or aeronautical engineer are in this category. Even the civil engineer, building stationary structures, cannot altogether remain aloof from dynamics, as was demonstrated one sad day in 1940 when the great suspension bridge near Tacoma, Washington, got into a violent flutter, broke to pieces, and fell into the water—a purely dynamical failure. The design of earthquake-resistant buildings requires a knowledge of dynamics by the civil engineer. But these are exceptions to the general rule, and it is mainly the mechanical engineer who has to deal with dynamical questions, such as the stability of governing systems, the smooth, non-vibrating operation of turbines, the balancing of internal-combustion engines, the application of gyroscopes to a wide variety of instruments, and a host of other problems.

2. Forces. Statics is the science of equilibrium of bodies subjected to the action of forces. It is appropriate, therefore, to be clear about what we mean by the words "equilibrium" and "force." **A body is said to be in equilibrium when it does not move.**

"Force" is defined as that which (a) pushes or pulls by direct mechanical contact, or (b) is the "force of gravity," otherwise called "weight," and other similar "field" forces, such as are caused by electric or magnetic attraction.

We note that this definition excludes "inertia" force, "centrifugal" force, "centripetal" force, or other "forces" with special names that appear in the printed literature.

The most obvious example of a pull or a push on a body or machine is when a stretched rope or a compressed strut is seen to be attached to the body. When a book rests on the table or an engine sits on its foundation, there is a push force, pushing up from the table on the book and down from the book on the table. Less obvious cases of mechanical contact forces occur when a fluid or gas is in the picture. There is a push-force between the hull of a ship and the surrounding water, and similarly there is such a force between an airplane wing in flight and the surrounding air.

If there is any doubt as to whether there is a mechanical contact force between two bodies, we may imagine them to be separated by a small distance and a small mechanical spring to be inserted between them, with the ends of the spring attached to the bodies. If this spring were to be elongated or shortened in our imaginary experiment, there would be a direct contact force. All forces that we will deal with in mechanics are direct contact forces with the exception of gravity (and of electric and magnetic forces). The mental experiment of the inserted spring fails with gravity. Imagine a body at rest suspended from above by a string. According to our definition there are two forces acting on the body: an upward one from the stretched string, and a downward one from gravity. We can mentally cut the string and insert a spring between the two pieces. This spring will be stretched by the pull in the string. We cannot imagine an operation whereby we "cut" the force of gravity between the body and the earth and patch it up again by a spring.

The unit of force used in engineering is the *pound*, which is the weight or gravity force of a standard piece of platinum at a specific location on earth. The weight of this standard piece varies slightly

from place to place, being about 0.5 per cent greater near the North
Pole than at the equator. This difference is too small to be considered
for practical calculations in engineering statics, but it is sufficiently
important for some effects in physics. (A method for exploring for oil
deposits is based on these very slight variations in gravity from place
to place.)

For the practical measurement of force, springs are often used.
In a spring the elongation (or compression) is definitely related to the
force on it so that a spring can be calibrated against the standard
pound and then becomes a "dynamometer" or force meter.

A force is characterized not only by its magnitude but also by the
direction in space in which it acts; it is a "vector" quantity, and not
a "scalar" quantity. The line along which the force acts is called its
"line of action." Thus, in order to specify a force completely, we
have to specify its line of action and its magnitude. By making this
magnitude positive or negative, we determine the direction of the
force along the line of action.

A very important property of forces is expressed by the **first axiom
of statics,** also known as Newton's third law (page 175), which **states
that action equals reaction.** Contact forces are always exerted *by*
one body *on* another body, and the axiom states that the force by the
first body on the second one is equal and opposite to the force by
the second body on the first one. For example, the push down on the
table by a book is equal to the push up on the book by the table.
Thinking about our imaginary experiment of inserting a thin "dynamometer" spring between the book and the table, the proposition looks
to be quite obvious: there is only one force in the spring. Newton's
third law, however, states that it is true not only for contact forces
but also for gravity (and similar) forces. The earth pulls down on a
flying airplane with a force equal to the one with which the airplane
pulls up on the earth. This is less obvious, and in fact the proposition
is of the nature of an axiom that cannot be proved by logical deduction
from previous knowledge. It appeals to the intuition, and the logical
deductions made *from* it (the entire theory of statics) conform well
with experiment.

Another proposition about forces, which appeals to our intuition
but which cannot be proved by logic, is the **second axiom of statics,**
or the **principle of transmissibility:**

**The state of equilibrium of a body is not changed when the point
of action of a force is displaced to another point on its line of action.**
This means in practice that a force can be shifted along its own line

without changing the state of equilibrium of a body. For ropes or struts in contact with the body, the proposition looks obvious; it should make no difference whether we pull on a short piece of rope close to the body or at the end of a long rope far away, provided that the short rope coincides in space with a piece of the longer one. For gravity forces the proposition is not so obvious.

It is noted that the equivalence of two forces acting at different points along their own line extends only to the state of equilibrium of a body, not to other properties. For instance, the stress in the body is definitely changed by the location of the point of action. Imagine a bar of considerable weight located vertically in space. Let the bar in

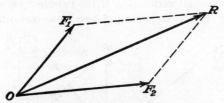

FIG. 1.   The parallelogram of forces.

case *a* be supported by a rope from the top and in case *b*, by a bearing at the bottom. The supporting force in both cases equals the weight and is directed upward along the bar; in case *a* it acts at the top, and in case *b*, at the bottom. By the axiom this should not make any difference in the state of equilibrium of the body, but in case *a* the bar is in tension, and in case *b*, in compression.

3. Parallelogram of Forces. The statements of action equals reaction and of transmissibility are not the only axioms about forces which are made. The **third axiom** in statics is that of the parallelogram of forces:

If on a body two forces are acting, whose lines of action intersect, then the equilibrium of the body is not changed by replacing these two forces by a single force whose vector is the diagonal of the parallelogram constructed on the two original forces.

This is illustrated in Fig. 1. The two forces, F_1 and F_2, have lines of action intersecting at *O*; they are said to be *concurrent* forces, to distinguish them from forces of which the lines of action do not intersect. The single force **R**, called the *resultant*, is equivalent to the combined action of the two forces, F_1 and F_2. Although this construction is now familiar to almost everyone, it is emphasized that it is an axiom, not provable by logic from known facts. It is based on experiment only, and the ancient Greeks and Romans did not know it, although Archimedes *was* familiar with the equilibrium of levers. The statement is about 350 years old and was formulated less than a century before the great days of Newton.

Many simple experiments can be devised to verify the axiom of the parallelogram of forces. They all employ for their interpretation two more statements which are sometimes also called axioms but which are so fundamental that they hardly deserve the honor. They are

Fourth axiom: A body in equilibrium remains in equilibrium when no forces are acting on it.

Fifth axiom: Two forces having the same line of action and having equal and opposite magnitudes cancel each other.

For an experimental verification of the parallelogram law, arrange the apparatus of Fig. 2, consisting of two freely rotating frictionless

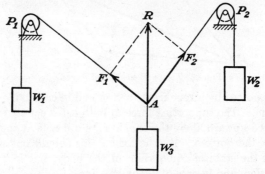

FIG. 2. Experimental verification of the parallelogram of forces.

pulleys, P_1 and P_2, of which the axles are rigidly mounted, and a completely flexible string or rope strung over them. Three different weights, W_1, W_2, and W_3 are hung on the string, and if the string is so long that the weights are kept clear of the pulleys, they will find a position of equilibrium, as the experiment shows. We observe the geometry of this position and reason by means of the five axioms. In this reasoning we employ a device, called **isolation of the body,** that is used in practically every problem in mechanics and that is of **utmost importance.** The first "body" we "isolate" consists of the weight W_1 and a short piece of its vertical string attached to it. The "isolation" is performed by making an imaginary cut in the string just above W_1 and by considering only what is below that cut. We observe from the experiment that W_1 is in equilibrium and notice that two forces are acting on it: the pull of the string up and the weight W_1 down. The lines of action of the two forces are the same so that **by** the fourth and fifth axioms combined, we deduce that the force or tension in the string is equal to the weight W_1. Next we look at or "isolate" the pulley P_1. The tensions in the two sections of string over a frictionless pulley are the same so that the tension in the string between P_1 and A is still W_1. This is not obvious, and the proof of it will be given much later, on page 21.

By entirely similar reasoning we conclude from the fourth and fifth axioms that the tension in the string between P_2 and A is W_2 and that the tension in the vertical piece of string between A and W_3 is equal to W_3.

Now we once more isolate a body, and choose for it the knot A and three short pieces of string emanating from it. This body is in equilibrium by experiment, and we notice that there are three forces acting on it, the string tensions W_1, W_2, and W_3, whose lines of action intersect at A. Now by the third axiom of the parallelogram of forces, the tensions F_1 and F_2 of Fig. 2 (which we have seen are equal to W_1 and W_2) add up to the resultant R. By the fourth and fifth axioms, it is concluded that R must be equal (and opposite) to W_3. By hanging various weights on the strings we can form parallelograms of all sorts of shapes and so verify the third axiom experimentally.

For example, if $W_1 = W_2 = 1$ lb and $W_3 = \sqrt{2} = 1.41$ lb, then the angles of the parallelogram will be 45 and 90 deg.

In constructing the parallelogram of forces, not all the lines have to be drawn in the figure. In Fig. 1 it is seen that the distance F_2R is equal (and parallel) to OF_1. In order to find the resultant OR in Fig. 1, it suffices to lay off the force OF_2, and then starting at the end point F_2 to lay off the other force $F_2R = OF_1$. The resultant OR is then the closing line of the triangle OF_2R. The lines OF_1 and F_1R do not necessarily have to be drawn. The construction then is called the **triangle of forces.**

If more than two forces are to be added together, this triangle construction leads to a much simpler figure than the parallelogram construction.

In Fig. 3a there are three forces, O_1F_1, O_2F_2, and O_3F_3, whose lines of action lie in one plane but do not meet in one point. We want to find the resultant of these three forces. To do this we first slide F_1 and F_2 along their lines of action to give them the common origin O_4 so that $O_4P = O_1F_1$ and $O_4Q = O_2F_2$. Then we construct the parallelogram, of which O_4S is the diagonal, and therefore O_4S is the resultant of O_1F_1 and O_2F_2. To add the third force O_3F_3 to this, we have to slide both O_4S and O_3F_3 along their respective lines of action to give them the common origin O_5 so that $O_5T = O_3F_3$ and $O_5R_{12} = O_4S$. Now complete the second parallelogram with O_5R_{123} as diagonal. Then O_5R_{123} is the resultant of the three forces.

Just below, in Fig. 3b, the construction has been repeated in the triangle form. We start at O and lay off one force OF_1; then from the end point F_1 we lay off the second force $F_1F_2 = O_2F_2$. The closing line of that triangle gives the intermediate result R_{12}: the resultant of the first two forces. Finally we lay off $F_2F_3 = O_3F_3$ starting at the end

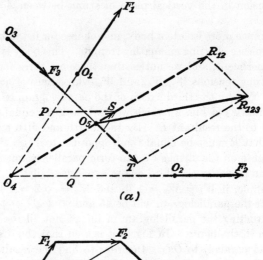

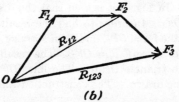

FIG. 3.   The compounding of three forces.

point F_2 of the diagram. The closing line OF_3 is the desired resultant R_{123}. It is seen that every line of Fig. 3b is parallel to a corresponding line in Fig. 3a and that Fig. 3b is a condensed and simplified version of Fig. 3a. However, Fig. 3a not only gives the magnitude and direction of the resultant R_{123}, but also gives the correct location of its line of action in relation to the three individual forces, whereas in Fig. 3b this location is not given—the resultant R_{123} is displaced parallel with respect to its true location in Fig. 3a. Later (page 20) we will see how the true location of R_{123} can be found as well.

The figure (3b) is known as the **polygon of forces,** and its usefulness is not limited to three forces but can be extended to any number of forces. The greater the number of forces, the more obvious becomes the simplification of the polygon figure (3*b*) as compared to the parallelogram figure (3*a*).

Problems 1 *to* 8.

4. Cartesian Components. Two intersecting or concurrent forces can be added to form a single resultant by the parallelogram construction. By the reverse procedure any force can be resolved into two components in arbitrary directions. In order to determine those components the directions must be specified. For instance, the force **F** of Fig. 4 can be resolved into the horizontal and vertical components **H** and **V** as shown. But the force **F** can just as well be resolved into the components **P** and **Q**. For any two chosen directions the force has its appropriate components, and since we can choose an infinite number of directions, the resolution of a single force into two components can be accomplished in an infinite number of ways.

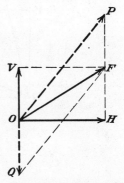

Fᵢɢ. 4. A force *F* can be resolved into two components in many different ways.

For many problems it is of practical advantage to resolve every force in the problem into its Cartesian *x* and *y* or horizontal and vertical components and then work with these components instead of with the forces themselves. Since all *x* components of the various forces lie in the same *x* or horizontal direction, their addition is an algebraic process instead of a geometrical parallelogram process. It can thus be stated that

The resultant of any number of forces can be found by first resolving the individual forces into their Cartesian *x* and *y* components, then by forming the algebraic sum of all the *x* components and similarly of the *y* components, and finally by compounding the *x* resultant with the *y* resultant by the parallelogram process.

This is illustrated in Fig. 5, where **R** is the resultant of the five forces $F_1, \ldots, F_5$, formed by the polygon method. The *x* or horizontal components of the five *F* forces are represented by the lengths OX_1, X_1X_2, X_2X_3, X_3X_4, and X_4X_5. It is seen that the first three forces have positive *x* components (to the right) and the last two have negative *x* components (to the left). The distance $OX_5 = OX_3 - X_3X_5$ represents the algebraic sum of the five *x* components and is the *x* com-

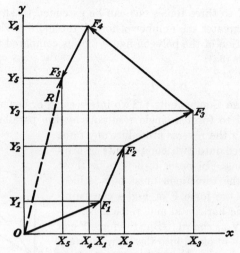

FIG. 5. The polygon of forces.

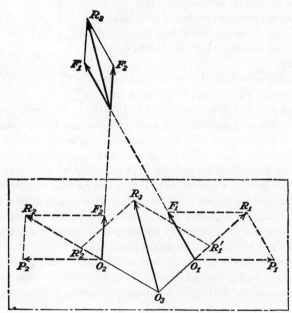

FIG. 6. Construction of the resultant of two forces when their point of intersection lies outside the paper.

ponent of the resultant OR. Similarly, OY_5 is the y component of the resultant, made up of four positive (upward) contributions and one negative (downward) contribution.

The method of components, therefore, is useful when compounding or resolving forces by numerical computation rather than by graphical construction. In the graphical construction of the parallelogram, a practical difficulty sometimes appears in that the paper on which we work is too small to contain all the lines. This occurs when the two components F_1 and F_2 to be added are nearly parallel and at some distance from each other, so that their point of intersection falls far

off the paper, as in Fig. 6. A trick that helps us out in such cases is to **add nothing to the system,** the "nothing" consisting of two equal and opposite forces, O_1P_1 and O_2P_2. These two forces are compounded with the original forces, F_1 and F_2, into the resultants O_1R_1 and O_2R_2, the sum of which two must be the same as the sum of O_1F_1 and O_2F_2, since "nothing" was added. The forces R_1 and R_2 intersect at O_3, nicely on the paper, and by making $O_3R_2' = O_2R_2$, and completing the parallelogram, the desired resultant O_3R_3 is constructed.

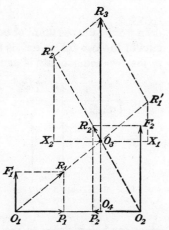

This trick of "adding nothing" even enables us to find the resultant

Fig. 7. Resultant of two parallel forces.

of two parallel forces whose point of intersection is infinitely far away. Clearly, in this case, the parallelogram construction breaks down altogether. Figure 7 is a repetition of Fig. 6 with the same letters, except that this time the two forces F_1 and F_2 are parallel, and the "nothing" (O_1P_1 to the right = O_2P_2 to the left) is taken perpendicular to F_1 and F_2. Again the resultant of O_1F_1 and O_2F_2 must be the same as the resultant of O_1R_1 and O_2R_2 because nothing was added. But O_1R_1 and O_2R_2 intersect at O_3, and now we slide the two forces along their lines of action to bring their origins to O_3 and then add them by the parallelogram of forces. The resultant is vertical (parallel to F_1 and F_2), which becomes clear at once if we consider the horizontal and vertical components of O_3R_1' and O_3R_2'. The horizontal components are equal and opposite (O_1P_1 and O_2P_2) and thus add to zero. The vertical components are F_1 and F_2 and thus add to $F_1 + F_2 = R_3$.

One more result can be deduced from Fig. 7 by geometry, and that is the location of the resultant, between $\mathbf{F}_1$ and $\mathbf{F}_2$. Consider the similar triangles $O_1P_1R_1$ and $O_1O_4O_3$. We have

$$O_3O_4 : O_1O_4 = \mathbf{F}_1 : \mathbf{H}$$

when $\mathbf{H}$ is the horizontal force $O_1P_1 = O_2P_2$.
Similarly we find from the triangles $O_2P_2R_2$ and $O_2O_4O_3$

$$O_3O_4 : O_2O_4 = \mathbf{F}_2 : \mathbf{H}$$

Dividing these two equations by each other, we find

$$O_1O_4 : O_2O_4 = \mathbf{F}_2 : \mathbf{F}_1$$

or in words, **the resultant of two parallel forces is equal to the algebraic sum of the two forces and is located so that the ratio of the distances to the two components is equal to the inverse ratio of the forces,** the

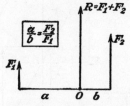

FIG. 8. Location of the resultant of two parallel forces.

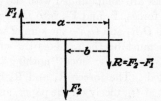

FIG. 9. Resultant of parallel forces in opposite directions.

resultant being closer to the larger of the two forces (Fig. 8). This result is true also when the two parallel forces are in opposite directions, in which case one of them is considered negative, and the formula of Fig. 8 shows that one of the two distances a or b must be negative. Since the resultant lies close to the larger force, b must be negative (Fig. 9). This is the relation of the lever, which was known to Archimedes. A better and clearer way of deriving and understanding these results is by means of moments, as explained in the next chapter.

Problems 9 *and* 10.

CHAPTER II

CONDITIONS OF EQUILIBRIUM

5. Moments. The concept of "moment," as it is used in mechanics, is the scientific formulation of what is everybody's daily experience of the "turning effect" of a force. Consider the wheel of Fig. 10a, which can turn with difficulty on a rusty axle. If we want to turn it, we know that we have to apply a force to the wheel away from its center, the farther away the better. Also we know that the

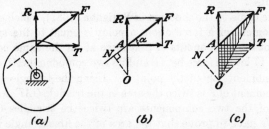

FIG. 10. Moments, showing a special case of Varignon's theorem.

force should be applied roughly in the tangential direction; a radial pull has no effect. Thus, in Fig. 10a, the turning effect of the force **F** is caused by its component **T**, while the component **R** is ineffective.

Abstracting ourselves from the wheel and directing our attention to the diagram of Fig. 10b, the moment is defined as follows:

The moment of a force F about a point *O* is equal to the product of the magnitude of the force F and the normal distance *ON* between the point *O* and the line of action of the force.

Thus a moment is measured in foot-pounds or inch-pounds. In Fig. 10b, let the distance *OA* be denoted by r, and let α be the angle between the force **F** and the tangential direction. Then the distance *ON*, the "moment arm," is $r \cos \alpha$, and the moment is $Fr \cos \alpha$. Applying the definition to the tangential component **T**, the moment of **T** is $T \cdot r = F \cos \alpha \cdot r$, equal to the moment of **F**. The moment of the radial component **R** is zero because its moment arm is zero. Thus we see that the moment of the force **F** about *O* is equal to the sum of the moments of the two components of **F** about the same point *O*. This is a special case of a general theorem, the *theorem of Varignon,* which we shall presently prove. In Fig. 10c look at the triangle OAF and consider the force AF as the base of that triangle and ON its height. The area of the triangle is half the base times height, or $\frac{1}{2}AF$ (measured in pounds) times ON (measured in inches). This is half the moment. Thus it can be said that **the moment of a force equals twice the area of the triangle made up of the force and radii drawn from the moment center O to the extremities of the force.** Thus the vertically shaded triangle OAF represents half the moment of the force **F** about O, while similarly the horizontally shaded triangle OAT represents half the moment of the force **T** about O. We have seen previously that those two moments are the same, and this is verified by the areas of the triangles, which are seen to be the same if we now consider the common side OA to be their base and $AT = RF$ to be their height.

Now we come to the **theorem of Varignon** (1687), which states that

The moment of a force about a point is equal to the sum of the moments of the components of that force about the same point.

In Fig. 11 let the force be AF and its two components AC_1 and AC_2. Consider moments about the point O. Using the triangle representation, the moment of F is twice the area of the triangle OAF, whereas the moments of the two components are twice the triangles OAC_1 and OAC_2. We have to prove that the area of the first triangle is the sum of the areas of the other two. We note that all three triangles have the common base OA. The three heights are FN_F, C_1N_1, and C_2N_2. Drop the perpendicular C_1P to FN_F. Then

$$FN_F = N_FP + PF = C_1N_1 + C_2N_2,$$

because C_1F being parallel and equal to AC_2, the triangles C_1PF and AN_2C_2 are equal. Thus the height FN_F of the triangle OAF equals the sum of the heights of the triangles OAC_1 and OAC_2, and since all three triangles have a common base, the area of AOF is the sum of the two other areas. This proves Varignon's theorem. As an exercise, the reader should repeat this proof for another location of the moment center O, such as O_2 in Fig. 11, for which the moments of the two components C_1 and C_2 have different signs, one turning clockwise, the other counterclockwise.

Another proof of this theorem, employing the method of resolution into Cartesian coordinates, is as follows: In Fig. 11, take a coordinate system with the moment center O as origin and with the line OA as y axis, the x axis being perpendicular thereto. Now resolve all three forces **F**, **C₁**, and **C₂** into their x and y components. The three y components have no moments about O. The three x components all pass through A and have the same moment arm OA. But the x component of F is the sum of the two other x components. With the same moment arm OA, the moment of the x component of F equals the sum of the moments of the two other x components. Finally, by Fig. 10b, it

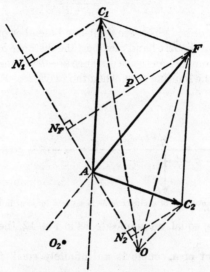

Fig. 11. Proof of Varignon's theorem.

was proved that the moment of any x component equals the moment of the force itself, which proves Varignon's theorem.

Problems 11 *to* 13.

6. Couples. Now let us return to the problem of composition of parallel forces (see Fig. 8). The force **R** is supposed to be the resultant of **F**₁ and **F**₂. Clearly **R** has no moment about the point O through which it passes, and if **R** is to be completely equivalent to the sum of **F**₁ and **F**₂, the sum of the moments of **F**₁ and of **F**₂ about O must also be zero by Varignon's theorem. Thus $F_1a - F_2b = 0$, or the clockwise moment F_1a equals the counterclockwise moment F_2b. The same relation holds for Fig. 9.

Consider two parallel forces in opposite directions, as in Fig. 9, in which **F**₁ and **F**₂ are nearly alike so that the resultant **F**₂ − **F**₁ is small. Then the distances a and b become large, which can be seen as follows:

$$F_1 a = F_2 b, \qquad a = \frac{bF_2}{F_1}$$

$$a - b = \frac{bF_2}{F_1} - b = b\left(\frac{F_2}{F_1} - 1\right) = \frac{b(F_2 - F_1)}{F_1}$$

so that

$$\frac{b}{a - b} = \frac{F_1}{F_2 - F_1}$$

If the forces are nearly alike, the right-hand member of this expression becomes large; in the left-hand member, the distance b of the resultant then is large in comparison to the distance $a - b$ between the two forces F_1 and F_2. In the limiting case, when the two forces F_1 and F_2 become

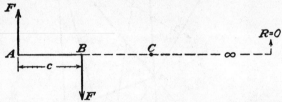

Fig. 12. The resultant of a pure couple is a "zero force at infinite distance."

completely alike, equal, and opposite, as in Fig. 12, the combination is called a couple.

The resultant of a couple is an infinitely small force at infinite distance.

The moment of the couple is the sum of the moments of the two forces, or the moment of the resultant. The latter is not useful to us as it gives the moment in the form of $0 \times \infty$, which may have any value. The moment in Fig. 12 about the point A is Fc clockwise; about the point B it is again Fc clockwise; and about the arbitrary point C it is again Fc clockwise.

The moment of a couple is the same about every point in the plane. The moment of a *force* varies, of course, with the choice of moment center.

We will now show that the only important property of a couple in a plane is its moment. The fact that the forces in Fig. 12 are drawn vertically, for instance, will be shown to have no significance for the determination of equilibrium. By "adding nothing" to a couple, we can change its appearance completely. Let in Fig. 13a the couple be **F, F,** and let us add to it "nothing" in the form of the forces **P, P** in the mid-point and parallel to **F**. Now let us combine one force **P** with each of the forces **F** to form the resultants **R**. The two resultants

R, R form a new couple with larger forces, a smaller distance between them, and with the same moment.

In Fig. 13*b*, the "nothing" **P, P** is in a direction perpendicular to **F, F**; the resultants **R, R** are larger and their normal distance smaller than the original couple **FF**, but the moment is again the same by Varignon's theorem. We do not distinguish at all between these various appearances of the couple and call **FF** or **RR** in Fig. 13*a* or 13*b* one and the same couple, designating it sometimes by a curved arrow as in Fig. 14.

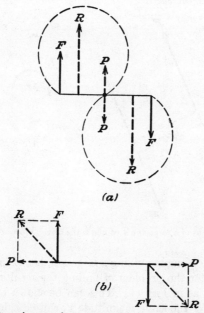

Fɪɢ. 13.   A couple in a plane can be represented by two forces in many different ways.

Therefore, **a couple in a plane is completely defined by its moment only, irrespective of the direction or magnitude of its two constituent forces.** A couple in a plane is thus an algebraic scalar quantity, whereas a force in a plane is a vector quantity. The magnitude of the couple can be expressed by a single number (inch-pounds, positive or negative), but for a force through a point in a plane, we have to specify two numbers: for instance, the x and y components in pounds.

A homely example of this occurs when a man alone in a rowboat near a dock wants to turn the boat. He can do this if he will grab two points of the dock with his two hands as far apart as possible and then push-pull.   The couple thus exerted by him on the dock has a reactive or counter couple from the dock on him and through his feet on the boat.   It is completely immaterial to the turning effort on the boat whether the man stands at the bow and push-pulls fore and aft or whether he stands at the stern and push-pulls sidewise; the couple is the same.   If, however, he attempts to turn the boat by a *force, i.e.,* by one hand only, his position in the boat is of crucial importance; when he pushes on the dock, the rotation of the boat will be in one direction when he stands in the bow and in the opposite direction when he stands in the stern.

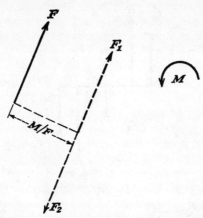

Fig. 14.   The resultant of a force and a couple is the original force displaced parallel to itself.

In the process of compounding a large number of forces on a body, we often encounter the problem of finding the resultant of a force F and a couple M, as in Fig. 14.   This can be shown to be equivalent to a single force of the same magnitude F, shifted sidewise through a distance M/F.   To understand, we use the trick of adding "nothing" in the form of the pair of forces F_1, F_2, each of which is equal to F.   Now consider the combination of F and F_2, which is a couple equal and opposite the original couple M, and thus the two cancel each other. All that is left is the force F_1.   Thus in Fig. 14 the combination of F and M, both printed in heavy line, is equivalent to the single force F_1, printed in heavy dashes.

The sum of a force and a couple in the same plane is a force equal in magnitude and parallel to the original force shifted sidewise through a distance equal to the moment of the couple divided by the force.

Problems 14 *to* 16.

7. Equations of Equilibrium. By the parallelogram construction
any number of forces in a plane can be combined into a single resultant
force, with the notable exception that if we end up with a pure couple,
consisting of two equal and opposite forces some distance apart, it is
impossible to combine these further into a single resultant.   (Another
way of expressing the word "impossible" is by saying that the resultant is infinitely small and located infinitely far away.)

The resultant force tries to push the body on which it acts in its
own direction and tends to accelerate it (page 175).   A resultant couple
tries to turn the body on which it acts and tends to accelerate it (page
215).   The fourth axiom (page 6) stated that a body in equilibrium
remains so if no force at all is acting on it.   If there are several forces
acting on it, it will still remain in equilibrium if these forces have no
resultant.   Thus we arrive at the **first statement** of the condition for
equilibrium in a plane:

**If, and only if, the total resultant of all forces acting on a body
in a single plane is zero, the body once in equilibrium will remain in
equilibrium.**   The qualification "once in equilibrium" in this definition calls for an explanation.   Later, on page 175, we will see that a
body on which no forces act may be moving at constant speed, and
hence is not "in equilibrium" by the definition of page 3.   In the
remaining chapters on statics in this book, up to page 156, we will
assume that the body is always in equilibrium to start with so that the
qualification will not be repeated.

If we work with x and y components in a coordinate system, the
sum of the x components of all individual forces is the x component of
the resultant, and similarly for the y components.   Thus, if the sums
of the x and y components are both zero, there is no resultant force,
and the only thing left may be a couple.   Then, if the sum of the
moments of all x and y components of the forces about one arbitrary
point is zero, we know that the couple has zero moment because the
resultant force is already zero and can have no moment.   Thus we
have the **second statement** of the conditions for equilibrium in a plane:

**A body is in equilibrium under the influence of forces in a plane if
and only if the three conditions below are satisfied:**

a. **The sum of the x components of all forces acting on the body
must be zero.**

b. **The sum of the y components of all forces acting on it must be
zero.**

c. **The sum of the moments of the x and y components of all forces
about one arbitrary point must be zero.**

Expressed in a formula this condition is

$$\Sigma X = 0, \qquad \Sigma Y = 0, \qquad \Sigma M = 0 \qquad (1)$$

In a third form in which the equilibrium conditions appear, only moments are considered, and no force resultant is computed at all. We have seen that any force system is equivalent to a single resultant force or to a couple. In either case, if we compute the moment about an arbitrary point and find that moment to be zero, the chances are that there is equilibrium. The only possibility of reaching a wrong conclusion is when the moment center happens to be chosen on the line of action of the resultant force. Then there is zero moment but no equilibrium. Thus the zero moment about one arbitrary point is a necessary condition for equilibrium, but it is not a sufficient condition. Now let us choose a second point and compute the moments about it. If there is equilibrium, the moment is found to be zero, and conversely, if we find a zero moment, the chances are very great that we have equilibrium. However, it is still not sure, because even that second point may lie on the line of action of the resultant force. To be completely sure, we compute the moments about a third point, not lying on the straight line connecting the first two moment centers. If the moment about this third point is still zero, we conclude that the resultant itself must be zero because the resultant must have a moment arm about at least one of the three points. In addition, there cannot be a couple because the moment is zero. Thus we arrive at the **third statement** of the conditions for equilibrium in a plane:

A body subjected to forces in a plane is in equilibrium if the moments of all forces taken about three different moment centers, not lying in a straight line, are zero.

It is useful to know all three forms of the equilibrium conditions because, depending on the character of the particular problem, one form will give a simpler solution than another.

Problems 17 to 25.

8. Applications. The foregoing principles of statics suffice to explain the operation of and the forces involved in many objects and mechanisms of daily use. In this section eight of these will be explained in some detail:

    a. The single pulley wheel
    b. The boom hoist
    c. The hinged arch
    d. The beam on two supports

 e. The crank mechanism
 f. The differential hoist
 g. The locomotive equalizer gear
 h. Sailing against the wind

Others are left as exercises in Problems **26** to **52**.

 a. The Pulley Wheel. On page 7 the statement was made that the tension forces in the two ends of a flexible rope slung over a (non-rotating) pulley with a frictionless bearing are the same. In order to understand this with the help of Fig. 15, we start with the first essential

step in any analysis of statics: the choice and isolation of the "body." We choose the wheel and the rope cut off at some distance from the pulley at both ends, excluding the journal or shaft on which the pulley can rotate. The friction between the rope and the wheel is supposed to be sufficiently large to prevent slipping, and we shall neglect the weight of the wheel and of the rope. Then there are three forces acting on the body: the two rope forces and the force from the shaft on the wheel. The body is in equilibrium; therefore, the sum of the moments of all forces about any point

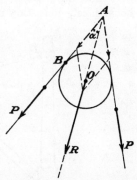

Fig. 15. Forces on a friction-less pulley.

must be zero. We take moments about the shaft center O and conclude (in the absence of bearing friction) that the two rope forces must be equal. Calling them **P**, their resultant **R** is constructed graphically in Fig. 15. It is seen to pass through the center O and to be of magnitude $R = 2P \cos \alpha$. Now we apply the first statement of the conditions of equilibrium (page 19). The moments are zero about O, and the resultant of all forces must be zero also. Hence, the third force on the body, *i.e.*, the force from the shaft on the wheel, must be equal and opposite to R. The force R, by the axiom of action equals reaction, is the force from the wheel on the shaft, which is then further transmitted to the supporting structure. In other words, R is the load of the pulley wheel on the supporting structure.

 In case there is friction in the bearing, that bearing can support a couple, and the two rope tensions are no longer necessarily equal. The difference between their two moments about O must be equal to the friction couple or friction "torque."

 This is the simplest way of treating the problem. It can be solved also by applying the third criterion of equilibrium (page 20), but at the expense of greater complication.   First we take moments about O and conclude that $P = P$.   Then we take moments about A, which automatically gives zero and leads to no new result.   Finally we take moments about B.   The moment arms are not too simple, and the reader should verify for himself that

$$P \frac{r \sin 2\alpha}{\tan \alpha} = Rr \cos \alpha$$

from which the previously obtained relation between R and P follows.

b. *The Boom Hoist.*   Consider in Fig. 16 a simplified version of a boom hoist, consisting of a mast AD, a boom AB, hinged without

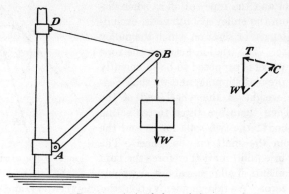

FIG. 16.   A simple boom hoist.

friction at A, held by a cable between B and the mast at D, and carrying a dead load W at B.   Neglecting the weight of the boom, what are the forces in the boom, the cable, and the mast?

First we must isolate a body, and we choose for it the boom.   At A there is a force on the boom from the hinge axis, but since there is no friction, there can be no couple at that end, and the force must pass through A.   At B there are two forces on the boom, from the two cables BW and BD.   Cables can transmit only tensile forces along their center lines; thus the two forces pass through B, and if we knew their magnitudes, we could construct their resultant.   This resultant, together with the force at A, must reduce to zero, since the boom is in equilibrium.   Hence the force at A must be along the center line of the boom and the resultant of the two forces at B must be equal and opposite to it.   In future applications we usually omit this long story and recognize at the start that **a bar, hinged without friction at both**

ends, and not supporting lateral loads along its length (in our boom
the weight was neglected), **can have only a compressive or tensile
force along its center line.**

The next body we isolate is the load W with a short piece of rope.
There are two forces: the weight or force of gravity W and the rope
tension. Hence the rope tension is W. In future applications, we
usually omit this analysis and state the conclusion immediately.

The third body we isolate is the crux of the problem. It is the
upper piece of boom at B with two short pieces of rope attached to it.
On this body there are three forces acting, all along the respective
center lines; one of the forces is known to be W, the other two are
unknown. Constructing a (closed) triangle of these three forces, as
in Fig. 16, leads to the values of the two other forces, such that the
resultant of all three forces on the body is zero. The directions shown
in the triangle of Fig. 16 are the forces *on the body;* therefore, the
lower part of the boom pushes up on the body (the boom is in compression with the force C), and the cable force, being to the left, pulls
on the body so that the cable is in tension with the force T. Finally,
the forces on the mast are a pull to the right at D and a push down to
the left at A. These forces bend the mast and are transmitted to its
foundation. The calculation of the stresses in the mast caused by
these forces is a question in "strength of materials," which we are not
yet ready to consider.

Now consider the more realistic picture of Fig. 17, where the load
can be hoisted up by means of a (frictionless) pulley at the end of the
boom and a winch drum. We do not repeat the analyses for the boom
and for the weight just given, but we immediately apply our equilibrium conditions to the body consisting of the pulley and the piece
of rope slung over it. We conclude from Fig. 15 that the tension in
the rope between the pulley and the winch is again W, and Fig. 17a
shows the force F exerted by the pulley on its shaft, *i.e.*, on the upper
end of the boom, as in Fig. 15. This force F is what the boom feels,
so that from now on the analysis is the same as in Fig. 16, only F
replaces W as the weight load on the boom. The triangle of forces
for the isolated body consisting of the top end of the boom is shown in
Fig. 17b. It is concluded that the angle α between the boom and the
winch rope had better be made smaller than the angle β between the
boom and the vertical; otherwise the force in the boom-supporting
cable reverses and becomes compression. Thus with the arrangement of Fig. 17, if the boom is too high (small β), the winch will pull
it still higher, and the system collapses. In actual constructions the

boom support cable is made of variable length, and the boom itself can be lifted or lowered by means of another winch.

c. The Hinged Arch. In some civil-engineering applications, the structure of Fig. 18 is encountered, consisting of an arch, hinged

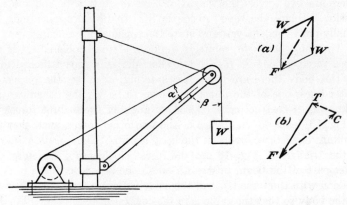

FIG. 17. A boom hoist with hoisting drum.

at its two points of support A and B, and having another hinge at the apex C. Certain loads are placed on the structure, and we are required to find the reactions at the supports A and B. We note that the half arch between A and C is a rigid body, hinged at both ends, carrying certain forces, and as long as the ends are hinged and the forces remain the same, the reactions remain the same whether the arch is painted green or red or whether it is made of steel or of wood or whether it changes its shape in any

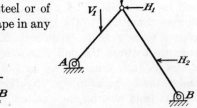

FIG. 18. A hinged arch.

FIG. 19. Another form of hinged arch.

way. Therefore, the problem of Fig. 19 is again a "hinged arch," although the "arches" are now straight bars. In order to solve the problem of Fig. 19 by statics, we subdivide the system into three isolated bodies: the two bars, each cut off just below the top hinge, and the top hinge itself (Fig. 20). At the cuts and at the supports unknown reaction forces are acting, which are denoted by X_1, X_2, X_3, and X_4 on the left bar, and by X_5, X_6, X_7, and X_8 on the right bar. Then, by the law of action and reaction, the opposites of the forces X_3, X_4, X_7, and X_8

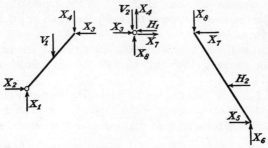

Fig. 20. Forces acting on the system of Fig. 19.

act on the hinge. For the left bar we can write three equations of equilibrium ($\Sigma V = 0$, $\Sigma H = 0$, $\Sigma M = 0$) and similarly for the right bar.

For the hinge body we can write $\Sigma H = 0$ and $\Sigma V = 0$, while the third or moment equation becomes meaningless ($0 = 0$) because all the forces pass through the same point. We thus have $3 + 3 + 2 = 8$ equations for the 8 unknown reactions. This is a great deal of work, and it can be greatly simplified by a different approach.

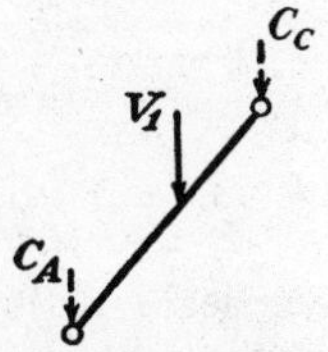

Fig. 21. Resolution of a force in the middle of a beam into components at its end hinges.

First, in Fig. 19, we take each load and resolve it into two parallel components C at the nearest hinges, as shown in Fig. 21. After this has been done, the hinged arch is loaded only by vertical and horizontal loads at each of the three hinges A, B, and C. The loads at A and B are directly passed on to the foundation so that what is left is the arch of Fig. 22, loaded only by a single force **F** at the top hinge. Then the arch consists of two bars, each of which is hinged at both ends and has no lateral loadings along its length, so that by page 23 each bar transmits only a force along its own center line. Now consider as the isolated body the top hinge only, subjected to three forces, the load **F** and the two bar forces $\mathbf{F}_A$ and $\mathbf{F}_B$, and construct the force triangle (Fig. 22). The directions of the forces indicated are those *from* the bars *on* the

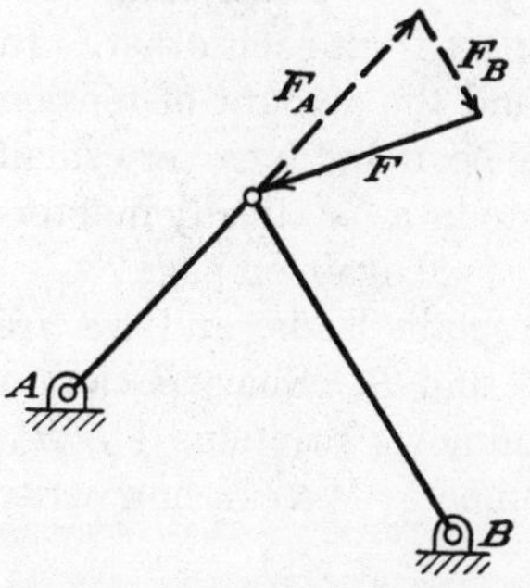

Fig. 22. Resolution of a force F into components along two hinged members.

hinge; thus we see that bar A is in compression and bar B is in tension. These two bar forces are transmitted to the bottom hinges A and B and to the foundation. The total force on the foundation is made up of these plus the contribution of the components C_A of Fig. 21 that are directly transmitted.

Problems with *two* hinges between the supports, known sometimes as three-bar linkages (Problems 50 and 51) are treated in a similar manner.

d. The Beam on Two Supports. Consider the beam of Fig. 23 on two supports. Note that one of the supports is indicated as a hinge, which is a symbol for a support that can give a horizontal as well as a

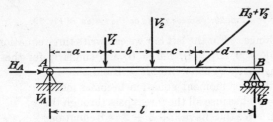

FIG. 23.   To find the support reactions of a beam.

vertical reaction force to the beam, while the other support is shown on rollers, which means that no horizontal reaction is possible. If we had not done this and had put the beam on two hinge supports, we could have installed the beam on its supports while pulling it lengthwise from both ends, or we could have just put it in plainly and then moved the hinge supports a little farther apart from each other. In either case the beam would be in tension, and the amount of tension would be determined not by the loads on the beam but by a very small stretch of the beam. The beam is then said to be a "statically indeterminate structure." This subject will be discussed again on page 73.

The beam of Fig. 23 is subjected to certain loads, and we are required to find the bearing reactions at A and B. Since B cannot have a horizontal reaction, we name the unknown reactions V_A, H_A, and V_B, and note that they are three in number. We can now write the three equations of equilibrium:

$$\Sigma V = 0, \qquad V_1 + V_2 + V_3 = V_A + V_B$$
$$\Sigma H = 0, \qquad H_3 = H_A$$
$$\Sigma M_A = 0, \qquad V_1 a + V_2(a + b) + V_3(a + b + c) = V_B l$$

and we can solve these for V_A, V_B, and H_A.

Another method of solving the problem is by writing three moment equations about three non-collinear points (page 20); for instance, the points A, B, and a third point at a distance p above A. The first of these equations is the same as the third of the previous set. The second equation is

$$\Sigma M_B = 0, \qquad V_A l = V_1(b + c + d) + V_2(c + d) + V_3 d$$

The third equation contains first of all the same terms as the ΣM_A equation (which all together are zero and thus can be eliminated) and in addition the terms $H_A p - H_3 p$. Setting this equal to zero gives the middle equation of the above set. This second method in this case leads to more algebraic work than the first method.

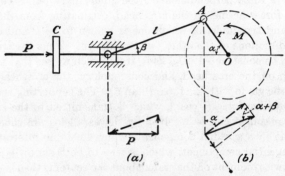

Fig. 24. The crank mechanism.

Still another method is to split up each vertical force into two parallel components at A and B. For instance, V_1 is replaced by a force $V_1(b + c + d)/l$ at A and by a force $V_1 a/l$ at B. When this has been done with all the forces, we have a beam loaded at the bearings only, and then the loads are directly transmitted to the ground as bearing reactions.

It is interesting to note the state of longitudinal tension of the beam. The portion between A and $H_3 V_3$ has a compressive force H_3 in it, while the right-hand part of the beam is free from compression. In case the roller support had been placed at A and the hinge support at B, the right-hand part of the beam would have been in tension H_3, and the left-hand part would have had no longitudinal force. In case of two hinge supports, all we could say is that the tension in the right-hand part is greater by the amount H_3 than the tension in the left-hand part, but the longitudinal force in each part of the beam would have been indeterminate ("statically indeterminate").

e. The Crank Mechanism. This mechanism, shown in Fig. 24, consists of the crank OA, the connecting rod AB, hinged at A to the crankpin and at B to the crosshead, which slides without friction between parallel guides, the piston rod BC, and the piston C. When a force P is exerted on the piston, the machine can be kept in equilibrium by a counter torque M on the crank. We want to know the relation between P and M for a certain position designated by the angle α. First we note that the piston force P is transmitted without change to the crosshead B. Next we isolate the crosshead and stubs of the piston and connecting rods. The connecting rod, being a double-hinged bar without side loads, has a force along its center line only (page 23). The guides, being frictionless, exert on the crosshead only a force perpendicular to the guide (page 84). Thus there are three forces acting on the crosshead, originating from the piston, the rod, and the guide, of which one is known to be P and of which only the directions of the other two are known. Thus the triangle of forces can be constructed (Fig. 24a), from which we see that the guide pushes up on the crosshead, and consequently, the crosshead pushes down on the guide with the force $P \tan \beta$. The connecting rod carries the compressive force $P/\cos \beta$, which is transmitted to the crankpin. At the crankpin it can be resolved by the parallelogram construction of Fig. 24b into a radial component (which holds no interest for us) and a tangential component, which is seen to be $P \sin (\alpha + \beta)/\cos \beta$. The clockwise moment on the crank about the center O then is

$$Pr \sin (\alpha + \beta)/\cos \beta,$$

and that moment must be held in equilibrium by a counterclockwise couple of the same magnitude. We note that at the two dead centers ($\beta = 0°$, $\alpha = 0$ or $180°$) the couple becomes zero. We will return to this problem on pages 170 and 199.

f. The Differential Hoist. This mechanism, shown in Fig. 25, consists of an upper wheel carrying two pulleys of slightly different radii, r_o and r_i, constructed together as one integral piece. An endless rope or chain is slung twice around the upper wheel and once around the lower floating pulley, as shown. The upper wheel is hung from a solid ceiling or crane; the lower pulley carries the load, and the operator pulls at the chain with a force P. The rope or chain does not slip around either wheel owing to sufficient friction or other means, and the friction in the upper and lower wheel bearings is made as small as possible. We assume it to be zero in the analysis. We want to find the relation between the load W and the pull P at equilibrium. The chains are so long that they are supposed to be vertical with sufficient accuracy.

We start the analysis with the lower pulley, which from example *a* (page 21) is in equilibrium if the two chain tensions T_1 and T_2 are $W/2$ each. Next we isolate the upper wheel with four chain stubs and the supporting hook stub protruding from it. The chain tension forces are $W/2$, $W/2$, zero, and the unknown P; the hook force is the unknown V. The equation of vertical equilibrium tells us that the hook supports a force $W + P$, which is not particularly interesting. The equation of horizontal equilibrium reduces to $0 = 0$, unless we pull sidewise on our chain, when we recognize that the hook supporting force must have a sidewise component equal to the sidewise component of the pull P. But the interesting equation is the moment equation about the center of the upper disk. It states that

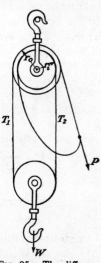

$$\frac{W}{2} r_o = \frac{W}{2} r_i + P r_o$$

or

$$P = \frac{W}{2} \frac{r_o - r_i}{r_o} = \frac{W}{2}\left(1 - \frac{r_i}{r_o}\right)$$

By making the outer radius r_o nearly equal to the inner radius r_i, we can make the pull P very small with respect to the load W and thus obtain a great "advantage." It is noted that in the absence of friction the load P just calculated holds the load in equilibrium. A slightly larger pull

Fig. 25. The differential chain hoist.

will pull the load up, and a slightly smaller pull will let the load down. In actual practice the effect of friction is very important. We will return to this problem on page 94.

g. The Locomotive Equalizer Gear. Anybody who has lived long enough has at one time or another been annoyed by the wiggly four-legged restaurant table and has tried to stop the wiggle by putting something under the short leg. The distribution of the load on the top of a three-legged table among the legs is a problem in statics (page 117), but for a table of four or more legs the distribution of the load depends greatly on very small variations in the length of the individual legs or on small variations in the flatness of the floor on which it stands. A large locomotive has twelve or more wheels on which the load rests, and if no particular precautions were taken, it could happen that if three of the twelve wheels were to strike high spots simultaneously on the rails, these three would carry all of the

load, and the other nine wheels would act as the short leg of the restaurant table.   Of course the springs between the wheels and the
frame make this statement not quite as drastic as it sounds, but springs
in locomotives are usually very stiff.   Thus designers have introduced
the "equalizer gear," consisting of a system of levers that ensures
proper three-point support of the weight of the locomotive.   Figure
26 shows three driving wheels on the same rail.   Their axles turn in
bearing boxes, and on each box the locomotive load presses from above.
In the figure this is done by two levers, one each between each two
wheels.   These levers in turn are pressed down by the ends of a nest
of leaf springs, which carry the locomotive load P in their center.

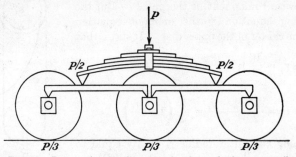

FIG. 26.   Locomotive equalizer gear for three wheels on one rail.

The static equilibrium of the leaf springs requires a division of P
into two equal halves $P/2$.   By placing the end of the spring at the
one-third-length point of the next lever, this $P/2$ is divided into $P/3$
and $P/6$ at the bearing boxes.   On the center wheel there are two
loads $P/6$ so that ultimately the load P is equally divided over the
three wheels by this equalizer gear.   The reader should convince
himself that if one of the three wheels is set on a 1-in.-high spot, then
still the load is equally divided between the three wheels.

Figure 27 shows an equalizer gear to divide a load into four equal
parts over four wheels on two axles on both rails.

A good locomotive construction consists of a combination of two
Figs. 26 and one Fig. 27, with the total weight supported in three
points, two as in Fig. 26 on the two sides of the frame comprising the
forward driving wheels, and one, as in Fig. 27, in the center of the
frame near the aft end close to the firebox.

h. Sailing against the Wind.   This is one of the major inventions
that ushered in our modern age.   The Greeks, Romans, and other
ancient peoples could not do it, and the art was developed, primarily
by the Portuguese, in the century before the great voyages of dis-

covery. Without being able to sail against the wind Columbus would hardly have opened up a new world, and when he did it, he was using a newly developed modern invention.

In Fig. 28 let β be the angle between the sail and the ship, and let α be the angle of attack between the wind and the sail. A sail acts

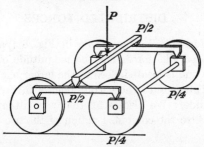

Fɪɢ. 27. Equalizer gear for four wheels on two rails.

very much like an airfoil or an airplane wing; the force F exerted by the wind is just about perpendicular to the sail. This force F can be resolved into the forward and sidewise components F_f and F_s, of which in "sharp" sailing, as in the figure, the sidewise force F_s is considerably greater than the forward force F_f.

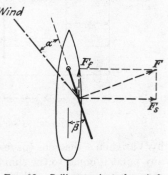

The ship is constructed with a large keel, which gives it great resistance against sideslip, while the resistance against forward motion is small. Therefore, even with a small forward force F_f, the ship will still move more in the forward direction than sidewise. From this simple explanation it would seem as if the ship could sail almost dead against the wind, with $\alpha = \beta = 1°$ or so. In practice this is not so; the wind force on the sail is

Fɪɢ. 28. Sailing against the wind.

roughly proportional to α, so that for $\alpha = 0$ the wind force F on the sail is zero, and for zero sail angle β the forward component of F becomes zero. In practice, therefore, both α and β must have definite values, and the ship cannot sail dead against the wind. Later (page 432) we will return to this problem to clear up some points; in particular, the wind direction shown in Fig. 28 is that of the wind relative to the (moving) ship, which is not the same as the direction of the wind with respect to the (still) water.

Problems 26 *to* 52.

CHAPTER III

DISTRIBUTED FORCES

9. Parallel Forces. We have seen in Fig. 8 (page 12) that the resultant of two parallel forces has the magnitude of the sum of the two components and is located close to the larger one of the two components, with distances to these components inversely proportional to their magnitude. We also recall that a resultant force is "statically equivalent" to the combined action of the two components and

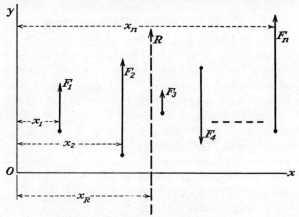

FIG. 29.   Resultant of many parallel forces.

by Varignon's theorem (page 14) the moment of the resultant about any point is equal to the sum of the moments of the components. In case we have three parallel forces, we can first combine two of these into a resultant and then add that resultant to the third force, and for more than three forces this process can be repeated over and over again, so that we can construct the resultant of a large number of parallel forces. In this manner we recognize that the value or magnitude of the resultant is the sum of the magnitudes of all components, taken algebraically. Also, the moment of the resultant about any point is equal to the sum of the moments of all components.

Thus in Fig. 29, taking moments about the origin O of the coordinate system, we have

$$R = \sum_n F_n \quad \text{and} \quad Rx_R = \sum_n F_n x_n$$

or combined

$$x_R = \frac{\Sigma F_n x_n}{\Sigma F_n}$$

These sums are understood in an algebraic sense—for instance, the force F_4 of Fig. 29 gives a negative contribution to the ΣF_n sum as well as to the $\Sigma F_n x_n$ sum.

The most frequently occurring parallel forces are the gravity or weight forces acting on the parts of a large body. We can imagine the body to be subdivided into small elements, and then we can

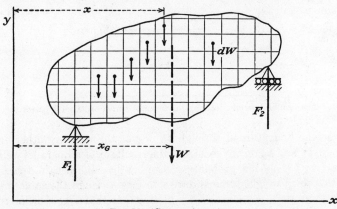

Fig. 30.   Center of gravity.

imagine invisible strings between each such element and the earth; each (imaginary) string having a tensile force dW in it. Thus the body is acted upon (Fig. 30) by a large number of small forces dW, all vertical and hence parallel. These forces are held in equilibrium by the support reaction forces F_1 and F_2. In order to calculate these reactions we replace the large number of weight forces by their resultant W, being the sum of all the little weights and hence the weight of the whole body. The location of this resultant W is found from the moment equation just written down, but following the usual convention, we replace the Greek letter s, written Σ, by the German letter s, written $\int$, because the number of elements involved is infinitely large. Thus

$$x_G = \frac{\int x \, dW}{\int dW} \tag{2}$$

Now we can turn the body through 90 deg, support it properly, and let gravity act again. We do not need to redraw the figure, but imagine in Fig. 30 the weight forces to be acting parallel to the x axis to the right. The resultant weight force W will act to the right at distance y_G from the x axis, and

$$y_G = \frac{\int y \, dW}{\int dW} \tag{2}$$

The intersection point of the two weight forces is called the *center of gravity* of the body, hence the subscripts G in Eq. (2). Now we

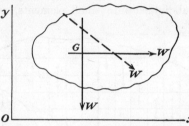

FIG. 31. The resultant weight force acting in many directions always passes through the same point G.

imagine the body turned through 45 deg or through any other arbitrary angle, and again we construct the resultant of all parallel weight forces (Fig. 31). We probably remember having seen in the past that this third weight force is drawn to pass through the same intersection G so that Fig. 31 is in error. Indeed it is, but this is by no means obvious and has to be proved, as follows:

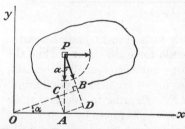

FIG. 32. Toward the proof of the theorem of Fig. 31.

Consider in Fig. 32 a single element dW of a body, and let the weight force act on this element in three different directions, 0, 90 deg, α deg, consecutively, by turning the body suitably. We call $OA = x$, $AP = y$, as usual, and we call $OB = p$. The quantities x, y, and p are the moment arms of the weight force dW in the three directions shown. The location of the point P is determined by x and y so that p is expressible in x, y, and the angle α.

$$p = OB = OC + BC = OA \cos \alpha + AD = OA \cos \alpha + AP \sin \alpha$$
$$= x \cos \alpha + y \sin \alpha$$

Now form the total weight resultant of all the elements of the body. In the vertical and horizontal directions this gives us Eq. (2) above, and similarly in the α direction, we have

$$p_W = \frac{\int p \, dW}{\int dW}$$

But, substituting the general value of p of each element into the numerator, that integral becomes

$$p_W W = \int p \, dW = \cos \alpha \int x \, dW + \sin \alpha \int y \, dW$$
$$= \cos \alpha W x_G + \sin \alpha W y_G$$

or

$$p_W = x_G \cos \alpha + y_G \sin \alpha$$

because in the integrations the angle α is a constant, being the same for every element of the body. Thus we come to the result that the relation between p_W, x_G, and y_G is the same as the relation between p, x, and y of a single element (Fig. 32). Since the three weight lines of a single point obviously all pass through that point, so must the three resultant weight lines pass through the point G. (As Fig. 31 is drawn, the p_W would be larger than the sum $x_G \cos \alpha + y_G \sin \alpha$.) Thus it is proved that

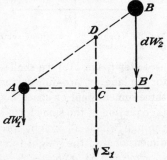

Fig. 33. Second proof of the theorem of Fig. 31.

The weight force W of a large body, being the resultant of all the elemental forces dW, passes through the same point G (the center of gravity) for all possible directions of the weight force relative to the body.

For another proof of this important proposition, we consider two elemental weight points only. In Fig. 33 let these two points be A and B with weights dW_1 and dW_2, not necessarily equal. First let the weight act downward, the resultant being Σ_1, which divides the distance AB' into two parts AC and $B'C$, inversely proportional to the weights dW_1 and dW_2. For simplicity we imagine that dW_2 is twice dW_1; then $AC = 2B'C$. It is clear then that this force line Σ_1 cuts the connecting line between the two weights AB into parts AD and BD in the same ratio (again $AD = 2BD$). Now we repeat this entire argument word for word in a different direction, the horizontal one or that of angle α. We always end up with a Σ line passing through point D. Thus D is the center of gravity for the two mass points, through which the resultant passes independent of the direction of the weight, and the two weights dW_1 and dW_2 can be replaced by their sum at point D. If there is a third weight point dW_3 at E, we can repeat the entire argument, this time with the two masses dW_3 at E and $(dW_1 + dW_2)$ at D, and conclude that the resultant of the weight forces of the three weights dW_1, dW_2, and dW_3 passes through a single point, the center of gravity, irrespective of direction. The proof can then be extended to four, five, and to an indefinite number of weights.

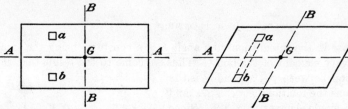

FIG. 34. Center of gravity of symmetrical flat plates. FIG. 35. Center of gravity of skew symmetrical figures.

Later (page 119) we shall see that Eqs. (2), which were proved here for a plane or two-dimensional body, will continue to be true for a three-dimensional or space body and are to be completed by a third (z) equation of the same form. Again there will be a center of gravity through which all weight forces pass for any direction (in all three dimensions) of the weight force.

Problems 53 *to* 55.

10. Centers of Gravity. In the practical numerical computation of centers of gravity, the first thing to look for is *symmetry*. When there is an axis of symmetry, either of the rectangular or skew type, the center of gravity must be on that axis. Consider the rectangular plate of Fig. 34, of uniform thickness and hence of uniform weight per unit area. Consider two equal weight elements at a and b. Their center of gravity lies on the axis of symmetry, AA, and the whole plate can be cut up into similar pairs of elements a and b. Thus the center of gravity must lie on AA. For the same reason, it must lie on the other axis of symmetry, BB.

The same argument, word for word, can be repeated for the case of skew symmetry, as in the parallelogram of Fig. 35.

The second thing to look for, after symmetry, is the possibility of *subdivision into plates of simpler shape*. Consider the L-shap d flat uniform plate of Fig. 36. This plate can be cut into two rectangular plates for instance, as indicated. Each rectangle has its center of gravity in its center of symmetry. The two rectangles have weights in ratio of their areas, at for the upper one and $(b - t)t$ for the lower one. We note that **for finding the center of gravity, it is not necessary to compute the weights of components in detail; their ratio suffices.** The combined center of gravity G is that of two weight points G_1 and G_2, and by Fig. 8 (page 12) it lies on the connecting line, so that

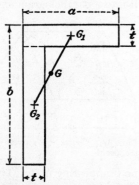

$$\frac{GG_2}{GG_1} = \frac{at}{(b - t)t} = \frac{a}{b - t}$$

Next consider the triangle of Fig. 37. The median, or line connecting a vertex with the middle of the opposing side, is an axis of skew symmetry and hence contains the center of gravity. We can concentrate the weight of each horizontal strip into its own center of gravity and thus replace the triangle by a "heavy center line" (Fig.

FIG. 36. Center of gravity of a figure compounded of two rectangles.

37). The weight along this line is not uniform, the lower part weighing more per unit length than the upper part. In fact, the weight increases linearly with the distance x from the top O and is C_1x per unit length. Thus the weight of a piece dx is $C_1x\,dx$, and its moment

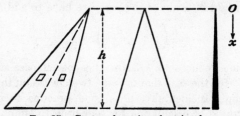

FIG. 37. Center of gravity of a triangle.

about O is $C_1x\,dx \cdot x = C_1x^2\,dx$. By Eqs. (2) the distance x of the center of gravity from the vertex is

$$x_G = \frac{\int_0^h C_1x^2\,dx}{\int_0^h C_1x\,dx} = \frac{C_1h^3/3}{C_1h^2/2} = \frac{2}{3}h$$

Thus the center of gravity of a triangle is at one-third height from its

base and at two-thirds height from a vertex. We note again that the
constant C_1 has dropped out of the above integration without our
having computed its value, for the same reason that we did not compute the individual weights of the parts of the previous example
(Fig. 36).

Now we are ready to consider a trapezoidal plate by subdividing
it into a parallelogram and a triangle (Fig. 38). The center of gravity

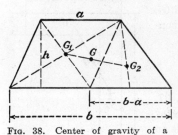

FIG. 38. Center of gravity of a
trapezoid.

G_1 of the parallelogram is at height $h/2$
above the base, and G_2 of the triangle
by itself is at height $h/3$ above the
base. The total weight of the parallelogram is Cah and that of the triangle, $C(b - a)h/2$, where the constant
C, being the weight per unit area of the
plate, does not need to be calculated,
since it will drop out later. The ratio
of the two weights is as $a : (b - a)/2$,
no C appearing in it. The center of gravity of the combination, point
G, lies on the connecting line of G_1 and G_2 at distances inversely proportional to the weights (Fig. 8). Let us calculate the height of G
above the base. The difference in height between G_1 and G_2 is $h/6$,
and thus the difference in height between G and G_2 is

$$\frac{a}{a + (b - a)/2} \cdot \frac{h}{6} = \frac{2a}{a + b} \cdot \frac{h}{6} = \frac{a}{a + b} \cdot \frac{h}{3}$$

To find the height of G above the base, we have to add $h/3$ to this or

$$h_G = \frac{h}{3}\left(1 + \frac{a}{a + b}\right) = \frac{h}{3}\frac{2a + b}{a + b}$$

This expression reduces to $h/3$ for the special case $a = 0$, and it
reduces to $h/2$ for the case that $b = a$—two checks on the accuracy of
our calculation. Incidentally, it is good practice to check a final
result of any kind of calculation for reduction to simplified cases for
the purpose of discovering possible errors.

Another way of finding G graphically in Fig. 38 is by remarking
that the line joining the mid-points of the base and top of the trapezoid is an axis of skew symmetry, and thus its intersection with the
line G_1G_2 is the desired point G.

The next example is a uniform plate in the shape of a sector of a
circle of central angle 2α (Fig. 39). First we subdivide the sector into
many very small pie-shaped pieces, so small that the base of each little

triangle, being infinitesimally short, is also straight. The center of gravity of each little triangle lies at distance $r/3$ from the periphery, and thus the center of gravity of the sector plate of radius r is the same as the center of gravity of a heavy circular arc of uniform weight and of radius $r_1 = 2r/3$. Let φ be the variable angle measured from the center, and consider an element of arc of length $r_1\,d\varphi$. Its moment arm for horizontal forces about O is $r_1 \cos \varphi$, and thus by formula (2) (page 33) the distance of the center of gravity from O is

$$r_G = \frac{\int_{-\alpha}^{\alpha} r_1 \cos \varphi r_1 \, d\varphi}{\int_{-\alpha}^{\alpha} r_1 \, d\varphi} = \frac{2r_1^2 \int_{0}^{\alpha} \cos \varphi \, d\varphi}{2r_1 \int_{0}^{\alpha} d\varphi}$$

$$= r_1 \frac{\sin \varphi}{\varphi} \bigg|_0^{\alpha} = r_1 \frac{\sin \alpha}{\alpha} = \frac{2}{3} r \frac{\sin \alpha}{\alpha} = r_G$$

Besides being at this distance from O, the point r_G, of course, also lies on the center line of the figure, this being an axis of symmetry.

Let us check our last result against the few simplified cases we already know. For $\alpha = 0$, the figure reduces to a thin triangle, and the distance becomes $2r/3$ as it should. This is so because although $\sin \alpha/\alpha$ reduces to the form $0/0$, we can develop $\sin \alpha$ into a power series $= \alpha - \alpha^3/6 \cdots$,

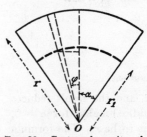

FIG. 39. Center of gravity of a sector of a circle.

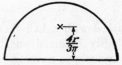

FIG. 40. Center of gravity of a uniform flat semicircle.

which for small α reduces to α so that the ratio $\sin \alpha/\alpha$ becomes unity for small α. On the other hand, for $\alpha = 180$ deg, the figure becomes a full circle, and the formula gives zero for the distance from the center, which it should.

A particular case of some interest is the semicircular plate (Fig. 40), for which the above formula shows the radius of the center of gravity to be

$$\frac{2}{3} r \frac{\sin \pi/2}{\pi/2} = \frac{2}{3} r \frac{2}{\pi} = \frac{4}{3\pi} r$$

A segment of a circle is a sector from which a triangle is subtracted, as shown in Fig. 41. G_1 is the center of gravity of the entire sector of area $r^2\alpha$ and weight $Cr^2\alpha$; G_2 is the center of gravity of the triangle of base $2r \sin \alpha$ and height $r \cos \alpha$; G_3 is the center of gravity of the shaded segment. We now write that the moment about O of the sector is the sum of the moments of the segment and triangle:

$$Cr^2\alpha \frac{2}{3} r \frac{\sin \alpha}{\alpha} = Cr^2(\alpha - \sin \alpha \cos \alpha)x + Cr^2 \sin \alpha \cos \alpha \frac{2}{3} r \cos \alpha$$

Again the weight per unit area, C, drops out, and we can solve for x with the result

$$x = r_{G_3} = \frac{2}{3} r \frac{\sin^3 \alpha}{\alpha - \sin \alpha \cos \alpha}$$

For the special case $\alpha = 180$ deg, the full circle, this distance is easily seen to be zero, as it should. Also for $\alpha = 90°$, the half-circular seg-

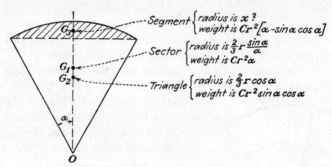

Fig. 41. Center of gravity of a segment.

ment is identical with a half-circular sector, and the result reduces to the previously obtained one, as it should. Finally, for $\alpha = 0$, the segment reduces to nothing, and x ought to be r. The formula goes to the form $0/0$ in a complicated way. The reader should check that everything is all right by developing $\sin \alpha$ and $\cos \alpha$ into power series, retaining terms up to and including α^3, and working out the value of $0/0$ in that way.

Although all problems treated so far have been in two dimensions only, many three-dimensional bodies can be reduced to flat plates or even heavy curved lines by elementary considerations. For instance, a solid rectangular block has three planes of symmetry intersecting in one point, which is the center of gravity. A circular rod can be cut up into thin circular disks with the center of gravity in the center of each disk, and thus the rod is reduced to a heavy center line with its center of gravity in the middle of that line. The same holds for rods of cross sections other than a circle.

Consider the solid cone of Fig. 42, and cut it into circular slices, each with its center of gravity in the center. Thus the cone is reduced to a heavy center line of varying weight density, similar to the flat triangle of Fig. 37. Only now the radius of the circle is proportional to x, and the area of the circle to x^2, so that the weight per unit length of the rod is Cx^2, where again we do not bother to be precise about C. Then the weight of an element dx is $Cx^2\,dx$, and its moment about the vertex O is $Cx^3\,dx$, so that

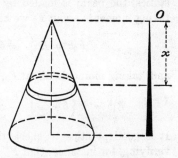

$$x_G = \frac{\int_0^h Cx^3\,dx}{\int_0^h Cx^2\,dx} = \frac{h^4/4}{h^3/3} = \frac{3}{4}\,h$$

The center of gravity of a solid cone is at one-quarter height above the base. This result is true for cones or pyramids of any other cross section as well, by exactly the same derivation. Truncated cones or pyramids are treated as the complete object less the truncated top, just as was the segment of the circle.

Fig. 42. The center of gravity of a solid cone is at one-quarter height above the base.

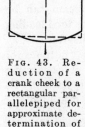

Fig. 43. Reduction of a crank cheek to a rectangular parallelepiped for approximate determination of of the center of gravity.

A half sphere or a spherical segment can be cut up into circular disks, then replaced by a heavy center line of variable weight, and integrated along the length of that center line.

Most objects in practical machine design consist of combinations of pieces of plate, cylinder, parallelepiped, cone, and the like, or can be approximated by these. For example, Fig. 43 represents a crank cheek with curved edges. To find the center of gravity we take the blueprint and sketch in by eye the dotted lines, cutting off and adding to the cheek approximately equal areas, *i.e.*, equal weights. Then the cheek becomes a parallelepiped of which the gravity location is known. In performing computations of this character, it is useful to estimate the possible error by a quick calculation (see Problem 62).

Problems 56 to 67.

11. Distributed Loadings. Consider the beam of Fig. 44 on two supports, loaded by a uniformly distributed load of q lb per running inch, and weighing itself w_1 lb per running inch. What are the reactions on the supports? The distributed loading $w + q$ per unit length has a resultant $R = (w_1 + q)l$, located in the center of the beam, which is at distance $l/2$ from the right support. Thus, for purposes of statics, the beam is loaded by three forces only: R, F_1, and F_2, and taking moments about F_1 we have

$$F_2 b = R\,\frac{l}{2} \quad\text{or}\quad F_2 = \frac{l^2}{2b}\,(w_1 + q)$$

and, taking moments about F_2,

$$F_1 b = R\left(b - \frac{l}{2}\right) \quad\text{or}\quad F_1 = \left(l - \frac{l^2}{2b}\right)(w_1 + q)$$

It is good practice to check this result by some other method, by verifying, for instance, that $F_1 + F_2 = R$, which is seen to be correct. Thus we have the general rule that

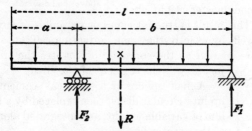

FIG. 44.   Uniformly loaded beam.

A distributed loading of parallel forces along a beam can be replaced for purposes of statics by a resultant equal to the total loading and passing through the center of gravity of the loading diagram.

We note that the above statement is restricted to "purposes of statics." This is to emphasize that for the purpose of calculating the stresses in the beam, the distributed load cannot be replaced by its resultant. The reader is referred to the argument in the middle of page 79.

In Fig. 45 let the weight of the beam itself be negligible, and let the loading grow linearly from nothing to 500 lb/ft as shown. The loading can be replaced by a resultant of

$$\tfrac{1}{2} \times 500 \text{ lb/ft} \times 12 \text{ ft} = 3,000 \text{ lb,}$$

located at the center of gravity 4 ft from the right support. Thus the right reaction is 2,000 lb, and the left reaction is 1,000 lb.

Now we turn to the more complicated application shown in Fig.

46a, which is a stiff (unbendable) beam of weight w_1 lb per running inch, loaded by a single eccentric load of P lb and supported by the ground along its entire length. The ground has more or less the property that it sags down locally in proportion to the load it supports locally. This is expressed by a constant k, measured in pounds per running inch loading on the ground per inch deflection of the ground, or lb/in./in., or lb/in.^{2}. The ground does not quite do this exactly, but it is a sufficiently approximate description of the facts

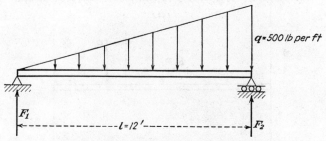

FIG. 45. Beam with triangular loading diagram.

to be useful. The question is: What are the deflections δ_1 and δ_2 at the two ends of the beam under the loading P and w_1?

To solve this, we first combine the two parts of the downward loading. The resultant weight is $w_1 l$, and it acts in the center of gravity of the weight loading diagram, *i.e.*, the center of the beam. That load combined with P gives a single resultant R, equal to $P + w_1 l$ and located between P and $w_1 l$ at distance c from the right, where

$$c = b + \frac{w_1 l}{P + w_1 l}\left(a - \frac{l}{2}\right)$$

This result is found by application of moments as in Fig. 8, and it is to be verified by the reader. In order not to obscure the further analysis with complicated algebra, we will from now on use the symbols R and c, instead of P, a, b, and w_1. As is shown in Fig. 49c, the beam will sag into the ground as a straight line because it is unbendable, and the reaction from the ground is proportional to the local deflection. Thus the reaction load diagram is represented by the trapezoidal figure $ABCD$, and, for equilibrium, it must be so constituted as to have a resultant equal and opposite to the load R. The intensity of the (upward) ground reaction loading at the left end is $k\delta_1$ lb/in. (k has the dimension lb/in./in., and δ_1 is expressed in inches, so that the product is pounds per inch). To find the total

reaction upward we need the center of gravity of the trapezoid. That
we find by subdividing it into a rectangle δ_1 and a triangle $\delta_2 - \delta_1$, as
in Fig. 46d. The reaction Q_1 due to the rectangular part is $k\delta_1 l$,
and the reaction Q_2 from the triangle is $\frac{1}{2}k(\delta_2 - \delta_1)l$. The com-

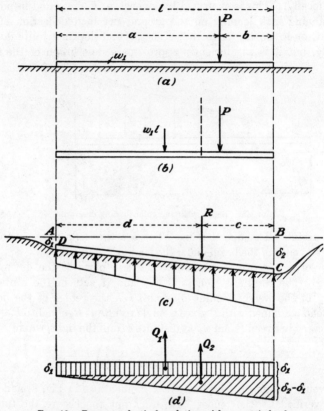

FIG. 46. Beam on elastic foundation with eccentric load.

bination of Q_1 and Q_2 must be statically equivalent to R, so that we
write, taking moments about the right-hand end,

$$Q_1 + Q_2 = k\delta_1 l + \frac{1}{2}k(\delta_2 - \delta_1)l = R$$

$$Q_1 \frac{l}{2} + Q_2 \frac{l}{3} = k\delta_1 \frac{l^2}{2} + k(\delta_2 - \delta_1)\frac{l^2}{6} = Rc$$

These are two algebraic equations in the unknowns δ_1 and δ_2 and
should be solved by the reader with the result

$$\delta_1 = \frac{R}{kl}\left(6\,\frac{c}{l} - 2\right)$$

$$\delta_2 = \frac{R}{kl}\left(4 - 6\,\frac{c}{l}\right)$$

which is the solution to the problem. When a solution is found, we should *always do two things: first, check for dimensions, and second, check for reduction to simplified special cases.*

The left-hand sides of the solution are deflections in inches; therefore, the right-hand sides must be in inches also; they could not possibly be pounds. R is pounds, k is pounds per square inch, l is inches, and the parentheses are pure numbers. Thus we verify that the right-hand side is also expressed in inches, as it should be.

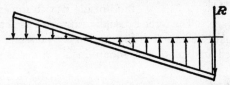

Fig. 47. A beam on elastic foundation will lift out of the ground if the load is too far from the center.

Now suppose R to be in the center of the beam: $c = l/2$. Then we see that $\delta_1 = \delta_2 = R/kl$ as it should. Next place R at the right end of the beam: $c = 0$. The formula shows that δ_1 becomes negative and half as large as δ_2. This means (Fig. 47) that instead of pushing up, the ground pulls down on the left-hand end, because our definition of k specified a ground loading proportional to the deflection. The left-hand end must be pulled down, because the right-hand end to the left of the load R pushes up, and without a downward pull at the left extremity, would have a resultant to the left of R, which could never be in line with R. The actual ground hardly ever pulls down on a beam, so that this assumption for k breaks down when the load is placed too far eccentrically.

Problems 68 *to* 73.

12. Hydrostatics. This is the name given to the subject of the statical equilibrium of objects in connection with water or other fluids. A fundamental property of a fluid at rest is that it can sustain or transmit only compressive forces, while it is incapable of taking tensile forces or shear forces. This means that if a piston in a cylinder containing a fluid is pressed into the cylinder, the forces exerted by the fluid are perpendicular to the surfaces of the piston and the cylinder walls, as shown in Fig. 48a. The condition of Fig. 48b, where the force between the fluid and the wall is oblique, thus having a normal or "pressure" component as well as a tangential or "shear" component, occurs only when the fluid is moving at some speed along the wall.   For fluids at rest, in statics problems, only the normal or

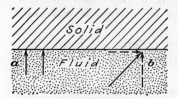

pressure component exists.   This is true not only between the fluid and a solid wall but also between two adjacent particles of the fluid itself. Imagine in Fig. 49 a certain portion of the fluid in the cylinder, bounded by an arbitrary surface.   Then the fluid outside that surface presses on the fluid inside it with forces perpendicu-

FIG. 48.   Distinction between pressure and shear in a fluid.

lar to that surface.   The force is usually expressed as acting on a unit area, say a square inch, and is then called a "pressure," denoted by p and measured in pounds per square inch.   Force is thus pressure times area.

Fluid pressures in many practical applications, such as hydraulic presses, are so great that the weight forces are negligible with respect to them.   Let us, therefore, consider first a weightless fluid.   Then the statement can be made that in a weightless fluid at rest, the pressure is everywhere the same.   To understand this, imagine a small cube of the fluid, and consider the equilibrium of this cube in a direction perpendicular to two of its six faces and hence parallel to the other four faces.   The pressures on the other four faces are perpendicular to those faces and hence have no component in the direction we are investigating. The only forces in that direction are caused by the pressures on the first two faces.   For equilibrium the forces must be equal and opposite, and hence the

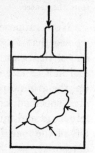

FIG. 49.   Fluid pressure on an arbitrarily bounded surface inside the fluid.

pressures must be equal on the two opposite faces of the cube.   A larger volume of fluid can be thought of as subdivided into a large number of small cubes, and by proceeding from cube to cube, passing from one to the next by means of the first axiom of page 4, we can show that the pressure is everywhere the same.

This holds true for all possible directions at one point as well. To see this, consider in Fig. 50a an infinitesimally small triangular prism of angle α and of unit thickness perpendicular to the drawing.

Imagine that the pressures on the three faces of this prism are not necessarily equal but are expressed by p_1, p_2, and p_3. The forces on the prism are then these pressures multiplied by the areas as shown. These three forces must be in equilibrium, and hence their force triangle (Fig. 50b) must be a closed figure. The directions of the

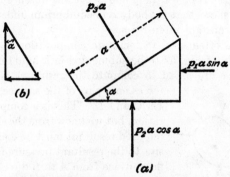

Fig. 50. Equilibrium of fluid pressure forces on a triangular prism.

forces are known; they are perpendicular to their respective faces. Then the lengths of the three forces in Fig. 50b must be in ratio of $1 : \sin \alpha : \cos \alpha$, which can be only if $p_1 = p_2 = p_3$.

Returning to Fig. 49, and considering that the arbitrarily bounded figure in the center obviously is in equilibrium, we can now imagine the interior of that figure to be frozen or solidified and state that

The resultant of all pressure forces exerted by a weightless fluid (at rest) on an arbitrary, completely immersed rigid body is zero.

To see that this statement is compatible with the one of the constancy of the pressure throughout the fluid, consider Fig. 51, and, since we have limited ourselves to two-dimensional or plane cases only, let the body of Fig. 51 be a cylindrical one, bounded

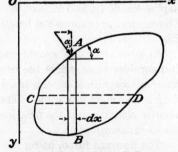

Fig. 51. The resultant of all pressure forces is zero on a body immersed in a fluid of equal pressure everywhere.

by straight lines in the z direction perpendicular to the paper. Let us imagine a strip of width dx cut out between A and B, and let us look at point A. For unit width in the z direction, the area of the strip at A is $dx/\cos \alpha$, and the force is $p \, dx/\cos \alpha$. Resolve this force into x and y components. The x component does not interest us, but we note that the y component is $p\,dx/\cos\alpha$ multiplied by $\cos\alpha$ giving $p\,dx$, a result independent of the angle α. Repeat the argument at B, and note that the y components at A and B cancel each other. This will be true for all other vertical strips of width dx, and thus the entire body will have no y resultant force. We then repeat the argument for horizontal strips CD of width dy and prove the absence of the x component. This shows that the body is in equilibrium under the hydraulic pressure of constant intensity p.

This fact is often used to facilitate computations. Suppose the cylindrical body of Fig. 52 to be submerged in a liquid of pressure p, and suppose we are asked to compute the resultant of all pressure

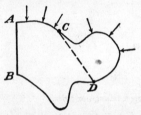

forces on a part of the surface, from A to B via C and D. This is a complicated integration, but we know from the above that the required resultant must be equal and opposite to the resultant pressure force over the flat surface from A to B directly, for which the answer can be written down immediately. Even if no such simple piece of straight boundary as AB actually exists, the reasoning can be used. If we ask for the resultant force on the surface from C to D along the right-hand side, we imagine the

Fig. 52. The resultant of the pressures on a curved contour is the same as the resultant on the straight line joining the ends, for a body in a fluid of the same pressure everywhere.

body cut along the straight line CD, unfreeze or liquefy the left-hand part of the body, and answer the question for the right-hand part only. The answer is p times the length CD in a direction perpendicular to CD.

Now we drop the simplifying assumption of the weightless fluid and consider cases where the pressure forces are smaller, of the same order of magnitude as the weight forces, so that the weight of the fluid is no longer negligible. The main problem here is that of *buoyancy* or *flotation*, and the principal **theorem** is that **of Archimedes** (the $\epsilon\upsilon\rho\eta\kappa\alpha$ theorem):

The buoyant force, being the resultant of all fluid pressure forces on a body partially or totally submerged, is equal and opposite to the weight of the fluid displaced by that body and hence passes through the center of gravity of the displaced fluid.

Two proofs of this important proposition will be given. The first and best proof is taken from a statics text by Simon Stevin, printed in the year 1585. The proof is shown in Fig. 53. The first figure a is the ocean at rest, obviously in equilibrium. In figure b an imaginary surface is drawn in the ocean. In figure *c* the water inside the surface is imagined to be frozen into a solid kind of ice, which has the same weight as water. The ocean is still unaware that anything has happened. In figure *d* the ice is replaced by a thin steel shell of the same shape with a weight in it, so placed that the center of gravity has not changed. In *e* the weight is removed and replaced by other material,

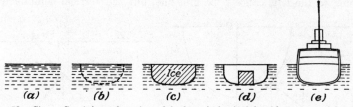

(a)   *(b)*   *(c)*   *(d)*   *(e)*

FIG. 53. Simon Stevin's explanation of Archimedes' principle of buoyancy of floating bodies.

leaving the total weight and the center of gravity undisturbed. In none of these steps has the previously existing equilibrium been disturbed in any way. We end up, in Fig. 53*e*, with a ship on which the downward gravity force is its weight, and the upward buoyancy force is the same as that on the body 53*b*, which sustains the displaced water. Thus Archimedes' theorem is proved.

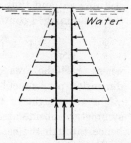

FIG. 54. Pressure distribution in still water under the influence of gravity.

For the second proof of Archimedes' law, we first deduce by means of Fig. 54 that the pressure of the water at a depth *h* below the surface is γh, where γ is the weight of water per unit volume.

The value of γ is 62.4 lb/cu ft, and thus the pressure at 33-ft depth, for instance, is $62.4 \times 33 = 2{,}060$ lb/sq ft = 14.3 lb/sq in., roughly 1 atmosphere.

Consider in Fig. 54 an imaginary cylinder of cross-sectional area *A* and depth *h*. This cylinder has a volume *hA* and a weight $\gamma h A$, which is kept in vertical equilibrium by the pressure force on the bottom, *pA*. Thus $p = \gamma h$, which is seen to increase linearly with the depth, as indicated on the sides of the column of Fig. 54.

Now consider the submerged part of a symmetrical and cylindrical ship's hull (Fig. 55). Take an element *ds* at a depth *x* below the water surface. Considering a slice of unit length perpendicular to the paper, the pressure force $pA = \gamma x\,ds$, perpendicular to *ds*. The line element *ds* has components *dx* and *dy*, and the total pressure force can be resolved into horizontal dH (to the right) and vertical (upward) dV components, which can be seen to be

$$dH = \gamma x \, dx \quad \text{and} \quad dV = \gamma x \, dy$$

Now consider the symmetrical line element on the other side of the hull at the same depth x.   Its horizontal component of pressure force, being to the left, cancels the first one, but the vertical component is

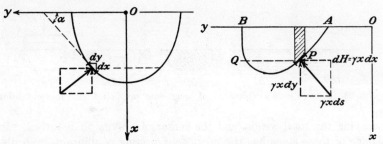

FIG. 55.   Toward the proof of the law of Archimedes.

FIG. 56.   Archimedes' law applied to a non-symmetrical ship.

still upward and thus adds.   Integrating the vertical force along the entire submerged contour of the hull, we have

$$V = \int dV = \int \gamma x \, dy = \gamma \int x \, dy = \gamma A$$

where A is the area of the ship's cross section under water.   The dimension perpendicular to the paper was taken to be unity, so that γA is the weight of the submerged water.   The resultant, from a symmetry consideration, passes through the vertical center line and hence through the center of gravity.

For a non-symmetrical submerged ship (Fig. 56), the argument is the same up to the two components of the pressure force dH and dV.   The total horizontal resultant is

$$H = \int dH = \gamma \int x \, dx$$

integrated from A to B or from $x = 0$ to $x = 0$, which is $\gamma x^2/2 \big|_0^0 = 0$. The vertical resultant is

$$V = \int dV = \int \gamma x \, dy = \gamma \int d \text{ area} = \int dW = W,$$

the weight of the displaced water.

In order to find the location of this vertical resultant W, we take moments about O

$$d \text{ moment} = y \, dV - x \, dH$$

First we integrate the second term

$$\int x\, dH = \gamma \int x^2\, dx = \frac{\gamma x^3}{3}$$

which between limits 0 and 0 equals zero. This is so because the horizontal components at points P and Q are equal and opposite, and their moments cancel. Then we integrate the first term

Moment $= \int y\, dV = \int \gamma xy\, dy = \int y(\gamma x\, dy) = \int y(\gamma\, d\text{ area})$
$$= \int y\, d\text{ weight} = y_G \cdot \text{weight}$$

Thus the resultant passes through the center of gravity of the displaced water.

Problems occurring in practice, which can be solved by the methods just discussed, are shown in Figs. 57 and 58, illustrating submerged

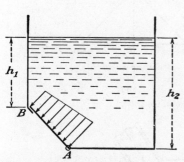

Fig. 57. Pressure distribution on a flat submerged gate.   Fig. 58. Pressures on a curved submerged gate.

sluice gates, hinged at one end A and freely supported without friction on the other end B. It is usually required to determine the load on the gate and the reactions at the hinge A and the support B. The case of the flat gate is in the same category as the beams of Figs. 44 and 45, but the arched gate of Fig. 58 is somewhat more complicated. It has to be solved by considering the force on an element of the gate and resolving it into horizontal and vertical components. Each of these then must be integrated. The total resultant must pass through the center C of the circle, because all elemental forces are radially directed. Then this resultant of all pressures has to be carried by the two reactions at A and B, both of which must be radial, through C. This is so because the reaction at B, being a frictionless support, must be perpendicular to the support, *i.e.*, radial through C. Then, since the B reaction and the total pressure force pass through C, so must their difference, the reaction through A.

Problems 74 *to* 80.

CHAPTER IV

TRUSSES AND CABLES

13. Method of Sections. A truss is a structure frequently used in roofs, bridges (Fig. 59), and hoisting cranes, consisting of a combination of slender bars, joined together at their ends so as to form a rigid unit. The bars are attached to each other at their ends usually by a riveted construction with "gusset plates," as in Fig. 60, which

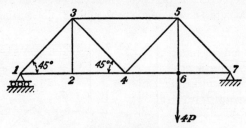

FIG. 59.   Simple bridge truss.

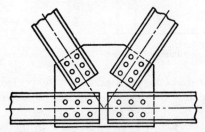

FIG. 60.   Detail of a "joint" with members riveted to a gusset plate.

represents a detail of joint No. 4 in the truss of Fig. 59.   It is seen that the center lines of the various bars of the joint all intersect in one point, which is supposed to be the case.   As a first approximation, we replace the riveted joint of Fig. 60 by a hinged joint, and also we assume the truss to be loaded at the hinge points or joints only, so that no forces are made to apply along the bars between joints. This is usual practice in these constructions, and the weight forces of the bars themselves are usually negligibly small in comparison to the loads applied at the joints.

Thus, all bars are supposedly hinged at both ends and carry no lateral loads along their lengths, so that by page 23 they are not subject to bending but transmit only longitudinal forces, either tensile or compressive. The calculation of these forces is the subject of this chapter. The stresses caused by these forces are known as the "primary" stresses of the structure. In addition, there will be some bending in the bars, because Fig. 60 is not a hinge joint, and these bending stresses are called "secondary" stresses. Because the bars are usually made very long in comparison to their width, these secondary stresses are much smaller than the primary ones; hence the assumption of hinged joints is well justified, because it leads to an answer that is close to the truth with one-tenth or less of the work required for the more exact solution.

In the first sentence of this chapter, a truss was said to be made "so as to form a rigid unit." This, of course, is not true of all combinations of hinged bars. For example, consider four bars of equal lengths hinged together to form a square. This square can be deformed in its own plane into a diamond or even into a double straight line without any resistance. On the other hand, three bars, pinned together to form a triangle, are a rigid structure.

The simplest way to construct a truss, *i.e.*, a rigid plane bar structure, is by starting with a triangle, like 1-2-3 of Fig. 59, and then going to a fourth point 4 by two bars, each starting from one of the vertices of the triangle. Each following joint is then rigidly attached to the structure by two bars starting from two previous joints (Fig. 59). Obviously in this manner we arrive at a rigid structure, but, although most trusses in practice are made in this manner, it is not the only way in which a rigid plane truss can be made. For further details on this subject, the reader is referred to Timoshenko and Young's "Engineering Mechanics." From the manner of building up a truss we can deduce a simple relation between the number of bars, b, and the number of joints, j. For each one joint added to the structure, two new bars appear. Thus $b = 2j + $ constant. The simplest truss is a triangle with $b = j = 3$. Thus we have

$$b = 2j - 3$$

a formula the reader should check on Fig. 59.

One more remark on trusses is useful before we start calculating the forces in the bars. Imagine that in Fig. 59 one more bar is inserted, say between joints 2 and 5, and imagine that the truss be without loads when this is done. Since the truss is rigid, it is clear that the new bar has to be made to an exact length in order to fit. If it is a quarter inch too long and it is forcibly inserted, the new bar will be in compression, and some other bars will be in tension, all of this without any external loads. Even if the new bar were just right in length and could be fitted in without force in the unloaded condition, the application of load would place us before a problem similar to that of the four-legged table of page 29: the problem is said to be "statically indeterminate," and the new bar 2-5 is said to be "redundant."

In what follows in this chapter we will consider only statically determinate trusses, *i.e.*, trusses without redundant members.

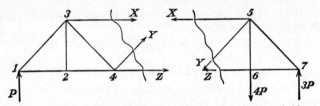

FIG. 61. The method of sections for finding bar forces applied to a bridge truss.

In order to calculate the forces in the individual bars, we first determine the reactions at the supports. The truss, being a rigid body in its own plane, can be treated as such and the reactions determined as if it were a simple beam. In Fig. 59, for example, the reactions to the single load $4P$ at joint 6 are $3P$ up at support 7 and P up at support 1. After the reactions have been determined, we have a choice between two different methods for the further procedure:

a. The *method of joints* is useful when we want to know the forces in all the bars. It will be discussed in the next article.

b. The *method of sections* leads to a quicker result if we want to know only the force in one arbitrarily selected bar. We now proceed with this method.

Suppose we want to know the force in the top member 3-5 of Fig. 59. Then we make a section through that bar and further across the entire truss, as in Fig. 61, replacing the cut bars by their unknown forces X, Y, and Z, which are shown in the figure as tensions. If, after the calculation, we find positive answers for X, Y, and Z, then they are really tensions, but in case of a negative X, the force is a negative tension or a compression. Now we can consider either the left half or the right half of the truss and note that either half is in equilibrium. Of course we take the simplest half, which is the left one, because there is only *one* force P acting on it, instead of two on the right. Then the three unknowns X, Y, and Z can be calculated from the three equilibrium conditions, by any of the methods of page 19. For the specific case that only the force X is wanted, while we do not care about Y or Z, we **write the moment equation about the intersection point of the two unwanted forces,** in this case about joint 4. Then the unwanted forces, having no moments about that point, do not appear in the equation. In this case only the reaction P and the force X have moments, the latter having half the moment arm of the former. Hence $X = -2P$, or the top bar is in compression with force $2P$. Similarly, if only Z in the bottom bar is wanted, we take moments about the intersection of X and Y, which is point 5. From that we conclude that $Z = 3P$ in tension. Finally, if only Y is wanted, the intersection of X and Z is at infinity, and the moment equation reduces to a statement that the vertical component of Y must balance the vertical component of P. Thus the vertical component of Y is P downward, and there must be sufficient

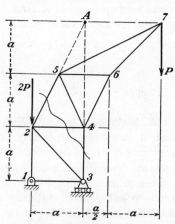

FIG. 62. The method of sections applied to a crane structure.

horizontal component to give Y the proper 45-deg direction. Thus we conclude that $Y = P\sqrt{2}$ in compression.

All of these results could have been obtained from the equilibrium of the right-hand part of Fig. 61 as well, and the reader should do this in detail.

It is noted that the section of Fig. 61 shows three cuts with three unknown forces and that there are also three equilibrium conditions for their calculation. This is no accident but is in consequence of the fact that the truss is a statically determinate one. In case there were an additional (redundant) bar 2-5, the section would show *four* unknowns, and it would be impossible to calculate the four forces from the three equations of statics.

Another example of a plane truss is the crane of Fig. 62, with a load P and a counterweight load $2P$ for the purpose of keeping the left wheel at joint 1 on the ground. We proceed from joint to joint in the order indicated to verify that this truss is a rigid one, without redundant members, as discussed on page 53. Now suppose we want to know

the force in bar 2-4. In this case it is not necessary first to calculate
the reactions at 1 and 3 because the section indicated in the figure
leaves an upper portion of the structure on which all forces are known
and in which no unknown reactions appear. This is an unusual case;
most truss cases are like Fig. 59, where each side of a section contains a
reaction, which therefore has to be calculated first. The section in
Fig. 62 passes through three bars; the point of intersection of the two
unwanted bar forces is A, and for the equilibrium of the upper part
we write moments about A.

$$F_{24} \cdot A4 + P \cdot A7 = 0 \qquad \text{or} \qquad F_{24} \cdot 2a = -P \cdot \tfrac{3}{2}a$$

or

$$F_{24} = -\tfrac{3}{4}P \qquad \text{(compression)}$$

Problems 81 *to* 85.

14. Method of Joints. With this method the equilibrium of each
joint is considered separately and consecutively. The joint is isolated as a body, *i.e.*, the joint itself and as many short bar stubs as
there are bars entering the joint. Thus the joint is subjected to a
number of intersecting forces, usually three or four (Fig. 59). It is
always possible (in a statically determinate truss) to find a joint in
which not more than two of the forces are unknown, and these two
can then be determined by a polygon construction. With two new
bar forces known after this construction, we can proceed to a contigu-
ous joint in which only two forces are unknown and so on throughout
the entire structure.

Figure 63 shows this for the bridge truss of Fig. 59. We start with
joint 1 on which are acting the known upward reaction P and the two
unknown bar forces 1-2 and 1-3. The triangle of forces of joint 1 is
started from the little circle by laying off P upward and then returning
to the little circle by two lines parallel to the bars 1-2 and 1-3. This is
possible only in the manner shown. The directions of the forces in
the diagram are as they act *on* the joint, the reaction force P pushing
up, the force 1-2 pulling to the right, and the force 1-3 pushing down
at 45 deg. Hence bar 1-2 is in tension, marked $+$, and bar 1-3 is in
compression, marked $-$, and of magnitude $P\sqrt{2}$. The next joint
to work on is either 2 or 3. We are not ready to go to joint 3, because
of the four bars entering that joint, three carry forces that are still
unknown, and the polygon can be constructed only with two unknown
forces. But in joint 2 only two forces are unknown: 2-3 and 2-4.
The force diagram for that joint reduces to a double line; the joint is
in equilibrium only if force 2-3 is zero and if force 2-4 is equal and

opposite to force 1-2. (If force 2-3 existed, it could not be balanced by 1-2 or 2-4, because neither of those has a vertical component.) Thus bar 2-4, like bar 1-2, is in tension with force P. Next we proceed to joint 3, in which two of the four forces are known. We start at the little circle in Fig. 63 and lay off force 1-3 = $P \sqrt{2}$, pushing up on the joint because the bar 1-3 is in compression. We add to this

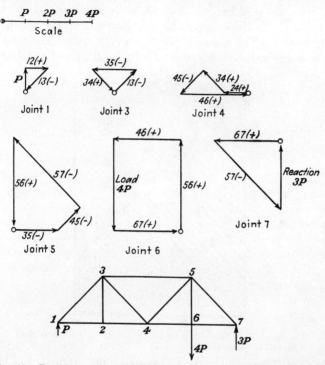

FIG. 63.   Equilibrium force diagrams at the various joints of a bridge truss.

the zero of bar 2-3 and now complete the polygon by returning to the little-circle origin in the two directions of bars 3-4 and 3-5. Again the directions shown are those *on* the joint; hence 3-4 is in tension and 3-5 in compression.

Next we proceed to joint 4, in which nothing new occurs, except possibly that the force vectors of bars 2-4 and 4-6 partly overlap, so that force in bar 4-6 equals $3P$ in tension. From now on we have a choice. We can proceed either to joint 5 or to joint 6. We do 5 first, 6 next, and end up with joint 7, on which all forces are known prior to the construction, so that the triangle diagram of joint 7

serves as a check only.   That check is a very potent one, because any
error in any of the previous polygons would have made the end reaction
force different from the correct value $3P$ up.

This is the method of joints, and it is seen to involve quite an
amount of work.   The number of lines to be drawn can be reduced
by 50 per cent when we remark that in Fig. 63 every force is drawn
twice, because every bar has two ends at two different joints.   Thus
the idea suggests itself to put all the diagrams of Fig. 63 into one
single figure in which each force is drawn only once.   The method of
doing this was invented independently by Luigi Cremona in Italy and

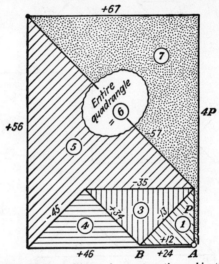

FIG. 64.   The Maxwell-Cremona diagram is a systematic combination of all the force
diagrams of Fig. 63 into a single figure.

by Clerk Maxwell in England, and the resulting diagram is known as
the **Maxwell-Cremona diagram.**   For its construction we have to be
neat and systematic.   **It is necessary to take the various bars in
each joint in the sequence in which they appear when proceeding
around a joint and to adopt the same sense of rotation for all the joints.**

An inspection of Fig. 63 shows that this procedure has not been
followed there; for example, joint 1 has been run around in a counterclockwise direction, and joint 3 has been run around in a clockwise
direction, while in joint 5, we started counterclockwise and then
skipped bar 5-6 until later.

Figure 64 shows the Maxwell-Cremona diagram for Figs. 59 and 63,
starting with joint 1 at the lower right-hand corner at A and proceeding

clockwise around each joint. The diagram for joint 1 starts at A, first by laying off the vertical bearing reaction P, then the force in bar 1-3, finally the force in bar 1-2, ending at the same point A. The polygon of joint 2 reduces to the line AB doubled up on itself, as in Fig. 63. Next we go to joint 3, in which forces 3-4 and 3-5 are unknown. Because we have to proceed around 3 in a clockwise direction and have to end up with the unknown bars, we are obliged to start with bar 2-3, which is at point B in Fig. 64. From B we lay off the zero length for bar 2-3, then follow the force 1-3, already drawn, strike out new with 3-5, and return home to B in the direction of bar 3-4. Proceeding to joint 4, we must start with bar 2-4, which means point A in Fig. 64. We follow 2-4 and 3-4, already drawn, and strike out anew with bars 4-5 and 4-6, ending up at the starting point A. Thus we see that force 4-6 is $3P$ in tension. In this manner we proceed further until we arrive at joint 7, of which we find all forces already drawn, constituting the check on the correctness of our construction.

We have drawn no arrows in the various bars as we did in Fig. 63, because each line in Fig. 64 is considered twice, at different joints. Thus each force in Fig. 64 would have two opposite arrow marks on it, which serves no purpose. The designation $+$ for tension and $-$ for compression has been retained. Each joint is represented in the figure by the area of a polygon, which has been crosshatched and marked by the number of the joint encircled. The reader should construct this diagram on paper line by line and understand every detail of it before attempting to work problems.

Problems 86 *to* 92.

15. Funicular Polygons. Figure 2 (page 6) shows the shape assumed by a flexible cable, supported between two points P_1 and P_2 and subjected to a single load W_3. In this chapter we propose to generalize that simple case and load the rope, first with several loads and ultimately with an infinite number of loads, or a "continguous load distribution."

Before we do that, we look once more at Fig. 2 and note that in the two sections of rope AP_1 and AP_2 the tensions F_1 and F_2 are different but that their horizontal components are the same, because the horizontal equilibrium of the isolated point A requires it. In general, **the horizontal component of the tension in a rope that is subjected only to vertical loads is constant along the length of the rope.**

Now consider in Fig. 65a three parallel forces of different magnitudes and at unequal distances from each other, and let these forces act on a flexible rope. What shape will the rope take? The question is indefinite and vague, because we have not specified at which places the rope is to be fastened to a solid support, nor have we specified how much longer the rope is than the distance between supports. The question ought rather to be: What possible shapes can a rope assume under the three forces? There exists a beautiful graphical solution to this problem, which is shown in Figs. 65b and c. First in Fig. 65b, we lay off the loads one after the other, hitching the tail end of each following load to the nose of the previous one. Then we take an arbitrary point O anywhere in the paper but outside of the force line

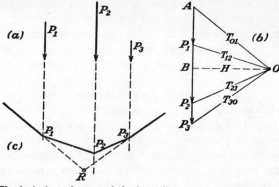

FIG. 65. The funicular polygon and the force diagram for parallel forces acting on a rope.

$AP_1P_2P_3$. Next we connect point O with the junction points and end points of all the forces, which practically completes Fig. 65b. Now we proceed to Fig. 65c and draw the four sections of heavy line parallel to the four rays through O in 65b, in such a way that the line OP_1 in 65b which connects O with the point joining forces P_1 and P_2, finds its parallel in 65c between the lines of action of the forces P_1 and P_2. Similarly, line P_2P_3 in 65c is drawn parallel to line OP_2 of 65b, which is the ray connecting O with the junction of forces P_2 and P_3. The ray OA in 65b running to the tail end of force P_1 is parallel to the line in 65c to the left of P_1, indefinite at the left-hand end.

We claim that 65c is a possible shape of the rope and that every line in 65b represents a definite force. Consider, for example, the triangle OP_1P_2 in 65b, made up of the force P_2 and the two rays from O to the end points of P_2. This triangle can be interpreted as the triangle of forces for the point in 65c where force P_2 rests on the rope. The rays OP_1 and OP_2 are parallel to the sections of rope, and if OP_1

represents the tension T_{12} in the rope between P_1 and P_2, and if further OP_2 in 65b represents the tension T_{23} in the section of rope between P_2 and P_3, then the point P_2 of the rope, as an isolated body, is in equilibrium under the three forces P_2, T_{12}, and T_{23}. Similarly, triangle OAP_1 in 65b is the triangle of forces for point P_1 in 65c, and triangle OP_2P_3 in 65b refers to point P_3 in 65c.

The scale of Fig. 65c is in inches or feet, and the scale of Fig. 65b is in pounds, tons, or kips (1 kip is a kilopound or 1,000 lb).

The dotted horizontal line OB in 65b, also measured in pounds, represents the horizontal component of tension in all sections of rope. For example, the force in the rope between P_1 and P_2 is T_{12} in Fig. 65b, which consists of a horizontal force OB and a vertical force BP_1.

Let us look again at 65c and note that the entire rope sustains five forces: P_1, P_2, P_3, and the two end tensile forces T_{01} and T_{30}. Obviously the resultant of P_1, P_2, and P_3 is downward and has the magnitude AP_3 in 65b. Similarly, the resultant of T_{01} and T_{30} is upward of equal magnitude P_3A. In Fig. 65c the resultant of the two end tensions must pass through their point of intersection R. Then also the *resultant of $\mathbf{P_1}$, $\mathbf{P_2}$, and $\mathbf{P_3}$ must pass through $\mathbf{R}$.*

The figure $P_1P_2P_3R$ in 65c is a closed polygon, and since every side is a piece of rope (or its extension), the figure is known as the *funicular polygon* (*funiculus* in Latin means rope). The diagram 65b is called the *force diagram.* The funicular polygon and its accompanying force diagram are of importance and are frequently used not only in connection with cables or ropes but also in beams (page 71) or for the graphical determination of centers of gravity (page 81), in which cases the rope in the funicular diagram is fictitious.

The diagrams are not limited to the case of parallel forces shown in Fig. 65. In Fig. 66 the three forces are not parallel, and the reader is advised to follow all the steps in this construction, if possible without referring to the text again. If a difficulty arises, the entire argument explaining Fig. 65 applies word for word to Fig. 66 with one exception. The difference between Fig. 65 and 66 concerns the horizontal component of rope tension. In Fig. 66 each of the loads P_1, P_2, and P_3 has a horizontal component itself; thus, in writing the horizontal equilibrium of a load point, we conclude that the difference between the horizontal components in the rope tensions to the left and to the right must balance the horizontal component of the applied load P. Only in Fig. 65 is this difference zero and the horizontal component H in the rope a constant. In Fig. 66c the resultant force R of $P_1P_2P_3$ is drawn parallel to AP_3 in the force diagram 66b.

The question asked in the beginning of the discussion of Figs. 65 and 66 was that of determining all the possible shapes the rope could assume under the given forces, and so far we have constructed one shape only. Where are the other possible shapes? It is recalled that the point O, the pole of the force diagram, was chosen arbitrarily. By choosing a point O closer to the forces, we make the horizontal pull H in Fig. 65 smaller; we increase the slopes of all the rays through O and make the funicular polygon sag through considerably more. If the horizontal pull is great, O is far to the right, the slopes of its rays are small, and the funicular shape is flat.

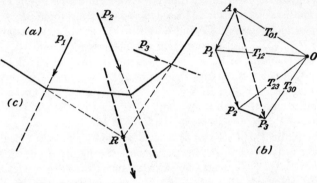

Fig. 66. The funicular diagram for non-parallel forces; construction of the resultant.

On the other hand, if we move point O up, we make the slope OA flat and OP_3 steep, and we get a rope of which the right support is higher than the left support. Conversely, lowering O raises the left support of the rope relative to the right end.

By considering all possible locations of point O in the plane of the force diagram, we find all the possible shapes the rope can assume. It is remarkable that among these many shapes which the rope can take, the resultant point R in 65c or 66c, being the intersection of the free rope ends extended, remains on the line of action of the resultant force.

The most important technical application of cables is the suspension bridge (Fig. 67), in which the bridge deck is suspended from a strong, heavy, main cable by means of many thinner vertical cables, usually equidistant and equally loaded for a uniform load on the bridge. The main cable is supported on the top of two towers and anchored to the ground at the shores.

The force diagram and the funicular diagram for such a case are shown in Fig. 67, and the pole O in the force diagram has been located in the middle vertically, so as to make the bridge symmetrical; also, O has been chosen horizontally so as to make the top of the cable assume a 45-deg angle. With this it is seen that the horizontal pull equals six vertical-cable pulls, and the total tension in the main cable at the top of the towers A and B is $6\sqrt{2}$ times one vertical-cable load. The bridge is usually built with the towers at one-quarter- and three-quarters-span distance, with side spans each a mirrored picture of half the center span. With this arrangement the horizontal

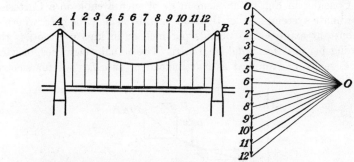

Fig. 67. Funicular and force diagrams for a suspension bridge.

tension components of the main cables at the top of the tower cancel each other, and the towers carry only a vertical compressive force. *Problems 93 to 96.*

16. Uniformly Loaded Cables. In Figs. 65 and 66 there are only a few discrete loads on the cable, which is straight between the loads and suffers locally concentrated changes in slope at the loads. In Fig. 67 there are many more loads, each one smaller than the loads of Figs. 65 and 66. Still the main cable is straight between loads and shows small changes in slope at the loads. The horizontal component of pull, H, is constant along the cable, and the vertical component of pull increases by one vertical-cable force P each time we pass a vertical cable. Thus the slope (more precisely, the tangent of the angle of the main cable with respect to the horizontal) increases in equal increments every time we pass a vertical cable, the increment being P/H, as can be seen from the force diagram in Fig. 67. In case we change the design of the bridge by adopting 24 verticals instead of 12 and thus carrying half the load on each vertical, the force diagram only gets additional rays between the old ones, the slope increments become half as large, and the funicular shape is practically unchanged. In the limit when the number of verticals becomes infinite and the individual loads approach zero, we have the case of the uniformly loaded cable, in which every point is subjected to an infinitely small change in slope, and the straight pieces in between approach zero length, so that the cable becomes curved, and the force-diagram triangle becomes uniformly full of radii. Practical applications of this case are found in electric transmission and telephone lines, under the influence of their own weight, of ice or sleet deposits, or of sidewise wind forces.

Consider in Fig. 68 an element ds of such a cable in a Cartesian coordinate system, and let the load on this element be $q\,dx$, where q, a constant expressed in pounds per inch or in kips per foot, is the loading per unit *horizontal* length dx. The actual weight, ice, or wind force, really is expressed as $q\,ds$ instead of as $q\,dx$, but for electric

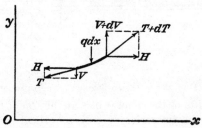

Fig. 68. Equilibrium of an element of cable.

cables with small slopes, the difference is not great. For the suspension bridge the load is the bridge deck, which is $q\,dx$ rather than $q\,ds$. The *real* reason for writing $q\,dx$ instead of $q\,ds$, however, is that the integration is very much simpler, as we shall soon see.

Further, in Fig. 68 let T be the tensile force in the cable (a function of x), let H be its horizontal component (constant along x, because the load has no x component), and let V be its vertical component (a function of x). Now we apply the equilibrium equations to the element dx. The horizontal equilibrium states that $H = H$; the vertical equilibrium equation is

$$(V + dV) - V = q\,dx \qquad \text{or} \qquad \frac{dV}{dx} = q$$

But from geometry, $dy/dx = V/H$, so that $V = H\,dy/dx$, and

$$\frac{d^2y}{dx^2} = \frac{-q}{H}$$

is the differential equation of the shape of the cable. This can be integrated twice to give

$$y = \frac{qx^2}{2H} + C_1 x + C_2$$

in which the integration constants are seen to be the ordinate and the slope of the curve for $x = 0$. Coordinate systems exist for the purpose of serving us and not the other way around, so that we now place the

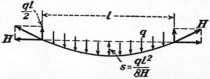

FIG. 69. A flexible rope loaded by a uniform loading per unit of horizontal length has the shape of a parabola.

origin O at the lowest point of the curve, where the ordinate and slope are both zero. The equation of the curve then is

$$y = \frac{q}{2H}\, x^2$$

a parabola. In particular, if the span is l and the sag in the center s, with respect to the end supports at equal height, then

$$s = \frac{q}{8H}\, l^2$$

which is illustrated by Fig. 69. This sag is seen to be proportional to the load and inversely proportional to the horizontal tensile force in the cable, as it should be.

The curve is always a parabola, even if the two end supports are at unequal heights. In that case the origin of coordinates, still at the lowest point of the curve, is no longer in the center of the span and may be even outside the span (Problem 97). In such cases we sketch a parabola and pick out such a portion of the curve as suits our end conditions.

Let us now consider the case of large sags and consequently large slopes in cables loaded by their own weight, so that in Fig. 68 the loading $q\, ds$ on an element can no longer be approximated by $q\, dx$. The differential equation for this new case is derived in exactly the same way as before, but for $q\, dx$ we substitute $q\, ds$, or for q we substitute $q(ds/dx)$ in the final result. Thus the differential equation becomes

$$\frac{d^2y}{dx^2} = \frac{q}{H}\, \frac{ds}{dx}$$

The line element $ds^2 = dx^2 + dy^2$, so that

$$\frac{d^2y}{dx^2} = \frac{q}{H} \sqrt{1 + \left(\frac{dy}{dx}\right)^2}$$

The integration of this equation is considerably more complicated than that of the parabola.

Let $dy/dx = p$ for the time being. Then

$$\frac{dp}{dx} = \frac{q}{H} \sqrt{1 + p^2}$$

and

$$\int \frac{dp}{\sqrt{1 + p^2}} = \int \frac{q}{H} dx$$

The left-hand integral we look up in tables and find that its value is

$$\log_e (p + \sqrt{1 + p^2}) = \frac{qx}{H} + C_1$$

in which C_1 is the first integration constant.

To save writing, let $(qx/H) + C_1 = A$. Then

$$p + \sqrt{1 + p^2} = e^A$$
$$\sqrt{1 + p^2} = -p + e^A$$

Square:

$$1 + p^2 = p^2 - 2pe^A + e^{2A}$$

Solve for p:

$$p = \frac{e^{2A} - 1}{2e^A} = \frac{e^A - e^{-A}}{2} = \sinh A$$

$$p = \frac{dy}{dx} = \sinh \left(\frac{qx}{H} + C_1\right)$$

$$y = \int \sinh \left(\frac{qx}{H} + C_1\right) dx = \frac{H}{q} \cosh \left(\frac{qx}{H} + C_1\right) + C_2$$

The two integration constants simply shift the curve bodily with respect to the origin of coordinates, C_2 up and down, C_1 horizontally. By making $C_1 = 0$ and $C_2 = -1$, we place the origin in the lowest point of the curve, and

$$\frac{q}{H} y = \cosh \frac{qx}{H} - 1$$

This is the equation of the *catenary*, or hanging-chain curve, from the Latin *catena*, chain (see Fig. 70).

For comparison, the equation of the parabola (for the case of loading proportional to the horizontal distance x instead of the curve length s) can be rewritten in the form

$$\frac{qy}{H} = \frac{1}{2}\left(\frac{qx}{H}\right)^2$$

and the hyperbolic cosine of the catenary can be developed into a Taylor power series:

$$\frac{qy}{H} = \frac{1}{2}\left(\frac{qx}{H}\right)^2 + \frac{1}{24}\left(\frac{qx}{H}\right)^4 + \frac{1}{720}\left(\frac{qx}{H}\right)^6 + \cdots$$

The first term of this development is the same as the formula of the parabola, and the other terms represent the difference between the

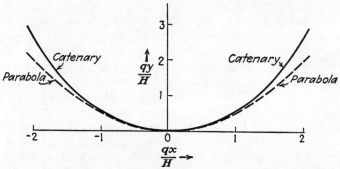

Fig. 70. A parabola is a good approximation to a catenary for small sags.

two, the catenary having the larger deflection. The difference between the two becomes significant only for large values of qx/H, that is, for large sags in comparison to the span. The table below gives some figures which the reader should verify from the formula.

qx/H	Sag/span	Error
2	½	34%
½	⅛	2%
⅛	⅟₃₂	0.1%

For the usual electric lines, therefore, the sag calculated by the parabolic formula is of the order of 1 per cent smaller than the true sag of a catenary. Such a small difference is without practical significance and is offset by a change in temperature of a few degrees in the cable.

Problems 97 *to* 100.

CHAPTER V

BEAMS

17. Bending Moments in Beams. The bars in trusses are subjected to tensile or compressive forces, as we have seen in Chap. IV, and only rarely does a bar in a truss construction have to support a sidewise load in the middle of its span. If an elongated structural member is subjected to lateral loads, and hence is subjected to bending, it is called a *beam*. Without any doubt, beams are the most important of all construction elements, and a good working knowledge of beam theory is indispensable to any civil, mechanical, or aeronautical engineer. Most beam theory is usually classified under the subject "strength of materials," but an important part of it, the determination of the bending moments in beams, can be carried out by the procedures of statics, and that is the subject of this chapter.

Consider in Fig. 71a a beam on two supports loaded by a single vertical load P, placed off center. The reactions to this load at the supports will be upward, of magnitude Pb/l at the left and Pa/l at the right. Now we make a section, an imaginary cut, through the beam and consider the portion of the beam to the left of the section (Fig. 71b). We know that that part of the beam is in equilibrium; also, we see the upward force Pb/l at the end L, and we suspect that some forces (as yet unknown) may be transmitted at the section by the right-hand portion of the beam acting on the left-hand portion. Then, from equilibrium, we conclude that these forces at the section must be equivalent to a downward force Pb/l and a counterclockwise couple Pbc/l. The vertical force at the section is called the *shear force*, and the couple is called the *bending moment* in the beam, both names being quite descriptive and appealing to our everyday experience. The manner in which the shear force and bending moment act and are distributed over the cross section is a subject in "strength of materials" and does not concern us at present. In Fig. 72 it is roughly indicated that when the beam is an I beam, the shear force is mainly taken by the vertical web of the I section, while the bending moment consists mainly of a compressive force in the fibers of the upper flange and of an equally large tensile force in the bottom flange.

The shear force Pb/l equals the left bearing reaction and is independent of c, being the same for all locations of the section between L and the load P.

A similar argument can be made for a section lying between the load P and the right-hand support R. The shear force then equals

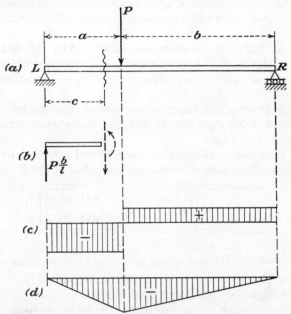

Fig. 71. Shear and bending-moment diagrams of a beam.

the right-hand reaction Pa/l and is likewise directed downward at the section. However, we are accustomed to consider this shear force as a negative one, while the shear force in the left-hand section is considered positive, and we now can plot the shear force as it varies along the beam in Fig. 71c. Why is it that of two shear forces, both acting downward at the section, one is considered positive and the other negative? Consider in Fig. 73 a small

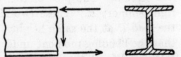

Fig. 72. Detail of how a shear force and bending moment is taken by an I beam.

length dx of the beam between two sections or cuts. The small piece is in equilibrium, so that, if the shear force S at the right is downward, there must be an upward force S at the left. Likewise if the bending moment M at the right is counterclockwise, as shown, there must be a clockwise moment at the left. We now state that the shear forces at

the right and left in Fig. 73 are the same, sign and all, and for the two
bending moments a similar statement is made. Thus, a shear force is
considered positive when the pair of shear forces acting on a small piece
dx of beam tend to rotate that piece in a counterclockwise direction.
A bending moment is considered positive when it causes a tensile
force in the top fibers and a compressive force in the bottom fibers of
the beam.

Now we return to the *shear-force diagram* (Fig. 71c) and check it
against the definition. Also we consider the vertical equilibrium of a
short section dx located just under the load P, and conclude from it
that:

**The shear force in a beam, when passing from the left side of a
vertical load to the right side of that load, is suddenly increased by
the amount of that load.**

Now we return to the bending moment and note that its value is
Pbc/l or the left-support reaction force multiplied by the distance from

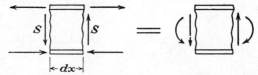

FIG. 73.   Definition of a positive shear force and of a positive bending moment.

it. When plotted in the *bending-moment diagram* of Fig. 71d, this
gives a straight line passing through the left support L. An equilibrium consideration of the right-hand portion of the beam for a cut
to the right of the load P leads to a bending moment in the right section equal to the right-hand reaction times the distance from it, also
giving compression at the top and tension at the bottom of the beam,
and hence of the same (negative) sign as the bending moment in the
left section. The plot of this is a sloped straight line through the
right support R. These two sloped lines are shown to meet under
the load P at the same height. The reader should check for himself
that the bending moment under the load gives the same answer when
calculated from the left and from the right. The bending moment
(unlike the shear force) retains the same value when passing under a
load for a short distance, but the slope of the bending-moment diagram
changes suddenly.

If the beam is subjected to several loads, instead of to a single load
as in Fig. 71, the shear force at any section equals the algebraic sum of
all the upward loads (including the bearing reaction) to the left of
the cut, or also it equals the algebraic sum of all the downward loads

(including the bearing reaction) to the right of the cut. Why do these two procedures give the same answer?

The bending moment at any section is $\Sigma_n P_n x_n$, summed over all downward loads and reactions to one side of the cut, where x_n is the distance between the nth load and the cut. Again, why does this lead to the same answer at a certain section, when calculated from first the left and then the right?

The funicular polygon explained in the previous chapter lends itself to a beautiful graphical interpretation in connection with bending

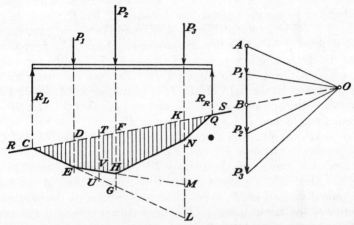

Fig. 74. The funicular diagram represents the bending moment.

moments, as is shown in Fig. 74 for the case of three loads on a beam. We start with the three loads P_1, P_2, and P_3 and do not as yet calculate the support reactions R_L and R_R, as they will be found by the graphical process. We lay off in the force diagram at the right the three forces AP_1, P_2, P_3 vertically, one hitched to the end of the other. Then we choose at random a pole O, but to obtain a drawing of nice proportions, we take O so that the two outside rays from it are not too far from 45 deg. We draw the four rays from O. Then we construct the funicular curve $CEHNQ$ as if the loads were supported by a cable, which, of course, is fictitious. We draw the closing line CQ of the funicular curve, and we draw in the force diagram its parallel OB. Then, as was explained previously, the line $RCEHNQS$ represents the shape of a cable subjected to the five forces P_1, P_2, P_3, R_L, and R_R, where R_L equals BA in the force diagram and R_R equals P_3B.

Now we make the statement that the **funicular diagram,** as shaded in the figure, **is the bending-moment diagram.**

In order to understand this important and useful relation we
consider a section in the beam between R_L and P_1. The bending
moment in that section is proportional to the distance from R_L (Fig.
71), and truly the funicular diagram represents this. The scale is
such that the distance DE represents as many foot-pounds bending
moment as the product of R_L in pounds and the distance R_LP_1 in
feet. The line CE can be extrapolated downward, and then KL to
the same scale represents the moment of the reaction R_L about the
point P_3 and is thus the contribution of R_L toward the bending moment.
Next, we consider a section T between P_1 and P_2. The bending
moment there equals R_L times the distance CT less P_1 times the distance DT. The first contribution is TU, and the second contribution
is UV, leaving TV as the bending moment. It has to be proved now
that the distance UV represents P_1 lb times DT ft to the same scale
that TU represents R_L lb times CT ft. This can be seen from the
force diagram to the right of Fig. 74. In triangle OBA, the line BA
represents the upward reaction R_L, and we note that OA and OB are
parallel to CE and CD in the funicular diagram. Likewise, in triangle OAP_1 the line AP_1 represents the force P_1, while OA and OP_1
are parallel to the lines EU and EV in the funicular diagram. The
distance UV grows with the distance DT if the section T shifts to the
right. Also, UV is proportional to the distance AP_1 in the force
diagram if we let force P_1 grow. Thus UV represents the product of
that force and the distance. TU is proportional both to the distance
CT and to the force $AB = R_L$, to the same scale.

Finally, we consider the point just under load P_3. The distance KL
represents the moment at K caused by the reaction R_L. From it we
subtract ML, representing the moment P_1 times DK. From that we
once more subtract MN, representing force P_2 times distance FK. To
get this clear the reader should repeat the same argument starting
from the right, from point Q, and understand the bending moment DE
under P as made up of contributions from R_R, P_3, and P_2.

The "base line" CQ of the bending-moment diagram, being the
closing line of the funicular polygon, is not horizontal in Fig. 74,
because we picked the pole O at random. The closing line could have
been made to come out horizontally by first calculating the end reactions, which gives point B in the force diagram, and then locating the
pole O at the same level, so that $OB//CQ$ is horizontal. This, however,
is not necessary, and a bending-moment diagram with a sloping base,
as in Fig. 74, is just as clear and instructive as one with a horizontal
base, as in Fig. 71d.

The determination of the bending-moment diagrams of the various beams in a structure is the first fundamental step towards their strength calculation and is, therefore, of utmost importance in design. Two more remarks will now be made to relate this discussion to what precedes in this book. The first is that the bars in trusses (pages 52 to 59) have no bending moments in them as long as the usual practice is followed of placing the loads at joints only, because then the loads, acting on the hinges at the ends of the bar, are along the bar and have no cross component to cause bending. For example, the roof truss of Fig. 75 sustains the rafters only at the joint points. The rafters are

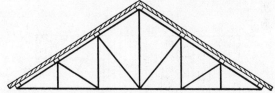

FIG. 75. A roof truss supports the roof at joints only so that no bending occurs in the bars of the truss.

beams in bending, which have to carry the snow or wind load more or less perpendicular to themselves and transmit this load by bearing reactions to the joints of the truss underneath. Thus the rafters are bent, but the bars in the truss are not.

Figure 75 suggests the second remark in connection with continuous beams that are supported in more than two points. Such beams are statically indeterminate (page 54). The resultant of any system of loading on a beam can be resolved into reactions at one hinge support and one roller support in one manner only (Fig. 23, page 26). However, if the beam is supported on three supports instead of on two, the resultant load can be resolved into these three reactions in an infinite number of ways. We can assign an arbitrary value to one of these reactions, consider it as a load on the beam, and then calculate the other two reactions by statics. This is what is meant by a beam that is statically indeterminate, because it has one or more redundant supports. The construction of a bending-moment diagram by the rules of statics, as in Fig. 71 or Fig. 74, can be accomplished only for statically determinate beams.

Not all statically determinate beams are on two supports; there are also cantilever beams, built in at one end and free at the other (Fig. 76). In such beams, the bending moment at the free end is zero (if in doubt about this statement, isolate the last piece dx of the beam and set up the equilibrium condition), while at the built-in end the wall may transmit forces, shear as well as bending, to the beam. The reader should go over the details of Fig. 76 and check the diagrams by calculating the shear forces and bending moments numerically in a few points.

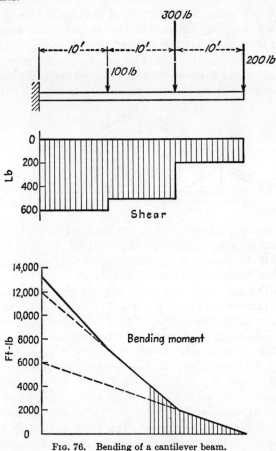

FIG. 76. Bending of a cantilever beam.

Figure 77 shows a plank on two supports *A* and *B* with an over-hang. A second plank is supported by the ground at *C* and is laid loosely on the overhang at *D*. In order to make the system statically determinate, it is necessary to assume a freely sliding contact at *D*; otherwise, the beam *CD* could be in tension or compression without load. The shear-force and bending-moment diagrams of this com-bination can be calculated and constructed by isolating various parts of the beams and writing the equilibrium conditions. In particular, the reader should reason out for himself that the bending moment must be zero at A, C, and also at D. The shear force at D is not zero. The loads P_1 and P_2 are drawn in full in the figure and the consequent bearing reactions in dashes.

Problems 101 *and* 102.

18. Distributed Beam Loadings. Consider the beam of Fig. 78, subjected to a uniform downward load of w lb per running foot, taken by two equal end reactions $R = wl/2$. What is the shear force and the

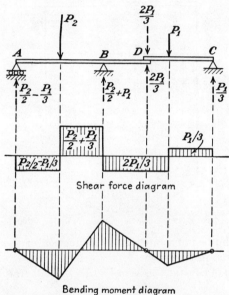

Shear force diagram

Bending moment diagram

Fig. 77. Two beams, hinged together, on three supports.

bending moment at distance x from the left support? We isolate the left portion of the beam and write the vertical equilibrium equation:

$$S(x) = R - \int_0^x w\,dx$$

or in words: the shear force equals the algebraic sum of all the upward forces and reactions to the left of the section (page 69) When the forces are discrete, the sum is written as Σ; when they are smoothly spread out, we write $\int$ instead. The above formula holds true whether w is constant or variable with x (Fig. 81). For the same left-hand portion we now write the moment or rotational equilibrium equation:

$$-M(x) = Rx - \int_{y=0}^{y=x} yw\,dy$$

In the latter integration, x is a constant, because we keep the section at a definite point x, and hence we have to write a new letter y for the variable, which is the distance from the load element to the cut. The load element then is dy. Again the above expression holds for constant as well as for variable w; in the latter case we have to express w as a function of y, because x is constant during the integration. Only after the integration is completed do we make x variable again

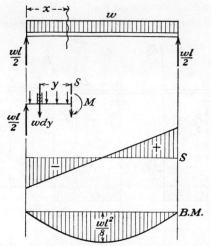

Fig. 78.   Beam on two supports with uniform loading.

and consider the bending moment as a function of the location x of the cut or section in the beam.

For the case of constant w, as in Fig. 78, the integrations are simple and lead to

$$-S(x) = R - wx = \frac{wl}{2} - wx$$

$$-M(x) = Rx - w\frac{y^2}{2}\Big|_0^x = \frac{wl}{2}x - \frac{wx^2}{2}$$

which results have been plotted in Fig. 78. First we notice that the bending-moment diagram is a parabola, and we remember that by Fig. 69 the shape of a cable loaded by a uniform load also is a parabola. This should not surprise us because we know from Fig. 74 that the bending-moment diagram is the same as the funicular diagram of a rope under the same loading, and of course this relation is just as true for distributed forces as it is for discrete ones.

The next thing we notice is that the above expression for the shear force can be derived from that of the bending moment by differentiation. This is an important relation, which will be shown to be generally true:

The ordinates of the shear-force diagram equal the slopes of the bending-moment diagram, or

$$S = \frac{dM}{dx}$$

To prove this consider an element of beam of length dx (Fig. 79), between locations x and $x + dx$, the coordinate increasing to the right. At location x we call the shear force S and the bending moment M; then at the other location we write $S + dS$ and $M + dM$. The beam section is loaded with $q\,dx$, and when dx is short enough, w can be considered constant. Also, we assume that there is no concentrated

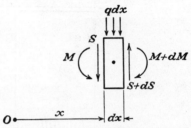

FIG. 79. Vertical equilibrium of a beam element requires that $w = dS/dx$; rotational equilibrium leads to $S = dM/dx$.

load P acting on this small piece of beam, so that $q\,dx$ is a small quantity. Then we write for the vertical equilibrium

$$dS = q\,dx \qquad \text{or} \qquad w = \frac{dS}{dx}$$

The ordinates of the loading diagram equal the slopes of the shear-force diagram. The rotational equilibrium or the moment equation about the center point of Fig. 79 gives

$$M - (M + dM) + S\frac{dx}{2} + (S + dS)\frac{dx}{2} = 0$$

or

$$dM = S\,dx + \frac{dS\,dx}{2}$$

The last term is small of second order and may be neglected, so that $S = dM/dx$, which proves our contention.

These relations between slopes and ordinates are generally true, and it is now useful to reexamine Figs. 71, 76, 77, and 78 with these relations in mind. Concentrated forces must be considered as locally infinite loadings w, which thus give locally infinite slopes in the shear diagram and sudden finite changes in slope in the bending-moment diagram.

As a next example, consider the beam of Fig. 80, loaded with a
uniform load along only half its length. First, for calculating the
reactions, we replace the distributed load by its resultant $wl/2$ at the
quarter-length point. This leads to bearing reactions $wl/8$ and
$3wl/8$. The shear-force diagram has zero slope in the left-hand part
and has slope w in the right-hand part; its end values must equal the
reactions. After the shear-force diagram has been thus constructed,

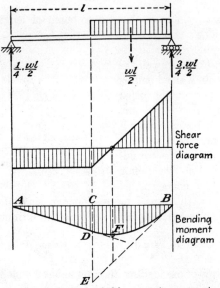

FIG. 80.   Partially loaded beam on two supports.

we proceed to determine the bending-moment diagram by using the
theorem $S = dM/dx$. In the left half, between A and D, the slope
of the bending-moment diagram is constant, equal to the shear force
$wl/8$, so that in the middle of the beam the bending moment is

$$\frac{wl}{8}\frac{l}{2} = \frac{wl^2}{16}$$

Next we start from the other end B, where the ordinate of the shear
diagram is three times that at A, with the opposite sign. The slope
of the bending-moment diagram then must be opposite in sign to the
slope at A and three times larger. Plot point E, making $CE = 3CD$,
and join B with E. This must be the slope at B. Since between B
and C the shear diagram is straight, the integral of it, or the bending
moment, must be a parabola or x^2 curve. Draw a parabola between

B and D, tangent to BE at B and tangent to AD at D. (Why is the curve smooth at D, and why does it not have a sudden change in slope there?) After the parabola is drawn we notice that it must have a zero slope at F, where the shear force is zero. Since F is such an important point, where the bending is at its worst, we will calculate the bending moment there. From the shear-force diagram we see that

$$x_F = \frac{l}{2} + \frac{1}{4}\frac{l}{2} = \frac{5}{8}l$$

Isolate the piece of beam BF of length $\frac{3}{8}l$. On it are acting the end reaction $3wl/8$ and a distributed loading $3wl/8$, of which the center of gravity is halfway or at $3l/16$ from F. Taking moments about F, the bending moment there is $9wl/128$, or 12.5 per cent greater than at C. The same result can be obtained by isolating the left section AF, which is left as a useful exercise to the reader.

An error into which most beginners fall at one time or another is the following. Suppose we ask for the bending moment at some point, say at F. In the top diagram of Fig. 80 we replace the distributed load by its resultant through the center of gravity as shown. Then the bending moment at the $5l/8$ point F is the left reaction $wl/8$ multiplied by the distance AF. Checking from the right, we isolate the piece FB, and the moment at F is the end reaction $3wl/8$ multiplied by its distance $3l/8$ less the load $wl/2$ times its distance $wl/8$. This checks the answer from the left, but both calculations are wrong. At which point in the argument has the error been made?

As another example, consider Fig. 81, a beam on two supports, loaded with a uniformly increasing load intensity. First let us try to construct the two diagrams with as little calculation as possible. The center of gravity of the loading diagram is at two-thirds distance, so that the end reactions are $P/3$ and $2P/3$ in which P is the total load of the triangle diagram. The shear forces at the two ends A and B then are $P/3$ and $2P/3$, and the intervening curve is the integral of a straight line, or a parabola. Near the left end A the distributed loading is small, practically zero; therefore, the shear force hardly changes; the tangent to the curve is horizontal at A. At the other end B the intensity of the loading q is $2P/l$ (why?), or written differently, $\frac{2}{3}P$ divided by $l/3$. The height BC is $\frac{2}{3}P$; take $BD = l/3$, and the connecting line CD must be the tangent to the shear diagram at C.

Next we try to find point E where the shear is zero. That point is so located that the triangular load from A to E just balances the left reaction $P/3$, or in other words, E is so located that the area of the partial triangle to the left is one-third of the area of the entire triangle. Areas of similar triangles are as the squares of their bases, so that $x^2/l^2 = \frac{1}{3}$ or $x = 0.578l$. With the three points A, E, and C and with two tangent directions, the curve can be nicely drawn.

Next we proceed to the bending-moment diagram, which must be zero at both ends. We lay off $GH = Pl/3$ at the right and join F with H. The slope of FH is $P/3$, and therefore FH is the tangent to

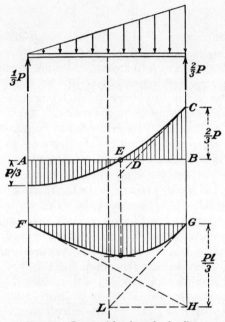

FIG. 81. Beam with triangular loading.

the diagram at F. The slope at G should be twice as large, and with the point L at mid-span we construct the tangent GL. Finally the intervening curve should have a horizontal tangent at E, which enables us to sketch it in quite nicely.

An algebraic calculation of the maximum bending moment at E proceeds as follows. Take a section at distance x from the left bearing. The area of the partial loading triangle at left is Px^2/l^2, and its center of gravity is at $2x/3$ from the left end or at $x/3$ from the cut. Thus the bending moment at x is

$$\frac{P}{3}x - \frac{x^2}{l^2}P\frac{x}{3} = \frac{Px}{3}\left(1 - \frac{x^2}{l^2}\right)$$

We notice that this correctly gives us zero bending moment at the other end $x = l$, and it is thus the equation of the bending-moment diagram. To find its maximum, we differentiate and set equal to zero

$$0 = \frac{P}{3}\left(1 - \frac{x^2}{l^2}\right) - \frac{Px}{3}\frac{2x}{l^2}$$

or

$$x = \frac{l}{\sqrt{3}}$$

Substituting this into the expression, we find for the maximum bending moment at E

$$M_E = \frac{Pl}{3\sqrt{3}}\left(1 - \frac{1}{3}\right) = \frac{2}{9\sqrt{3}}Pl = 0.128Pl$$

We end this chapter with an application of the funicular diagram to the graphical determination of the location of the center of gravity.

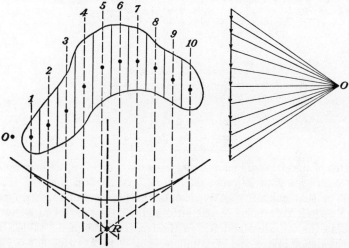

FIG. 82. The funicular diagram construction as a means for finding the center of gravity.

This construction is useful only when the body has a complicated shape that is not readily reducible to a combination of rectangles, triangles, and circles, and therefore is used but seldom in practice. It is mentioned here primarily because it is a good and instructive application of the principles. The (two-dimensional) body shown in Fig. 82 is first subdivided into a suitable number of strips. A few

strips give little work and poor accuracy; more strips give more work
and a correspondingly better return. In the example, ten strips are
chosen. We draw parallel dotted lines through the centers of gravity
of the individual strips and determine the weight of each strip by
estimating or by planimeter measurement or by any other suitable
method. The weight of the body is thus divided into ten partial
weights, and the center of gravity lies on the line of action of the
resultant of these ten forces. The funicular construction of this
resultant is done by the method discussed with Figs. 65 and 66 (page
62). To determine the center of gravity, a second similar construction
is required in a different direction, preferably perpendicular to the
first direction. In actual practice, it is usually simpler and more
accurate to make a table of the weights or areas of the ten strips of
Fig. 82, listing them in a vertical column, as shown below. If the

No.	Area	Product
1	4.10	4.1
2	5.25	10.5
3	6.30	18.9
4	7.20	28.8
5	7.10	35.5
6	6.85	41.1
7	6.45	45.2
8	5.90	47.2
9	5.40	48.6
10	4.35	43.5
Sums	58.90	323.4

widths of the strips are made equal, as is usual practice, we may multiply each area with its order number, as in the third column. Each
item in that column then represents the moment of the strip area
with respect to a point O, one strip width to the left of area No. 1.
Adding the ten areas and adding the ten moments in the table gives
us the denominator and the numerator of Eq. (2) (page 33). Thus
the center of gravity is to the right of point O by $323.4/58.90 = 5.50$
strip widths, which means that it is on the dividing line between strips
No. 5 and No. 6. This is an example of numerical integration of a
case in which the "functions" are given on a blueprint only and cannot readily or accurately be reduced to analytical form.

Problems 103 *to* 109.

CHAPTER VI

FRICTION

19. Definition. Friction is of far greater importance in statics than is often realized. Without it no ladder could stand against a wall, a two-legged animal could hardly remain upright, and a large number of structures would become unstable. Most simple one- or two-family houses are set loosely on their foundations and depend on their weight and friction to stay there. Without friction, such houses would be blown away by the slightest breeze. Worse than that, without friction, all houses, animals, and men who were not standing on solid rock, but on sand, clay, or dirt, would sink down in it until Archimedes' law restored equilibrium through buoyancy. This is so because the solidity we observe in sand or dirt is a consequence of the friction between the individual grains, which prevents these grains from sliding over each other freely. At some depth below the surface of the earth there is considerable pressure in the soil, because the weight of the upper layer has to be carried. With this great pressure the friction forces between the grains are large, and the conglomerate of grains acts like a solid body. This can be illustrated by a beautiful experiment, in which a small rubber bag is partially filled with sand and closed airtight with a screw cap. The bag is soft to the touch, like a tobacco pouch, because there is no pressure and hence no friction between the grains of sand. When the air is evacuated from the bag by a small hand pump, the atmosphere gets a chance to press on the bag from the outside (with 14.7 lb/sq in. pressure). This presses the sand grains together and causes friction between them, with the result that the bag becomes as hard as a rock and can be used spectacularly to drive a nail into wood.

Another illustration is the behavior of sand on the beach. We observe that dry sand is soft to the touch, and we can easily dig a foot into it. The same is true for the sand when completely submerged under water. But the just-dry beach close to the water's edge is hard, and we can ride a bicycle on it. This is caused by the fact that there is water between the sand grains just under the surface and that this water, by the action of capillarity, pulls the sand grains together. Thus the sand acts as if it were under the same pressure as at great

83

depth and becomes quite hard on account of the friction between the grains.

The clear understanding gained during the last thirty years of the relationship between capillarity and friction forces and its consequent application to conglomerates of grains of various sizes (gravel, coarse sand, fine sand, silt, clay) has given rise to a new subject in engineering, known as "soil mechanics."

So far we have talked about "friction" in the commonly understood sense of that term without being precise about it. Now we

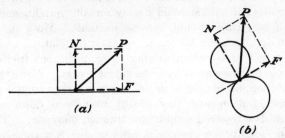

Fig. 83. The definition of friction force F, normal force N, and total force P.

turn to an exact definition. Consider in Fig. 83 two bodies in contact with each other, either along a flat area (Fig. 83a) or at a point of two curved surfaces (Fig. 83b). In both cases let P be the force exerted by one of the bodies (the lower one) on the other (the upper one), and let this force P be resolved into components, normal and tangential to the surface element of contact. Then N is called the *normal force*, F is called the *friction force*, and P is called the "total force" or the *contact force*. Thus the **friction force is defined as that component of the contact force between two bodies which lies in the tangent plane of the contact point.**

In the case of statics, *i.e.*, in the absence of motion, particularly of tangential sliding motion between the bodies, innumerable experiments during the past two centuries have shown that the maximum obtainable friction force for two given bodies is approximately proportional to the normal force, or in a formula,

$$F \leqslant fN \qquad\qquad (3)$$

where f is known as the *coefficient of friction*. We notice that the words "maximum obtainable" appear in the statement of the experimental law and that consequently the symbol $\leqslant$ appears in Eq. (3) instead of the symbol $=$. This is in accordance with common observation. In Fig. 83a, let the normal force N be held constant, say at 100 lb. Then the friction force in the condition of equilibrium at rest may be anything (including zero) up to a certain limit, say 30 lb. The law states that if the friction force F *surpasses* the limit $fN = 0.3N$, then no equilibrium is possible, and the bodies start sliding over each other. What happens to the friction force after slipping has started, during relative motion between the bodies, is another matter and is not of our concern when studying statics.

For the friction force during motion there are two principal laws, depending on whether we deal with "dry friction" or with "film lubrication." The law for dry friction is the same as Eq. (3), except that the $\leqq$ symbol is replaced by the $=$ symbol, or in words, with dry friction during relative motion, the friction force equals f times the normal force, independent of the velocity of slip. The law for film lubrication, such as applies to oil-lubricated bearings, is entirely different. In that case the friction force grows with the velocity of slipping and is more or less independent of the normal load N. The study of this law belongs to the field of fluid mechanics. The law of dry friction applies with tolerable accuracy wherever no oil film develops, even in the presence of a lubricant. Two metal surfaces with cup grease between them obey the law of dry friction, with a small value of the coefficient f.

The law of static friction, Eq. (3), was discovered and stated in the dim past; it can be found, for example, in a physics textbook printed in 1740 with elaborate tables of numerical values of f.[1] However, it is now usually referred to as **Coulomb's law,** and dry friction is usually called **Coulomb friction** (to distinguish it from viscous friction, which occurs in oil films), after a book on machine design[2] that had widespread influence in its day.

The numerical values of the friction coefficient f vary over a wide range and depend not only on the materials of the two bodies but also on the roughness of their surfaces and on the degree of cleanliness. The smallest practical coefficient of dry friction is about 0.02 (2 per cent) for smooth, well-greased metal surfaces, while on the other side of the range, an automobile tire on a dry concrete road has a friction coefficient of about 1.00 (100 per cent). Other intermediate values for all sorts of materials can be found in handbooks[3] but have to be

[1] "A Treatise on Natural Philofophy for the Ufe of Students in the Univerfity," by Pieter Van Musschenbroek, translated into English by John Colson, Lucafian Professor of Mathematics in the Univerfity of Cambridge, 2d ed., London, 1740.

[2] "Théorie des Machines Simples," by Coulomb, Paris, 1821.

[3] In Marks' "Mechanical Engineers' Handbook," 4th ed., p. 233, we read: "The static coefficient of friction between metal and oak (dry) is 0.62; for steel on steel 0.15." In John Colson's "Treatife" of 1740, cited above, we read: "A steel axis

considered only as average values, subject to considerable variation
depending on circumstances.

The simplest way to determine experimentally the value of a
coefficient of friction is by means of an inclined plane (Fig. 84). The

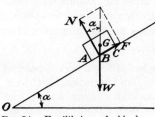

plane is lined with one substance and
the block on it with the other substance. The angle of inclination of the
plane is slowly increased until the block
starts to slide down. From the angle
α at which this occurs, the value of f
can be determined. On the block of
Fig. 84 there are three forces acting as

FIG. 84. Equilibrium of a block on
a rough inclined plane.

shown. The weight passes through the
center of gravity, and hence the reaction
force from the plane also passes through that point. It is resolved into
N and F, and from the equilibrium conditions, before sliding, we see
that

$$N = W \cos \alpha, \qquad F = W \sin \alpha, \qquad F = N \tan \alpha$$

The angle between the total contact force W and its normal component
N, when sliding just starts, is called the "friction angle," which in this
case equals the angle of inclination of the plane. Thus

$$f = \tan \alpha$$

The friction coefficient equals the tangent of the friction angle, which
again is equal to the tangent of the angle of inclination of the plane.
If $f = 1.00$ between a rubber tire and a concrete road, it means that a
car with completely locked brakes can stand on a 45-deg incline and
that when the incline becomes slightly steeper, the car will start sliding
down with locked wheels.

Problems 110 *and* 111.

20. Applications. *a. Tipping or Sliding?* Consider again the block
of Fig. 84. In the drawing it is shown low and squat, but suppose it
had been higher and with a shorter base. The question arises as to
whether with increasing angle α, it will not first tip over about the
forward edge A before it slides down. The answer to the question is
almost immediately visible from the figure. It is seen that the weight

being dry, and received into a bearing of guaiacum wood, the friction to the weight
is as 1 to 3⅓. The same axis received into brafs, anointed with olive oil, the
friction to the weight is as 1 to 7⁷⁄₁₅."

force W intersects the base at point B. This then must be the point through which the resultant of the contact forces must pass. If point B lies higher than A, or in other words, if B is in the base of the block, this is possible without trouble. However, if B lies below A, or outside the base of the block, the only way in which distributed normal forces on the base of the block can give a resultant at B is by being partially upward-pushing and partially downward-pulling. Thus, for equilibrium, it would be necessary for the inclined plane to pull down on the block near the rear contact point C. In the absence of this pull, no equilibrium is possible. Let b be the base length of the rectangular block, and let h be its height. Then the reader should derive that for $b > h \tan \alpha$, point B lies within the base, and for $b < h \tan \alpha$, it lies outside the base below A. Thus if $b/h > \tan \alpha$ or $b/h > f$, the body will

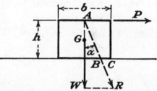

FIG. 85. A packing case will tip about C—instead of slide—if the resultant R intersects the ground to the right of C.

slide down first upon increasing the inclination α; but if $b/h < f$, it will tip over about point A before it slides.

A similar problem is shown in Fig. 85, where a packing case of dimensions b and h, with its center of gravity in the center, is subjected to a horizontal pull P at the top. When P is slowly increased, the resultant of P and W (which must be equal and opposite to the contact force) turns gradually about point A, increasing the angle α. Nothing will happen until *either* $\tan \alpha$ reaches the value f (when the case will start sliding) *or* the point B, where the resultant intersects the base, moves to the right of point C (when the case will tip forward).

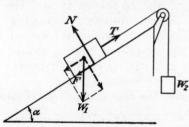

FIG. 86. A block sliding up an inclined plane.

Verify that for $f = 0.6$ or 60 per cent friction the base must be at least 1.2 times the height to prevent tipping.

b. Pulling a Block up the Inclined Plane. In Fig. 86 let a weight W_1 be held in equilibrium on an inclined plane by a second weight W_2, and let us ask for what value of W_2 the weight W_1 will just start sliding up and for what value' W_2 it will just start sliding down. Assuming no friction in the pulley wheel, the rope tension T equals

W_2 (page 21).   Other forces acting on the block are its weight W_1 and
the contact reaction from the incline.   We resolve W_1 into its normal
($W_1 \cos \alpha$) and tangential ($W_1 \sin \alpha$) components and resolve the contact force into N and F.   Then the equilibrium in a direction normal
to the plane requires that $N = W_1 \cos \alpha$, and the equilibrium tangential to the plane requires that

$$W_2 = W_1 \sin \alpha + F$$

where F is the friction force *from* the incline *on* the block, counted
positive down the plane.   If the block is just ready to slide up the
plane, we have $F = fN = fW_1 \cos \alpha$, so that

$$W_2 = W_1 \sin \alpha + fW_1 \cos \alpha \quad \text{(just going up)}$$

For downward motion the direction of the friction force is reversed,
so that

$$W_2 = W_1 \sin \alpha - fW_1 \cos \alpha \quad \text{(just going down)}$$

Now let us adjust either α or f so that without pull in the rope ($W_2 = 0$)
the block is just ready to slide down.   Then the two terms on the
right of the last expression cancel each other by being equal, and the
pull W_2 required to pull *up* thus becomes

$$W_2 = 2W_1 \sin \alpha$$

or twice as large as would be required with no friction at all.   The
friction force when just going up is directed downward, and the friction force when just going down is equally large and directed upward.
Thus when for $W_2 = 0$ the weight W_1 is just held by friction, we need
twice that friction force in rope tension to pull up, once to overcome
friction and once to overcome the gravity component $W_1 \sin \alpha$.

This long argument on a simple case has been made so elaborate
because it is the simplest manifestation of a general proposition as
follows:

**A continuously operating hoisting mechanism that is self-locking
by friction cannot be more than 50 per cent efficient when hoisting.**
The term "self-locking" means that the load on the hoist does not
slide down by itself when the upward pull is removed.   Efficiency is
defined as the ratio of the pull required to hoist without friction to the
actual pull with friction.   The phrase "continuously operating" is
inserted to exclude intermittently operating devices, containing
pawls or similar elements.   Other applications of this proposition
will be discussed presently: the screw jack and the differential pulley,

while in Fig. 92 (page 98) an intermittently operating hoist is shown in which the efficiency is better than 50 per cent.

   c. The Screw Jack.   The principle of operation of a screw jack can best be understood by considering a modification of Fig. 86. In that figure let the angle α be small, of the order of 1 deg, let the block be very long (100 in.) and low ($\frac{1}{4}$ in.), let the thickness of the block and the incline perpendicular to the paper be small, say $\frac{1}{4}$ in., and let them both be made of a plastic, easily bendable material. Now

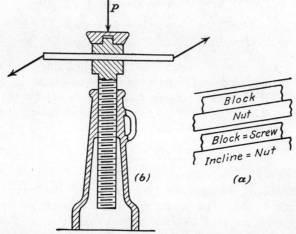

FIG. 87.   A screw jack operates on the principle of the inclined plane.

we cut and throw away most of the inclined plane, keeping only a strip of $\frac{1}{4}$-in. width and 100-in. length. Then we roll up the whole assembly to such a radius that the consecutive windings show a small clearance on the top side of the block, as illustrated in Fig. 87a. Finally, we cast the wound-up inclined plane into solid material on the outside, forming a nut, and we cast the wound-up block into solid material on the inside, forming a square threaded screw.

   Now we apply a torque of moment M to the screw and a sufficient counter moment to the nut to preserve equilibrium, and we also put a load P on the screw, as shown in Fig. 87b. This load P and torque M are transmitted by the screw to the nut, and Fig. 86 can be considered to be a small element of it. The forces transmitted by the block or screw element (Fig. 86) to the nut or inclined plane are W and T as shown, or $W - T \sin \alpha$ vertically and $T \cos \alpha$ horizontally or tangentially. The vertical forces of all elements in Fig. 86 add up to P in Fig. 87, so that

$$P = \Sigma(W_1 - T \sin \alpha)$$

The horizontal forces cancel each other, as they are wrapped around
several circles, but they all have a moment about the center line.
With r, the (pitch) radius of the threads, we have

$$M = \Sigma(Tr \cos \alpha)$$

From the previous example, we can now take the relations between T
(which is W_2) and W_1, and substitute.   This gives

$$P = \Sigma W_1(1 - \sin^2 \alpha \mp f \sin \alpha \cos \alpha)$$
$$M = \Sigma W_1 r(\sin \alpha \cos \alpha \pm f \cos^2 \alpha)$$

where the upper of the two signs belongs to "just going up" and the
lower to "just going down."   We divide these two equations together
to eliminate W_1 (which is only an auxiliary quantity for our screw)
and observe that α and f are the same for all elements of the screw,
so that the Σ signs can be left out.

$$\frac{M/r}{P} = \frac{\sin \alpha \pm f \cos \alpha}{\cos \alpha \mp f \sin \alpha}$$

again with the upper signs for "just going up," the lower ones for
"just going down."   The quantity M/r is the tangential force that
would have to be applied to pitch radius to hold the load P.   In the
absence of friction, this turning force M/r is considerably smaller
than P.   Thus, without friction a very large load can be lifted with a
small torque.   For self-locking, the "just going down" torque M
should be zero or f should be tan α.   Since tan α is only a few per cent
by the geometry of the screw and the friction coefficient in the threads
cannot be made so small, all practical screw jacks are self-locking.
Suppose we just have self-locking, $i.e.$, the jack is just ready to run
down by itself without torque.   What torque must we apply to just
push it up?   To find that we divide the above equation by cos α,

$$\frac{M/r}{P} = \frac{\tan \alpha \pm f}{1 \mp f \tan \alpha} = \tan (\alpha \pm \beta)$$

where $f = \tan \beta$, and β is the angle of friction.   For self-locking
$\alpha = \beta$, or the angle of friction equals the angle of inclination.   To
push up against this friction,

$$\frac{M/r}{P} = \tan 2\alpha$$

To push up (or down) without any friction,

$$\frac{M/r}{P} = \tan \alpha$$

which is less than half of $\tan 2\alpha$. Thus the torque necessary to hoist with self-locking friction is more than twice the torque to hoist without friction, or the efficiency is less than 50 per cent.

d. The Ladder against the Wall. In Fig. 88 let a ladder of length l rest against a vertical wall at angle α. Assume the friction coefficient on the ground equal to that on the wall. Let a man climb up the ladder. How far can he go up before the ladder slips? For simplicity, let the weight of the ladder be neglected, or rather let P represent the resultant of the man's weight and the ladder's weight. If we are just ready to slip, the friction forces are

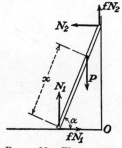

Fig. 88. The sliding ladder.

fN_1 and fN_2, where N_1 and N_2 are the unknown normal reactions. The direction of the friction forces is against that of impending motion. The equations of equilibrium are
Vertical forces:

$$N_1 + fN_2 = P$$

Horizontal forces:

$$fN_1 = N_2$$

Moments about bottom:

$$Px \cos \alpha = N_2 l \sin \alpha + fN_2 l \cos \alpha$$

If f is given, the three unknowns in these three equations are N_1, N_2, and the height x. Solving for x gives

$$\frac{x}{l} = \frac{f}{1 + f^2} (\tan \alpha + f)$$

which is the answer for all values of α and for all values of f. As an example, consider $f = 1$, which is reasonable for rough ground. Verify that if α is 45 deg, the ladder will slip when the load P just reaches the top.

e. The Buggy Wheel. Figure 89a represents one wheel and half the axle of a two-wheel horse-drawn buggy. The non-rotating axle is shown with a square cross section in the portion between the wheels, while the ends of that axle, which fit in the wheel bearings, are round.

The downward force W shown on the half axle in the figure equals half the weight of the loaded buggy less wheels, while the force P in the figure represents half the horse's pull. The wheel of weight w rests on the ground and feels an upward push, which is equal to $W + w$ because of vertical equilibrium of the combined wheel and half axle. It also feels a horizontal force, which must be equal to P for horizontal

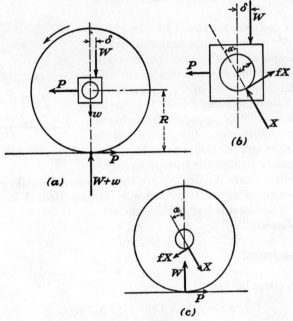

Fig. 89. The buggy wheel.

equilibrium. It is noted that the buggy load W is shown off center; this must be so for moment equilibrium: $W\delta = PR$. The force P on the ground is furnished by friction between the ground and the wheel rim; it will be seen later that P is very small with respect to W (the lucky horse!), so that P/W is certainly smaller than f, and the wheel will not slip on the ground. We want to calculate the horse's pull P when there is friction in the axle bearing. Figure 89b shows the axle by itself with the forces acting on it, and Fig. 89c, the wheel itself. The bearing forces X and fX require some explanation. First, the normal force X is unknown, and it is shown at an unknown angle α. (Try to establish equilibrium with a purely vertical X and find that it cannot be done.) Second, the friction force *equals* fX and is not smaller than fX, because we assume that the horse pulls hard enough to just start the buggy, so that there is impending relative motion in the axle bearing. Let the axle diameter be r and the wheel diameter R. We now write the equilibrium equations for the axle

$$\updownarrow \quad W = X \cos \alpha + fX \sin \alpha$$
$$\leftrightarrow \quad P = fX \cos \alpha - X \sin \alpha$$
$$(\quad W\delta = fXr$$

For the wheel the vertical and horizontal equilibrium equations are the same as for the axle, while the moment equation is

$$PR = fXr$$

Thus we have four equations, and, counting the unknowns, we also find four: X, P, δ, and α. The rest of the problem is algebra. We do not particularly care what X, δ, and α are, but we are interested in P. From the last equation we calculate X and substitute it into the first two equations, omitting the third equation. From the first two we calculate $\sin \alpha$ and $\cos \alpha$ separately, then eliminate α by squaring and adding, since $\sin^2 \alpha + \cos^2 \alpha = 1$. This leads to the result

$$\frac{P}{W} = \sqrt{\frac{f^2r^2}{(1 + f^2)R^2 - f^2r^2}} = f\frac{r}{R}\sqrt{\frac{1}{1 + f^2 - f^2r^2/R^2}}$$

Now, in a first-class buggy $R \gg r$, and $f \ll 1$, so that the square root hardly differs from unity, and

$$\frac{P}{W} = f\frac{r}{R} \qquad \text{(approximately)}$$

The moral: In order to keep the horse happy, we must grease the axle well and use a large-diameter wheel with a small-diameter axle. For a much shorter derivation of this result by the "method of work," see page 149.

The actual pull of the horse is larger than is indicated by the above formula. We have assumed a purely round wheel on a purely straight road, which can be true only for infinitely hard materials in wheel and road. Actually, the wheel will flatten locally, and the ground will be pushed in slightly, so that the wheel has to be "tipped" forward over the leading edge of its ground contact in the manner of Fig. 85. This effect is sometimes called "rolling friction," a bad misnomer, because it has nothing to do with friction. The coefficient of rolling friction is defined as the ratio of the tipping force to the weight. This tipping force has to be added to the above P in order to obtain the total horse's pull.

f. The Differential Hoist. In Fig. 25 (page 29) we studied this device, assuming frictionless bearings, and came to the conclusion that for a small difference between the radii r_o and r_i the pull P could be made a very small fraction of the load W. In the discussion that now follows we shall call the difference in diameters $\delta = 2r_o - 2r_i$, and we shall call $D = 2r_o$, the larger diameter, so that the smaller diameter $2r_i$ becomes $D - \delta$. With this notation the result of page 29 for the frictionless case is

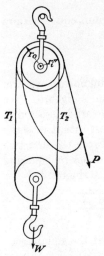

$$\frac{P}{W} = \frac{\delta}{2D}$$

Now let there be friction in the bearings; let the coefficient f be the same for the upper and lower sheaves; let the axle or journal diameter d also be the same for both; and let the diameter of the lower pulley wheel be D. Then, when pulling down on P (Fig. 25) and hoisting up on W, the two rope tensions T_1 and T_2 cannot be the same; T_1 must be larger than T_2 to overcome friction on the lower pulley axle. Let $T_1 = (W/2) + X$ and $T_2 = (W/2) - X$, thus satisfying vertical equilibrium of the lower pulley. The friction force there is fW, and its

FIG. 25, reprinted.

moment with respect to the center of that pulley is $fWd/2$. Then the moment equation of the lower sheave is

$$\left(\frac{W}{2} + X\right)\frac{D}{2} - \left(\frac{W}{2} - X\right)\frac{D}{2} = fW\frac{d}{2}$$

or, worked out,

$$X = \frac{fWd}{2D}$$

For simplicity we say that the normal force on the upper axle is W, thus neglecting P, which is small. The friction torque there, then, is the same as below, and the moment equation for the upper pulley is

$$\frac{PD}{2} + T_2\frac{D - \delta}{2} = T_1\frac{D}{2} + fW\frac{d}{2}$$

which after working out becomes

$$P = \frac{W\delta}{2D} + 2fW\frac{d}{D} - \frac{fW\delta d}{2D^2}$$

where the last term is negligibly small compared to the previous one. Thus,

$$P_{up} = \frac{W\delta}{2D} + 2fW\frac{d}{D}$$

which for $f = 0$ checks our previous result. In case we should give the pull P a small positive or a negative value so that the load goes down instead of up, the entire analysis would be the same, except that the friction would be reversed, which in the final answer results in a change in sign of f. (The reader should check this statement and never believe the Printed Word.) Thus, for letting down,

$$P_{down} = \frac{W\delta}{2D} - 2fW\frac{d}{D}$$

For self-locking P_{down} is zero, and the friction coefficient necessary to give this value is

$$f_{self-lock} = \frac{\delta}{4d}$$

In the usual chain hoists, δ is about $d/2$, so that 12.5 per cent friction will cause self-locking. Most differential chain hoists, even well-lubricated ones, are self-locking, except when specially provided with ball bearings top and bottom. Suppose we do have just self-locking; then the two terms on the right side of the formula for P_{down} (or for P_{up}) are equal: the first term pulls the load up (or down) without friction, the second overcomes friction, and we again verify the rule of not more than 50 per cent efficiency for a self-locking hoist.

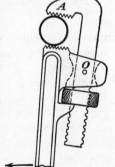

Fig. 90. The pipe wrench is a self-gripping device.

    g. The Pipe Wrench and the Automobile Brake. Figure 90 shows a pipe wrench, which is a tool for turning with considerable torque a smooth round pipe on which no ordinary wrench will take a bite. The only way to exert the torque is by friction, but friction requires normal force. The pipe wrench is one of those hen-and-egg devices in which only the deepest philosophy can decide which came first, the friction or the

normal force.   In Fig. 90 the upper jaw *A* loosely fits in the body
of the wrench, and it can be described with a little simplification
as being pivoted about point *O*.   The wrench is placed on the pipe
and turned clockwise.   When the wrench slips, the upper jaw at *A*
moves to the right with respect to the pipe.   If there is no friction
or normal pressure at *A*, nothing happens; the wrench slips and does
not bite.   But, if either the slightest bit of normal pressure or friction
appears, then the friction force on *A* will move *A* to the left in relation
to the lower jaw, and the pivot *O*
is so placed as to move the two
jaws closer together, increasing
the normal pressure and hence
the friction, which again moves
A farther to the left, thus leading
to an "infinite" friction torque,
limited only by the strength of
the plumber, the wrench, or the
pipe.   As soon as the wrench is
turned the other way, the sign of
the friction reverses; *A* moves to
the right relative to *O*, and the
wrench loosens up and slips.

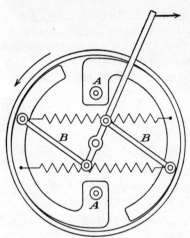

FIG. 91.   Self-gripping brake.

Another application of this
self-gripping operation is shown
in Fig. 91, illustrating an auto-
mobile wheel brake in use before the present hydraulically operated
brakes superseded them.   The two (non-rotating) brake shoes are
pivoted at the points *A* and are inside the rotating brake drum.
Two springs pull the shoes together and away from the drum.
When the brake lever is pulled to the right, the two compression
struts *B* press the shoes against the drum, stretching the springs.
Then friction appears, and the friction forces (for counterclockwise
rotation of the drum, as shown) are such as to rotate the shoes about
their pivots *A* so that more normal pressure and more friction are
generated.   Thus a very large braking effect is produced with a
small effort on the part of the driver.   The motion of the car really is
utilized to tighten the brakes.   In case the car is moving in reverse
and the driver wants to use the brake, the effect is reversed, and he
has to pull all the more, proving again that in this world one seldom
gets something for nothing.

h. The Automobile Friction Jack.   In this device, which is self-locking by friction an efficiency of much better than 50 per cent is obtained by intermittently relieving the friction lock, so that the mechanism during the hoisting period is really different from that during the locking period. The heart of the machine is shown in Fig. 92a, where a horizontal arm fits around a vertical stem, either round or square in cross section.   A load P is applied eccentrically at distance a from the center of the supporting rod.   This will cock the arm so that it bears on the upper left and lower right corners as shown.   The normal forces N and the friction forces fN (for impending slip) are shown as they apply from the vertical rod on the horizontal bar.   Horizontal equilibrium requires that N = N, which has already been assumed.   Vertical equilibrium requires that fN = P/2.   The moment equation about point O (or about any other point) is Pa = Nh. Then, substituting and eliminating the load P and the normal force N, we find

$$f = \frac{h}{2a}$$

for impending slip.   If the coefficient of friction is less than $h/2a$, the arm will slip down; for f greater than $h/2a$, it will be self-locking. For the usual case f is, say, 50 per cent, so that the device is self-locking when $h/2a$ is less than $\frac{1}{2}$ or when h is less than 4a.   Only when h is quite large with respect to the eccentricity a will the bar slip down; for most practical dimensions, it will stay put by self-locking under any magnitude of the load P.

Now, consider the jack of Fig. 92b, made of a steel casting about 6 in. high, loosely fitting about a ¾-in. round bar.   The load P (from an automobile bumper) is shown acting on the casting through a piece for convenient rotatory adjustment.   The load is transmitted through A to B and from there by friction lock through the vertical rod down to the ground G.   The bar AB is of self-locking dimensions, as explained above; on the other hand, the casting itself, between the points H and L, constitutes a beam (Fig. 92a) in which h/a is large, so that it can slip there.   The only way in which the load P can be lowered is by breaking the friction lock at B by loading the bar ABC so that the resultant load lies very close to the center of the bar (a < h/4).   This is done by pushing down on C with a force almost equal to the weight of the car P, by inserting a steel bar lever between C and the lip of the casting ¾ in. above it.   Then the whole casting suddenly slides down and lets the car down with it.   To hoist the car up, the lever E is used, which is pivoted at F around a pin held by the casting. The distance FE is about 16 in., and FD is about ½ in. When we pull up at E with unit force, we consequently push down on D with a force of 32 units. The point D is a fixed point by friction lock of the short horizontal bar D. The force at F is 33 units and is down on the lever

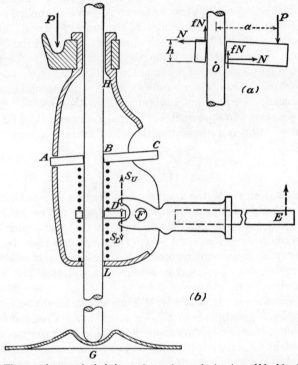

Fig. 92. The auto-bumper jack, being an intermittent device, is a self-locking hoist with better than 50 per cent efficiency.

from the casting or up on the casting from the lever. During the upward pull at E we thus exert a force, 33 times larger, upward on the casting at F, which will lift the whole casting with the automobile bumper P. The friction lock at D holds, but the friction lock at B is broken, because A moves up with respect to B and relieves the slight cocking angle shown. At the end of the upward stroke, the casting has been lifted, and D is still where it was before, so that the lower spring S_L is compressed, and the upper spring S_U is elongated. Letting down at the end E relieves the load at D. The friction lock at B takes over immediately, and the arm D, totally unloaded, is pushed up by the two springs S_L and S_U to its mid-position shown.

Thus, by intermittently carrying the load P by friction lock at B and at D, we avoid pushing against the locking friction during the hoisting period, and the efficiency of the device is close to 100 per cent.

i. The Rope and Hoisting Drum. Figure 93a shows a hoisting drum, rotating in a counterclockwise direction, over which a stationary rope is slung. On account of the friction between the moving drum and the steady rope the rope tensions T_0 and T_1 at the two ends are

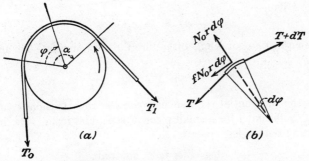

(a)   (b)

FIG. 93. The rope on the capstan.

different, and we now propose to study the relations involved. The rope tension T varies as a function of the angle φ from a minimum T_0 at $\varphi = 0$ to a maximum T_1 at $\varphi = \alpha$, and about all we can say of the manner in which T changes with φ is that it must be in a continuous way. Then it is appropriate to study the equilibrium of a short element of rope, an infinitesimal piece $r\,d\varphi$ at an arbitrary location φ, as shown in Fig. 93b. The forces acting on this element are the tensions T and $T + dT$ in the rope, and the normal and friction forces. The amount of normal force will be proportional to the length of the element and hence can be written $N_0 r\,d\varphi$, where N_0 is the normal pressure per inch of periphery. The friction force then is $f N_0 r\,d\varphi$, because there is relative motion. In order to write the equilibrium equations, we resolve these forces in two perpendicular directions, which we can choose freely. In order to write the equations as simply as possible, we choose the radial and tangential directions. The radial equilibrium requires that

$$T \sin \frac{d\varphi}{2} + (T + dT) \sin \frac{d\varphi}{2} = N_0 r\,d\varphi$$

Since $d\varphi$ is very small, we have $\sin d\varphi/2 = d\varphi/2$. Then of the four terms in this equation, three are small of the order of $d\varphi$, while the fourth one $dT\, d\varphi/2$ is small of the second order, and hence is neglected with respect to the other three. Thus, after dividing all terms by $d\varphi$, we have

$$T = N_0 r$$

In writing the equilibrium equation in the tangential direction, we encounter cosines of small angles, which differ from unity only by quantities small of the second order. (Readers who do not see this immediately are advised to look at the Taylor series development of a cosine.) Thus,

$$dT = fN_0 r\, d\varphi$$

Between the two equations of equilibrium we eliminate N_0 and find

$$\frac{dT}{T} = f\, d\varphi$$

This is the differential relation between the rope tension T and the friction coefficient f for a small piece of rope, and from it we find, by integrating both sides,

$$\log_e T = f\varphi + \text{constant}$$
$$T = e^{f\varphi + \text{constant}}$$
$$T = Ce^{f\varphi}$$

The constant of integration has a simple physical interpretation. At one end of the rope-drum contact we call $\varphi = 0$ and then $e^{f\varphi} = 1$, so that C is seen to be the rope tension at $\varphi = 0$, which we will call T_0. Thus

$$T = T_0 e^{f\varphi}$$

gives the tension in the rope at any angular distance φ from the first contact point where the tension is T_0. The tension rises like an e function with the angle, and this rise is extremely rapid. As an example, let $f = 0.5$; then the table below shows the ratio of the rope tensions on the two sides of the drum for various angles.

φ, radians	Times around drum	T/T_0
$\pi/2$	$\frac{1}{4}$	2.19
π	$\frac{1}{2}$	4.8
2π	1	23.1
4π	2	535.
6π	3	12,392.

The large magnifications T/T_0 that occur when the rope is slung two or three times around the drum are used for handling cargo on shipboard. A drum of about a foot diameter is rotated at constant speed always in the same direction by a steam or electric drive, and the end of a hoisting rope is wrapped around it several times. A man at the T_0 end of the rope can hoist very heavy loads by pulling on the rope by hand, and similar heavy loads can be let down by slacking off on the rope.

Problems **112** *to* **123.**

CHAPTER VII

SPACE FORCES

21. Composition of Forces and Couples. In Fig. 94 we see three forces F_1, F_2, F_3, drawn in full lines, all intersecting in a common point O, but not lying in the same plane, and with arbitrary angles (**not** equal to 90 deg) between the forces. The resultant of these three forces is the diagonal of the parallelepiped with the three forces as sides. This can be shown on the basis of the axioms of pages 4 to 6

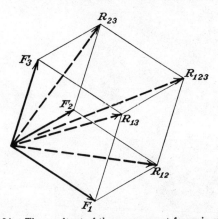

FIG. 94. The resultant of three concurrent forces in space.

as follows: Take two of the three forces, say F_1 and F_2, and as they lie in one plane, we can form their resultant R_{12} by the parallelogram construction. Then we consider the plane passing through R_{12} and through F_3 and find the resultant of R_{12} and F_3 by the parallelogram construction. Thus we arrive at R_{123}, being the diagonal of the parallelepiped. The order in which this composition is performed is arbitrary; we could just as well have performed either of the following sequences:

$$F_1 + F_3 = R_{13}, \qquad R_{13} + F_2 = R_{123}$$

or

$$F_2 + F_3 = R_{23}, \qquad R_{23} + F_1 = R_{123}$$

An important special case of this construction occurs when the angles between F_1, F_2, and F_3 are all 90 deg, as in a rectangular Cartesian coordinate system. Then the three forces are usually called X, Y, Z, and their resultant R is expressed by

$$R = \sqrt{X^2 + Y^2 + Z^2}$$

The resultant of more than three concurrent forces can be constructed by first forming the resultant of any three, then adding the fourth and fifth to it, etc.

From Fig. 94 it can be conversely concluded that **any single force in space,** say R_{123}, **can be resolved into components along three concurrent lines in arbitrary directions, provided the three lines do not lie in one plane.** On page 9, in discussing Fig. 4, it was said that a force could be resolved into *two* components in one plane with the force. If, in the more general three-dimensional case, we happen to pick two of our lines in the same plane with the force to be resolved, while the third direction must be outside that plane, we simply say that the component of our force in that third direction is zero.

On page 13 a definition was given of the moment of a force about a point O in its own plane. For space forces this definition has to be generalized appropriately. In the first place, the idea of "moment" is a turning effort, and we do not turn about a point but rather about an axis. In the previous definitions we really meant a rotation about an axis or line passing through O in a direction perpendicular to the plane about which we were talking. In three dimensions then, what is the moment of an arbitrary force about an arbitrary axis or line? The line of action of the force and the axis of rotation in general will not be perpendicular to each other, as they are in the plane case. But we always can resolve the force into a component parallel to the moment axis and into another component perpendicular to the moment axis. The first component, by our physical conception, does not try to rotate the body about the axis, and only the second component does. But this second component is situated just as is the force in the two-dimensional case of page 9. Thus we define

The moment of a force in space about a line equals the magnitude of the component of that force perpendicular to the line multiplied by the shortest, *i.e.,* **perpendicular, distance between the moment line and the line of action of the force.**

From this definition it follows that the statement of Varignon's theorem on page 14 is true also in three dimensions, provided that we read "line" instead of "point."

The moment of a force about a line in space is equal to the sum of the moments of the components of that force about the line.

For the proof of this statement, consider a plane through the point of action of the force, perpendicular to the moment axis. The force and all its components are now resolved into components parallel to the moment axis and into components lying in the afore-mentioned plane. The parallel components have no moments, and for the components lying in the plane, Varignon's two-dimensional theorem of page 14 holds, which proves the proposition. A further generalization of this theorem to non-intersecting force components is mentioned on page 136.

Next we turn our attention to *couples*. In Fig. 13 (page 17) it

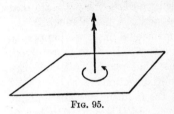

FIG. 95.

was seen that the two forces constituting a couple can be changed in magnitude and in direction without affecting the value of the couple; in other words, the turning effort or moment is the same whether we have two large forces close together or two small forces far apart, or whether these forces point east-west or north-south. The only thing that counts is the moment of the couple and the plane in which it acts. In Fig. 13c, the couple was represented as a curved arrow in the plane; for what follows it is more convenient to represent it by a straight arrow perpendicular to the plane (Fig. 95). This straight arrow is given two heads to distinguish it from a force; it is given a length measured in foot-pounds, proportional to the moment of the couple, and the direction of rotation of the couple is related to the sense of the straight arrow by a right-hand screw. From the discussion of page 17 we remember that the couple can act anywhere in the plane; therefore, **the double-headed (couple) arrow can be displaced parallel to itself to any location.** This cannot be done with a force or single-headed arrow. From Fig. 14 (page 18) we remember that if a force is displaced parallel to itself, we have to add a couple, or

A single-headed (force) arrow can be shifted parallel to itself only when a double-headed (couple) arrow is added to it, the double-headed arrow being perpendicular to the force vector as well as to the direction of the displacement.

We will now prove that double-headed (couple) arrows can be added by the parallelogram construction just like ordinary single-headed (force) vectors, or

The resultant of two couples, lying in different planes, is represented by the vector sum (parallelogram sum) of the vectors of the individual couples.

The proof is indicated in Fig. 96. Two planes are shown intersecting each other along the line AB at about 45 deg. In plane I a couple F_1F_1 is acting, and, for convenience, the forces F_1 have been drawn perpendicular to the intersection AB and at unit distance (1 ft) apart, which is always possible by virtue of Fig. 13. Likewise, in

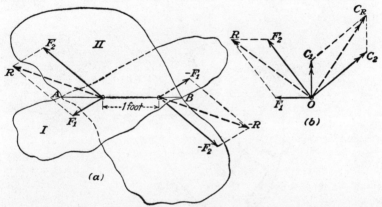

FIG. 96. The addition of couples lying in different planes.

plane II there is a couple F_2F_2, drawn in the same manner. It is now possible to lay a plane through F_1 and F_2 and another parallel plane through $-F_1$ and $-F_2$. These new planes will be perpendicular to AB and hence perpendicular to both plane I and plane II. The new planes are shown in Fig. 96b, F_1 and F_2 lying in the plane of the paper, and the line AB of Fig. 95a appearing as point O. Now we form R, the resultant of F_1 and F_2, and also $-R$, the resultant of $-F_1$ and $-F_2$. The resultants R and $-R$ again form a couple, lying in a new plane through R, $-R$, and AB. So far we have drawn no double-headed arrows. This can now be done in Fig. 95b. C_1 represents the couple F_1F_1. It is perpendicular to the plane I in which the couple operates, and since the moment arm of the couple is 1 ft, the length of the double-header C_1 equals the length of the single-header F_1. Similar properties hold for C_2 and for C_R. But now we observe that the entire figure $C_1C_2C_R$ is the same as the figure F_1F_2R, turned through 90 deg. This proves the statement.

Now we can proceed to the problem of finding the resultant of a large number of non-intersecting forces in space. Take one force and shift it parallel to itself until it intersects a second force. This introduces a couple. Combine the two forces by the parallelogram construction. Shift a third force parallel to itself to intersect with the resultant of the first two. This introduces another couple. Proceeding in this way we end up with one resultant force and a large number of couples in all directions in space. These double-headed couple arrows can be shifted parallel to themselves without punishment. We do this and make them all intersect; then we add them by repeated

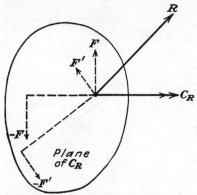

Fig. 97. A force and a couple is equivalent to a space cross of two forces.

parallelogram constructions. Thus, **any number of space forces can be reduced to a single resultant force and a single resultant couple.** We note that we say *a* force and *a* couple, not *the* force and *the* couple. Having found the resultant force and couple, we can shift the resultant force parallel to itself, introducing an additional couple, which can be compounded with the original resultant couple. Then we have again a force and a couple, different from the previous set.

There are two other simple forms to which an arbitrary system of forces in space can be reduced: the "space cross" and the "screw." The first of these is illustrated in Fig. 97, where the fully drawn vectors R and C_R constitute the resultant. We replace C_R by two equal and opposite forces F and $-F$ in the perpendicular plane, and choose F so as to intersect R as shown. Then the system is reduced to two non-intersecting forces, namely, $-F$ and the resultant of F and R. Two non-intersecting forces are called a "space cross." Again this can be done in millions of ways, because instead of F and $-F$, we could have chosen F' and $-F'$, still in the plane perpendicular to C, with a different, although equivalent, result.

Figure 98 shows the reduction to a *screw*, which is the combination of a force and a couple of which the vectors are along the same line, making the name quite descriptive. In Fig. 98 we resolve the resultant couple C_R into a longitudinal component C_l along F and a cross component C_c perpendicular to F. Then we use C_c to shift F parallel to itself until C_c is used up in the process. For example, in Fig. 98, F and C_c combined are equivalent to a force F located at a distance C_c/F below the paper. Then we shift C_l parallel to itself (without punishment) until it coincides with the new location of F. Thus a force F and a couple C_R with an angle α between them are equivalent to a screw of which the force has the value F and the moment has the value $C_R \cos \alpha$. The axis of the screw is found by moving F parallel to itself in a direction perpendicular to the plane defined

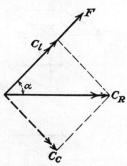

FIG. 98. Determination of the screw equivalent to a force and a couple.

by F and C_R through a distance $C_R \sin \alpha/F$. This construction is definite: **there is only one possible resultant screw.**

Summarizing, we can say that an arbitrary system of forces in space can be reduced to

    a. A force and a couple (in an infinite number of ways), or

    b. Two non-intersecting forces, *i.e.*, a space cross (in an infinite number of ways), or

    c. A screw (in only one way)

If the force and couple of the screw point in the same direction, the screw may be said to be a right-handed one; for opposite directions, it is left-handed. If the force is large and the couple small, the screw may be said to have a large pitch.

Now let us illustrate these theories with an example. Figure 99a shows a space cross consisting of two equal forces at right angles to each other. Where and how large is the equivalent screw? Let the forces be F and the distance between them a. First we make the forces intersect by moving the upper one down to meet the lower one. We do this by adding "nothing" in the form of $\pm F$ at OD and OE. Then we replace AB by $OD + (AB + OE)$. The latter term is a couple in plane OAD of value Fa, and with the right-hand-screw convention it is represented by a double-headed arrow pointing to the right, as shown in Fig. 99b. The force OD is compounded with OC to $F \sqrt{2}$ at 45 deg (Fig. 99b). The next move is to resolve the couple Fa into $Fa/\sqrt{2}$ along the force of Fig. 99b and $Fa/\sqrt{2}$ perpendicular to the force. The latter couple can be written $F \sqrt{2}(a/2)$, representing two forces $F \sqrt{2}$ with a moment arm $a/2$, and it can be interpreted as a force equal and opposite to $F \sqrt{2}$ of Fig. 99b plus a force parallel to $F \sqrt{2}$ at distance $a/2$ above it. Thus the force $F \sqrt{2}$ and the couple perpendicular to it together are equivalent to a force $F \sqrt{2}$ at mid-height. The other component of couple can then be shifted to mid-height, which leads to a right-hand screw of intensity $F \sqrt{2}$ (measured in pounds) and $Fa/\sqrt{2}$ (measured in foot-pounds), as shown in Fig. 99c.

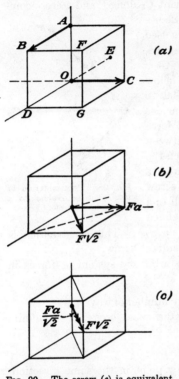

After this result appears, it looks quite plausible, and we might have derived it more simply as follows. We resolve force AB of 99a into components along AF and perpendicular to it. Similarly we resolve force OC along OG and perpendicular to it. The forces along AF and OG are alike and hence have a resultant at mid-height of $F \sqrt{2}$. The two components perpendicular to AF and OG form a couple in a plane perpendicular to AF and with moment $Fa/\sqrt{2}$.

Fig. 99. The screw (c) is equivalent to the space cross (a).

Problems 124 to 128.

22. Conditions of Equilibrium.

A three-dimensional body remains in equilibrium if, and only if, the total resultant of all forces acting on it is zero, and since this resultant takes the form of a screw, or of a space cross, or of a force and a couple with different directions in space, the condition for equilibrium can be worded in various ways.

For the two-dimensional case, three modifications of the conditions were given (page 19). Here we do likewise. The first statement is

A necessary and sufficient condition for the equilibrium of a body in space is that the resultant force vector (———>) and the resultant couple vector (———>>) both are zero.

Quite often we work with a rectangular or Cartesian coordinate system, and we resolve every force first into its x, y, and z components, which we denote by X, Y, and Z. Then we form the x component of the resultant, which is the sum of all individual x components of the individual forces, or in a formula

$$X_r = \sum_n X_n$$

or shorter

$$X_r = \Sigma X$$

and similarly for the other two directions leading to Y_r and Z_r. These three components of the resultant force can be compounded by the parallelepiped construction into the total resultant, but in this analysis we usually do not take that step.

Next we turn our attention to moments. We form the moments of all individual forces, first about the x axis. Let one of these forces (Fig. 100) be resolved into its three components X, Y, and Z. Of these, X has no moment about the x axis by the definition of page 13. The moment of the Y force is Yz, and the moment of the Z force is $-Zy$, the signs having been taken so as to call the double-headed moment arrow positive when directed toward increasing x. It is seen that force Z tends to screw (in right-handed direction) along the x axis toward the origin,

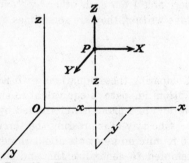

FIG. 100.   The moment of a space force.

i.e., toward decreasing x, and therefore the moment is negative, $-Zy$. Thus the total moment about the x axis of one force is $Yz - Zy$, and the moment of all the forces about the x axis, being the x component of the resultant double-headed arrow, is $\Sigma(Yz - Zy)$, summed over all the forces. We have equilibrium if the resultant force and the resultant couple are zero, hence if all their components are zero.

The **second statement of the equilibrium condition** then, expressed in the form of six formulas, is

$$\left.\begin{array}{l}
\Sigma X = 0 \\
\Sigma Y = 0 \\
\Sigma Z = 0 \\
\Sigma M_x = \Sigma(Yz - Zy) = 0 \\
\Sigma M_y = \Sigma(Zx - Xz) = 0 \\
\Sigma M_z = \Sigma(Xy - Yx) = 0
\end{array}\right\} \quad (4)$$

Before proceeding, these formulas should be clearly understood. First the reader should check the last two on Fig. 100. Then he should take a point P in another octant, where one or two x, y, or z values are negative and see that Eqs. (4) are still correct. Finally he should notice the symmetry in the moment equations. In the parentheses of the x equation, only the letters x and X are absent. The fifth equation can be derived from the fourth by replacing in it x by y, y by z, and z by x, or, when writing the letters xyz in a circle, 120 deg apart, by replacing each letter by the next one in the circle, either clockwise or counterclockwise, but always in the same direction. Starting from any one of the three moment formulas, this process twice repeated will give the other two, and of course the third repetition brings us right home again. The three formulas are said to be derivable from each other by "cyclic permutation," and in order to save writing, they are sometimes written

$$\left.\begin{array}{l}
\Sigma X = 0, \text{ cyclic} \\
\Sigma(Yz - Zy) = 0, \text{ cyclic}
\end{array}\right\} \quad (4)$$

Comparing these equations with Eq. (1) (page 20), we see that **equilibrium in space is equivalent to six algebraic conditions, while equilibrium in a plane is determined by three algebraic conditions.**

One way of expressing the three algebraic conditions in a plane is to require no moments about three points or axes (page 20). We are tempted to generalize this and state that a body in space is in equilibrium when the moments taken about six axes in space become zero. However, there is a limitation to this. In the plane the three moment points could not be on a common line, could not be collinear. In space we have something similar. Imagine the resultant of all forces on the body to be a single resultant force, without resultant couple. Then if we take six (or more) axes in space, all of which happen to intersect the line of action of the resultant force, the moments about these axes are zero, and still the system is not in equilibrium, having a resultant. Thus the six axes must be so taken that they do not have a common line of intersection. It is easy enough in a plane to see whether three points are collinear or not, but it is much more difficult to decide whether six arbitrarily chosen axes do or do not have a common line of intersection. Therefore, the equilibrium condition of no moments about six axes is of little value for practical purposes.

If a body is subjected to arbitrary forces in a plane, it is adequately supported by a hinge and by a roller support (Figs. 23, 71, and 101a).

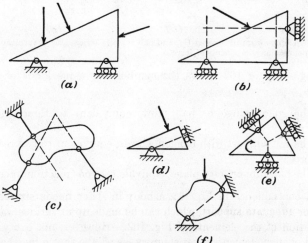

Fig. 101. Various manners of supporting a plane body.

A roller support can take a force perpendicular to the track only, and a hinge can take a general force in the plane having two components, say horizontal and vertical. Thus we have three supporting reactions, just the right number to satisfy the three equations (1). A hinge in a plane is thus equivalent to two reactions, and a roller support is equivalent to one reaction. Then a body in a plane can be supported in a statically determinate manner by three roller supports, as shown in Fig. 101b (provided, of course, that the reactions at all three supports are compressive). A bar support as shown in Fig. 101c is equivalent to a roller support, because the bar can take a single force only along its center line, which is the same as in the roller where the only possible force is perpendicular to the track. The only limitation in all of this is that the three possible reaction forces should not pass through a single point, because in that case they can be compounded into a single resultant, which is equivalent to only two com-

ponents, not three.   Thus the plane structures of Figs. 101*d* and *e* are
inadequately supported, whereas Fig. 101*f* is statically indeterminate,
having supports equivalent to four forces, one too many.

Now we are ready to consider the space generalization of the

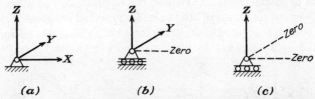

FIG. 102.   A hinge (*a*), rail support (*b*), and ball support (*c*) capable of furnishing 3, 2,
and 1 perpendicular components of force.

foregoing.   Figure 102 shows the symbols for three kinds of space
supports:

    a. The space hinge or ball joint, equivalent to three reaction
        forces
    b. The rail support with flanged wheels, equivalent to two reaction
        forces
    c. The ball support on a plane, equivalent to one reaction force

As a consequence of Eq. (4), **a body in space needs six reaction
forces for adequate support,** which can be made up in various ways by
combination of the elements of Fig. 102.   However, not every combination of such elements is satisfactory, as will be shown in example *d*
of the next article.

    Problems 129 *and* 130.

    23. Applications.   *a.   The Boom with Two Guy Wires.*   Figure 103*a*
shows (in elevation) a boom *OAB*, pivoted at *O*, loaded with a vertical
load *P* at the end *B* and supported by two guy wires *AC* and *AD*.
These latter can be seen better in the plan drawing 103*b*.   We want
to know the forces in the guy wires as well as the forces and bending
moments in the boom.   First we remark that the guy wires *AC* and
AD lie in one plane; hence their two forces must have a resultant in
that plane, acting through point *A*, and, on account of the symmetry,
that force must lie in the plane of elevation 103*a*.   Then the boom, as
an isolated body, feels three forces: the load *P*, the resultant of the
guy-wire forces, and the reaction at *O*.   The first two forces intersect
at *E*, and hence the third force must also pass through *E*.   Construct
the triangle of forces at *E*, consisting of the load *P*, of the ground

reaction force of the boom B, parallel to OE, and of the guy-wire resultant G. The force G is *from* the guy wires *on* the boom and is toward the left; hence the guy wires are in tension.

Then, in order to resolve the resultant between the two guy wires, we must have the triangle ACD in full size, not foreshortened by projection. This is shown in Fig. 103c, where $C'D' = CD$ of 103b, and

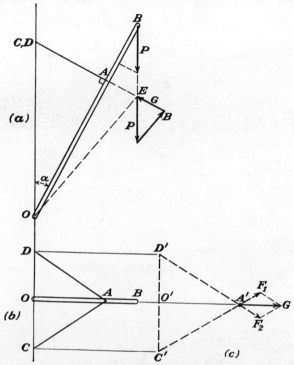

FIG. 103.   Plan and elevation of a boom with two symmetrical guy-wire supports.

$O'A' = CA$ of 103a. In 103c the resultant force G is drawn in full length, this time to the right, being now the force from the boom on the wires. The force is resolved into $A'F_1$ and $A'F_2$, representing the tensile forces in the two guy wires. The compressive force in the boom is $P \cos \alpha$, constant all along its length because in this example the plane of the guy wires is perpendicular to the boom. If this angle were different, the compressive force in the boom would change at A by an amount equal to the projection of force G on the boom. The shear force in the boom is $P \sin \alpha$ in the section AB; it becomes $P \sin \alpha - G$

in the section OA. The bending-moment diagram of the boom consequently is a triangle with its peak at A, the moment there being $P \sin \alpha \times AB$. This solves the problem.

How does this example connect with the theory of page 112, requiring six supports for a body in space? The boom plus guy wires are supported by a ball joint at O, equivalent to three reaction forces, and by wires at C and D, each equivalent to one force—a total of five.

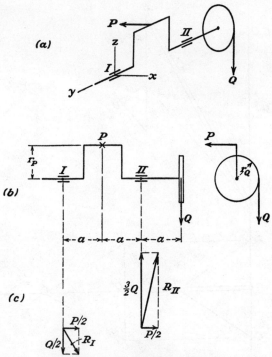

Fig. 104.   The equilibrium of a crank.

This is one less than required for adequate support, and on inspection, we notice that there is one kind of loading the boom cannot resist and that is a twisting couple or a double-headed arrow along its center line. Therefore, Fig. 103 is not capable of taking all possible loads, and in order to support it adequately, we would need one additional support in the form of a wire or strut between the end of a rigid cross-arm of the boom and a sidewall.

b. *The Crank and Pulley.* Figure 104a shows a crank in two bearings and a pulley attached to it. The dimensions can be seen in the projections of Fig. 104b. We want to know the bearing reactions. First we try counting to six to test whether the support is adequate. A bearing can take a sidewise or an up-and-down force, but the shaft is presumably free to slide along the bearing. Therefore, a bearing counts for two, and there are four supports only. We need two more, for instance, a thrust collar on the shaft to prevent motion along the y axis and a wire at Q attached to the ground to prevent rotation. If we want the system as shown to be in static equilibrium, we must take care not to load it in a sliding or in a rotational direction. There are no forces shown along the y axis, but there are two forces shown, P and Q, having moments about the y axis, and in general, the sum of these two moments will not be zero, and there will be no equilibrium. Only for the relation $Pr_P = Qr_Q$ will the total y moment be zero. In order to determine the bearing reactions, we take an xyz coordinate system with its origin in bearing I as shown. We write the six equilibrium equations of the shaft in terms of the forces P, Q, and the four unknown bearing reaction components, X_I, Z_I, X_{II}, Z_{II} (*from* the bearings *on* the shaft), as follows:

x forces:
$$X_I + X_{II} - P = 0$$

y forces:
$$0 = 0$$

z forces:
$$Z_I + Z_{II} - Q = 0$$

x moments:
$$Z_{II}2a - Q3a = 0$$

y moments:
$$Pr_P - Qr_Q = 0$$

z moments:
$$Pa - X_{II}2a = 0$$

Solving algebraically for the various unknowns from these equations gives the result

$$Q = \frac{Pr_P}{r_Q}$$
$$X_I = \frac{P}{2}, \qquad X_{II} = \frac{P}{2}$$
$$Z_I = -\frac{Q}{2}, \qquad Z_{II} = \frac{3Q}{2}$$

which solves the problem.

The same result can be derived (substantially in the same manner) by inspection. First we shift force P parallel to itself down through a distance r_P, so that it acts in the center line, and also shift force Q to the left until it is in the center line. The moments introduced by

these two parallel shifts cancel each other.   Then the force P gives two
horizontal bearing reactions $P/2$, both in the $+x$ direction.   Similarly
the force Q is equilibrized by two vertical bearing reactions: $3Q/2$
upward at II, and $Q/2$ downward at I.   The total bearing reactions
R_I and R_{II} are then found by a parallelogram construction, as shown in
Fig. 104c, representing views from the left (in the direction of the $-y$
axis) on bearings I and II.

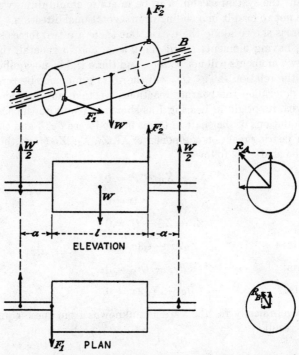

FIG. 105. Finding the bearing reactions of a rotor subjected to space forces.

c. *The Unbalanced Rotor.*   Figure 105 shows a rotor supported in
two bearings A and B acted upon by three forces W, F_1, and F_2,
as shown.   (The forces F_1 and F_2 may be interpreted later, on page 216,
as the centrifugal action of unbalances in the spinning rotor; for our
present problem this does not concern us, and F_1 and F_2 are just
forces of given magnitude.)   The problem consists in finding the bearing reactions at A and B.

As in the previous example, the two bearings are equivalent to
four reaction forces; two other reactions are necessary to support

the rotor adequately, but since there are no forces acting that tend to shift the rotor longitudinally or tend to rotate it, these two supports may be left out. The forces on the rotor all pass through the center line and are all either horizontal or vertical. Thus the problem falls apart into two plane problems. The elevation sketch shows the vertical forces. The load W has reactions $W/2$ at both bearings. The force F_2 has a reaction $F_2(a + l)/(2a + l)$ at the right and $F_2a/(2a + l)$ at the left. The plan sketch shows the horizontal force F_1 and its two reactions. These reactions are compounded in the sketches at the right of the figure, both seen from left to right along the rotor. The reader should verify that

$$R_B = \sqrt{\left[\frac{W}{2} - \frac{F_2(a + l)}{2a + l}\right]^2 + \left(\frac{F_1a}{2a + l}\right)^2}$$

and that the direction shown is that *from* the bearing *on* the rotor.

In case the forces F_1 and F_2 do not happen to lie in the horizontal and vertical planes, but at random angles, those forces can be resolved into components and the problem solved in the same manner.

d. The Table on Three or More Legs. Consider in Fig. 106 a table on three legs A, B, C, loaded with a vertical load P on the top. To make things general, the legs have been irregularly spaced, and the load is off center. How does the load distribute itself between the three legs? The problem can be solved by plane statics. Draw the

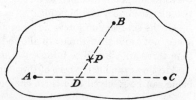

Fig. 106. A vertical load P on a table with three legs A, B, and C.

line BP, which intersects AC in point D. In the vertical plane BPD, resolve the load P into its components at B and at D. Then, in the vertical plane ADC, resolve the force D into components at A and C. Thus the force P is resolved into three components at A, B, and C, which are the compressive forces in the legs. Another way of finding the leg reactions is by taking moments about the line AC. It is seen that reaction B is smaller than P in the ratio of the distances from AC, *i.e.*, in ratio $PD:BD$. To find the reactions at A or C we take moments about the lines BC and AB, respectively. It is interesting to note that the location of P is the center of gravity of three weights at A, B, and C, each equal to the respective leg reaction.

The table on three legs standing on an icy floor has three reactions and therefore is not adequately supported, according to page 112. It can only take a purely vertical load and is powerless against sidewise forces (both in the x and y directions) as well as against a couple in the plane of its top. In order to support it adequately, one of the legs has to be made a "space hinge" equivalent to three reactions, and another leg has to be a "rail support," equivalent to two reactions. This would make the table capable of sustaining any kind of force or couple. However, not all possible combinations of six supports are satisfactory. If we place the table on four legs, one of which is a space hinge, the other three being "plane ball supports" (Fig. 102), the table is statically indeterminate under a vertical load, whereas it is not supported at all against a couple in its own plane, being free to rotate about the one hinged leg. Or if we place the table on six legs all on plane ball supports, it can sustain only vertical loads and has three legs too many for that function. The two last examples both have six supports.

The question of how to recognize when six supports are adequate and when they are not is too involved to be treated completely in this book. However, two unsatisfactory combinations we can understand fairly simply.

Of the six reaction forces necessary for the adequate support of a body in space, not more than three are allowed to lie in one plane, and also not more than three are allowed to intersect in one point.

First we examine the case of three lines in one plane. A force lying in that plane can be resolved into three components along the three prescribed lines

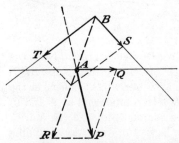

in a unique manner, as is shown in Fig. 107. The force there intersects one of our lines at A. Connect A with the intersection B of the other two lines, and resolve P along the first line (component AQ) and along line AB (component AR). The force AR can be shifted along its line to start at point B and again resolved into components BS and BT. Thus the force AP is resolved into components AQ, BS, and BT in a definite manner, or in other words, a plane structure, supported by

Fig. 107. A force P can be uniquely resolved into components along three arbitrary lines in its own plane.

three rods in arbitrary directions in that plane, can take any arbitrary force. In case there were more than three rods in that plane, the construction would break down, and the fourth rod would be redundant.

Next we examine the case of three or more lines intersecting at a point. A special case of this occurs when the point goes to infinity and the lines become parallel. In Fig. 106 we have seen how a force parallel to three other parallel lines can be resolved in a unique manner into components along those three lines. Later, in Fig. 111 (page 123), we shall see that a force in general can be uniquely resolved into components along three arbitrary directions intersecting with the force in one point. Both constructions, Fig. 106 and Fig. 111, break down for more than three directions. Mechanically speaking this means that if four or more rods, all intersecting in a point, support a body, then the fourth and additional rods are redundant.

e. The Center of Gravity of a Three-dimensional Body. On page 34, the center of gravity of a two-dimensional body was defined as the point through which passes the resultant of the weight forces of all small individual elements, irrespective of the direction in which these weight forces act. The location of this point G was expressed by Eq.

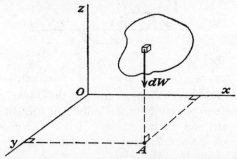

Fig. 108. The center of gravity of a three-dimensional body.

(2). Now let us generalize this to a three-dimensional body, placing it in an xyz coordinate system (Fig. 108) and letting gravity act in the negative z direction. The moment arm about the y axis of the weight dW of an element is Ay, which is equal to the coordinate x of the element. Thus the sum of the moments about the y axis of all the elements of the body is $\int x \, dW$, and this should be equal to the moment of the resultant force W about that axis, which is $x_G W$. Thus we find for x_G

$$x_G = \frac{\int x \, dW}{\int dW} \qquad (2a)$$

Next let the weight act in the x direction and take moments about the z axis, and after that let the weight act in the y direction and take moments about the x axis. This leads to the additional equations

$$y_G = \frac{\int y \, dW}{\int dW} \qquad (2b)$$

$$z_G = \frac{\int z \, dW}{\int dW} \qquad (2c)$$

The three coordinates x_G, y_G, and z_G determine the location of a point in space, which is called the "center of gravity," denoted usually as G. It is left to the reader to generalize the proof of page 34 to three dimensions, and thus to demonstrate that if the weight acts in a direction skew with respect to the three axes, the resultant force will still pass through the same point G.

As an example let us apply this to Fig. 109, showing an idealized two-throw crank, which is made up of nine pieces of heavy line, all of equal length a, but in which the three pieces along the x axis (the main

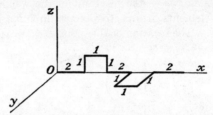

FIG. 109.   The center of gravity of a crankshaft.

journals of the crank) weigh twice as much as the crankpins and crank cheeks.   The centers of gravity of the individual straight stretches lie in their mid-points, and the integrations of Eq. (2) reduce to sums of nine terms, which are summarized in the table below.   The pieces are numbered starting from the origin.

No.	Weight	x	xW	y	yW	z	zW
1	2	$a/2$	a	0	0	0	0
2	1	a	a	0	0	$a/2$	$a/2$
3	1	$3a/2$	$3a/2$	0	0	a	a
4	1	$2a$	$2a$	0	0	$a/2$	$a/2$
5	2	$5a/2$	$5a$	0	0	0	0
6	1	$3a$	$3a$	$a/2$	$a/2$	0	0
7	1	$7a/2$	$7a/2$	a	a	0	0
8	1	$4a$	$4a$	$a/2$	$a/2$	0	0
9	2	$9a/2$	$9a$	0	0	0	0
Sums	12		$30a$		$2a$		$2a$

$$x_G = \frac{\Sigma xW}{\Sigma W} = \frac{30a}{12} = 2\frac{1}{2}a$$

$$y_G = \frac{\Sigma yW}{\Sigma W} = \frac{2a}{12} = \frac{a}{6}$$

$$z_G = \frac{\Sigma zW}{\Sigma W} = \frac{2a}{12} = \frac{a}{6}$$

This describes the location of the center of gravity, and the reader should visualize that location in Fig. 109.

Problems 131 to 136.

24. Space Frames. A space frame is a structure built up of hinged bars in space; it is the three-dimensional generalization of a truss (pages 52 to 59). Every statement made and every result found in connection with trusses can be appropriately generalized to space frames, and as long as we are content with these general statements and theorems, the discussion of space frames is fairly simple. However, we can no longer draw all the bars in one drawing on a sheet of paper, and the force triangles at each joint of the truss become space diagrams in the space frame. One such space construction is not particularly difficult, but it *is* more complicated than a triangle, and this, combined with the fact that the number of bars in space frames is considerable, makes the actual calculation of such a structure an arduous task. We will discuss here only some of the simpler aspects of it and work out in detail as an example the crane structure of Fig. 112.

On page 53 we saw that the simplest combination of bars that forms a stiff frame in a plane is a triangle, and the simplest way of building up a statically determinate truss is to add new triangles to the previous ones by putting in two new bars and one new joint at a time.

In space we start again with a plane triangle ABC (Fig. 110), and now we wish to fix a point D in space (outside the plane of the triangle) to the triangle in a rigid manner with a minimum number of bars. We need three bars to do it, because if we had only two, say AD and CD, then point D could still rotate with triangle ACD about the line AC. The third bar BD stops this rotation and fixes point D. Thus, a **tetrahedron** (Fig. 110) **is the simplest statically determined, stiff space frame.** From this funda-

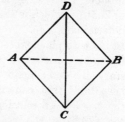

Fig. 110. Six bars, forming the sides of a tetrahedron, constitute the simplest rigid space frame.

mental frame we can start building. If a fifth point E is to be attached to it, we can choose one of the triangles of the tetrahedron, say BCD, and attach the new joint E to it by three new bars BE, CE, and DE. Thus we add three new bars and one new joint at a time. In such a structure, therefore, the relation between the number of bars, b, and the number of joints, j, must be $b = 3j +$ constant. In the tetrahedron we count $b = 6$ and $j = 4$, which, substituted in the formula, gives $6 = 12 +$ constant; hence, the constant must be -6. The relation between bars and joints thus is

$$b = 3j - 6$$

A space frame of this description with hinged joints, subjected to arbitrary forces at the joints only (and hence *not* to forces on the bars between joints), will have forces in the bars along their center lines only, so that the bars are without bending.

The method of joints in a truss (page 56) is based on the fact that there are two equilibrium equations at a joint: the horizontal and vertical force equations; there is no moment equation because all forces pass through the joint and hence have no moment arms.   Then, if all but two of the bar forces at a joint are known, these two can be calculated.   In space there are three equilibrium equations at a joint: the x, y, and z force equations; again there are no moment equations for the same reason as in the plane truss.   Thus, if in a space joint all but three bar forces are known, these three can be calculated.   By setting up equilibrium equations for all the joints of a space frame we obtain $3j$ equations.   The unknowns in these equations are the bar forces, b in number, and the reactions at the supports, which, as we saw on page 112, are six in number.   Thus we see that in principle the number of statics equations $3j$ is just sufficient to solve for the $b + 6$ unknowns, which is another way of saying that the frame is statically determinate.

As with plane trusses (page 53), the method of constructing a space frame by consecutive tetrahedrons is not the only one whereby a stiff, statically determinate frame can be obtained.   Many frames have been built that cannot be so constructed, although they do satisfy the $b = 3j - 6$ relation.   Further discussion of this subject leads us too far, and the reader is referred to Timoshenko and Young's "Engineering Mechanics" or to their "Theory of Structures."

Now we come to the fundamental construction that in the method of joints has to be performed over and over again: the determination of the forces in three bars coming together at a joint, subjected to a given external force at that joint.   The graphical construction, solving this problem, is performed in Fig. 111, where F is the force and a, b, and c are the three bars, drawn in plan and elevation.   Before going into details, the general procedure of the construction will be described.   There are four intersecting forces, F and the three unknown ones, A, B, and C.   We choose one of these three, say A, and compound it with F. The resultant must lie in the plane through A and F. For equilibrium this resultant must be equal and opposite to the resultant of the two remaining forces B and C, which lies in the plane through B and C. Thus both resultants lie along the line of intersection of the planes A,F and B,C. In the figure the points a, b, c, and f, in plan as well as in elevation, denote the intersections of the

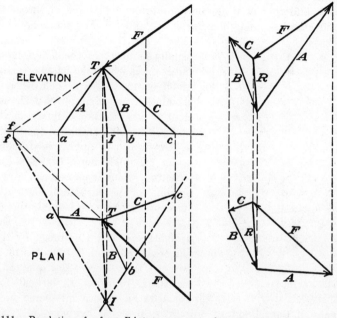

Fig. 111. Resolution of a force F into components along three lines intersecting in a point with the force.

corresponding forces with the base plane. Then the dot-dash lines af and bc in the plan drawing are the traces of the planes A,F and B,C with the base plane. These two traces intersect at I; hence TI (both in plan and elevation) is the line of intersection of the two planes and is the direction of the partial resultants $(A + F)$ and $(B + C)$. Now we proceed to the force diagrams to the right of Fig. 111. First the known force F is drawn, and a triangle is formed on it with directions parallel to A and to TI, both in plan and elevation. This leads to the vertical and horizontal projections of force A and of the resultant R. This resultant equals the sum of the unknowns B and C, so that other triangles with sides parallel to B and C are erected on R as a base. The arrows are drawn in the direction of the forces acting *on* the point *T*, and it is seen that, in both the elevation and plan projections of the force diagram, the polygon is a closed one with all the arrows in the same direction. The bars *A* and *C* are in compression, and *B* is in tension.

The same problem can be solved analytically by resolving *F*, as well as the unknown forces *A*, *B*, and *C*, into their *x*, *y*, and *z* components, and then writing the equilibrium equations in those three directions and solving them for *A*, *B*, and *C*. For the general case

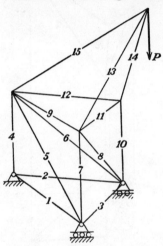

FIG. 112. A rigid and statically determined space frame.

(Fig. 111) the graphical solution is usually simpler than the analytical one, but in cases where the bars have angles of 90 or 45 deg with each other, the calculation may simplify itself sufficiently to make the analytical method preferable to the graphical one.

Now we will apply the foregoing to an example (Fig. 112) of a crane structure, which is chosen as simple as possible for the purpose of illustration. It consists of two horizontal triangles, 1,2,3 and 9,11,12, joined by three uprights 4,7, and 10, braced by three diagonals 5,6, and 8. On top of this box the three bars 13,14, and 15 fix the apex, carrying a vertical load. The bars have been numbered in the

sequence of their build-up process, as described on page 121. Note that there are 15 bars and 7 joints, satisfying the equation of page 122. The structure is shown supported on a hinge under 4, equivalent to three reactions; a rail under 10, equivalent to two reactions, and a plane ball support under 7. All of these are necessary to support a general loading at the apex, but in this case, for simplicity, the loading *P* has been assumed purely vertical. This gives only vertical reaction forces at the three supports, so that for this loading, three plane ball supports would have been sufficient.

In Fig. 113, the construction has been carried out in plan and elevation by repeated operations of the procedure of Fig. 111. First we search for a joint where only three bars exist, and the top appears to be a good place to start. (We could also start from joint 1,2,4, but in that case, we would have to find the reaction first, which is not

necessary at the top.) The two force diagrams just to the right of
the main figures in Fig. 113 are for the top joint and are started from
the circled point. First we lay off *P*, which in the plan appears as a
point only. Then we conclude from symmetry that the forces in
13 and 14 must be alike, so that the resultant of those two forces lies
in the (vertical) plane through *P* and 15. The elevation force diagram then consists of three forces in a plane (force *P*, force 15, and the
resultant 13,14), and it can be drawn immediately. In the plan dia-

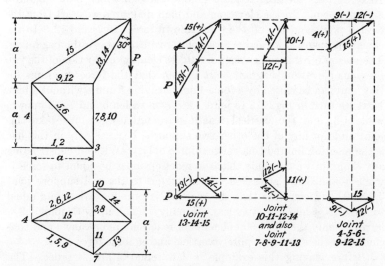

Fɪɢ. 113. Determination of the bar forces in the crane of Fig. 112.

gram we have three unknown forces: 13, 14, and 15, so that usually
we could not draw it. But we know that forces 13 and 14 have the
same magnitude, so that the diagram follows. The arrows are forces
on the top joint, so that we see that 15 is in tension and 13,14 are in
compression. Force 15 appears in its full size in the elevation diagram; 13 and 14 must be compounded properly from the two projections shown. (They are *not* to be compounded by the Pythagorean
theorem; it is left to the reader to visualize how these forces are
oriented in space and to find their magnitudes by graphical construction from the projections shown.) Thus we know forces 13,14,15,
and, looking at Fig. 112, we search for another joint with not more
than three unknown bars. We have only one such joint, and it is
10,11,12,14. The force diagrams for this joint are shown next in

Fig. 113, again starting from the circled points with the known force 14. The diagrams are simple because of the fortunate circumstance that one force reduces to a point in projection (force 11 in the elevation and force 10 in the plan). Note that the arrows of force 14 in these diagrams are the reverse of those for the top joint. (Why?)

The next joint we can approach (Fig. 112) is 7,8,9,11,13, and we conclude from symmetry that force 9 must equal force 12, force 7 must equal force 10, and consequently the diagonal 8 is without stress. Next we take the six-bar joint 4,5,6,9,12,15, in which only 4,5,and 6 are unknown. In the elevation and plan diagrams we start from the circled point and lay off the three known forces 15,9, and 12. In the elevation diagram we arrive at a point vertically above the beginning. The reason for this coincidence should be clear on inspection of the two previous elevation diagrams. It is seen that the force diagrams for this joint can be closed by a force 4 only, leaving 5 and 6 without stress. Next we turn in Fig. 112 to joint 1,3,5,7, in which 5 and 7 are known, while 1,3, and the vertical reaction are unknown. (We take this joint and not any of the other two supports, because only in this one is the reaction limited to a vertical force only.) Without constructing any diagram we conclude that the reaction must be equal to force 7, while 1 and 3 are stressless. Next we go to the rail support joint 3,2,8,10, and conclude that the reaction is purely vertical and equal to force 10, while 2 is stressless. Finally we end up with joint 1,2,4, in which only the three components of reaction are unknown and conclude that the reaction is purely vertical and equal to force 4.

Before leaving this example, a few remarks are in order. The entire bottom part of the structure, with the exception of the three vertical struts, is seen to be without force. Could then the diagonals 5,6,8 and the bottom bars 1,2,3 have been omitted without danger? The answer to this is yes, provided that the load is purely vertical as assumed. For a small horizontal component in the load P, all those bars will acquire forces, and then they are necessary.

On page 54 the method of sections was discussed for plane trusses, and was seen to be of advantage in case only a single bar force was to be found. This method of sections also can be generalized to space frames. For example, if in Fig. 112 only the force in bar 13 is wanted, we can cut off the top of the crane through bars 10,11,12,13, and 15. That top end then is loaded with those five bar forces and with P. Now we write the moment equation about bar 12 as axis. All moments are zero except those of P and of bar 13, so that the force in that bar can be determined immediately. If force 15 were wanted, we could make the same section and write the moment equation about bar 11 as axis.

Problems 137 *and* 138.

25. Straight and Curved Beams. On pages 68 to 80 we discussed shear-force and bending-moment diagrams in straight beams subjected to forces in one plane. This will now be generalized in two directions: first to straight beams subjected to forces and moments in several

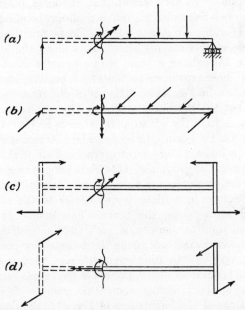

(a)

(b)

(c)

(d)

FIG. 114. Bending moments in two planes and the twisting moment in a beam.

planes, and second to curved "beams," such as arches in building construction, rings, helical springs, and other common machine elements.

In the bending-moment diagrams of Figs. 77 to 81, where the loading (and the reactions) were all in one plane, the bending moment at any section could be represented by a curved arrow in the plane of the loadings or by a double-headed straight arrow perpendicular to that plane (Fig. 95). This is illustrated in Fig. 114a, where all loads and reactions are in a vertical plane, and the double-headed bending-moment arrow at a section is horizontal, pointing into the paper. By turning that figure through 90 deg we arrive at Fig. 114b with horizontal loadings and reactions, and a bending-moment vector in the (vertical) plane of the paper, pointing downward. In case a beam is loaded with forces at arbitrary angles, neither horizontal nor vertical, these forces can be resolved into horizontal and vertical components, and we have a combination or "superposition" of the two cases 114a and 114b. At each section, then, the beam has a bending moment in the horizontal plane and another in the vertical plane. Plotting these along the beam we obtain two bending-moment diagrams: a horizontal and a vertical one. At each section it is, of course, possible to compound the two bending moments into a resultant bending moment in a skew plane (Fig. 96b), but as a rule this has no practical advantage, since the angle of this oblique plane varies along the length of the beam. Figure 114c shows a beam with crossarms at the ends, and the loading of these crossarms amounts to a bending moment in the vertical plane, as in Fig. 114a. The only difference between 114c and 114a is that the bending moment in 114a grows proportionally with the distance from the left support, whereas in 114c it is constant along the length of the beam. Of course, Fig. 114c can again be turned 90 deg about the beam center line, crossarms and all, to give horizontal bending moments. Finally, Fig. 114d shows something new. Here the forces on the crossarms are perpendicular to the beam, and they amount to a couple in a plane perpendicular to the beam with a double-headed arrow along the beam. This couple at the beam section is called the "torsion couple" or "twisting couple," which in 114d is of constant magnitude along the beam. For other loadings the twisting couple or "twisting moment" may vary from point to point along the beam, and the magnitude of the couple can be plotted along the beam as the "twisting-moment diagram'"

In the sections of the four beams of Fig. 114, the double-headed straight arrows are a much clearer indication of the moments involved than are the curved arrows. This becomes more pronounced in cases of greater complication, so that in what follows the curved plane arrows will be dropped altogether in favor of the double-headed straight ones.

Consider Fig. 115, showing an L-shaped beam built into a solid support at the bottom. In 115a this beam is loaded with a single load in the plane of the L. This being a plane problem, we expect only a bending moment in the plane of the beam with the moment vector perpendicular to the plane and no twist or bending in the other plane. In the upper leg of the L, the bending moment is Px, proportional to the distance x from the load. In the vertical leg the bending moment is Pa, independent of the location along the leg. Thus the bending-moment diagram has the shape shown, where the value of the moment is plotted in a direction perpendicular to the beam. Now we examine the space problem of Fig. 115b, where the beam is loaded by a force P perpendicular to the plane of the L. In the upper leg we have again a bending moment Px, as before, but this time the bending moment lies in a plane perpendicular to the paper with the moment vector lying in the plane of the L as shown. In the vertical leg there is a constant twisting couple Pa, which is plotted perpendicular to the beam in Fig. 115 and is cross-hatched diagonally. It is interesting to follow what happens to this moment when it turns the sharp corner of the L. For this purpose consider two sections: a

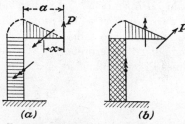

(a)   (b)

Fig. 115. Bending and twist in an L-shaped cantilever.

first one just to the right of the corner, and a second one just under that corner. Since the two sections are infinitely close to each other, the moment vector is the same in both sections, namely, Pa and pointed upward in the plane of the paper. But for the first section this vector is perpendicular to the beam and thus represents a bending moment, while for the second section it is along the beam and represents a twisting moment.

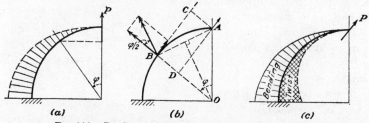

(a)   (b)   (c)

Fig. 116. Bending and twist in a quarter-circle cantilever.

A bending moment in a plane perpendicular to an L-shaped beam element becomes a twisting moment when the beam turns through a 90-deg corner.

Figure 116 shows a quarter-circle beam, built in at the bottom. When proceeding from one point on this beam to a next point, we do not turn through 90 deg but through a small angle. Let us investigate the moment situation. In Fig. 116a the force lies in the plane of the quarter circle; the problem is a plane one, and at the section φ indicated, the moment arm (dashed line) of force P is $R \sin \varphi$, so that the bending moment is $PR \sin \varphi$, with the arrow perpendicular to the paper pointing upward out of the paper. This value has been plotted perpendicular to the curved bar. In Fig. 116b, the end force is perpendicular to the paper, and at point B the moment arm of the force is $AB = 2R \sin (\varphi/2)$. The total moment vector shown has the value $2PR \sin (\varphi/2)$ and is directed perpendicular to AB, i.e., at angle $\varphi/2$ with respect to the radius through B. This moment vector can be resolved into radial and tangential components. The tangential component, interpreted as the local twisting moment, is $\sin (\varphi/2)$ times the total moment or

$$M_{\text{twist}} = 2PR \sin^2 \frac{\varphi}{2} = PR(1 - \cos \varphi)$$

The normal component, interpreted as the local bending moment, is $\cos (\varphi/2)$ as large or

$$M_{\text{bend}} = 2PR \sin \frac{\varphi}{2} \cos \frac{\varphi}{2} = PR \sin \varphi$$

These results can be derived somewhat differently. The moment arm of force P about the tangent BC is

$$AC = BD = OB - OD = R - R \cos \varphi = R(1 - \cos \varphi).$$

This is the twisting-moment arm, which checks the above result. The bending moment is the moment about the normal axis OB, which is AD in the figure; equal to $R \sin \varphi$ as before. In Fig. 115c the bending moment has been plotted to the outside of the circle and the twisting moment to the inside, both in a radial direction. At the built-in point the two moments have the same value; everywhere else the bending moment is the larger of the two.

One of the most useful applications of these relations is to coil springs. Figure 117 shows a coil spring in three conditions of loading:

  a. In tension: forces P at the two ends
  b. In twist: twisting couples M_t at the ends
  c. In bending: bending moments M_b at the ends

What are the reactions at a section through the spring wire in these cases? The spring is supposed to be tightly coiled, which means that the various circles of the coils practically lie in a plane and for this analysis are considered to be in a plane. At the ends the spring turns through 90 deg in the plane of the circle to continue along a radius to the center of the circle; there it turns once more through 90 deg and continues along the center line as shown. In case a the load is an upward force P. If somewhat more than half a turn is cut off at A as shown, vertical equilibrium of that piece requires that at A there is a downward shear force P at the section A. Then the two forces P form a couple in a plane perpendicular to the wire at section A, and this couple must be held in equilibrium by an equal and opposite couple at A. The direction of the double-headed arrow of this couple

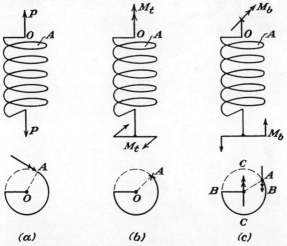

Fig. 117. A coil spring in tension (a), twist (b), and bending (c).

is tangential to the spring wire, independent of our choice of the location of A along that wire. Thus **the wire of a spring subjected to tensile forces is in twist.** If the forces P were reversed, all reactions would be reversed with it, including the twisting moment in the wire. It follows that the wire of a spring subjected to compressive forces likewise is in twist.

In Fig. 117b the ends of the spring are subjected to a twisting couple. Isolating the portion above the section A, we see that the equilibrium of this portion requires a moment vector at A equal and opposite to the one at O, because on page 104 it was seen that such vectors can be shifted parallel to themselves without penalty. Thus in the vertical projection of 117b we look on the point of the arrow at O, and we look on the tail or feather end of it at A. This, as is usual, is indicated by a dot for the point and by a cross for the feather end.

The vector at A is parallel to the center line of the spring, independent of our choice of point A; the moment is a bending moment in the wire. **A spring subjected to end twist has bending in its wire.**

Finally, Fig. 117c shows a spring subjected to bending moments at its ends. The moment at section A lies in the plane of the coil, and now its nature does depend on the location of A along the wire. For points B,B the moment is a twisting moment; for points C,C it is a bending moment; for any point in between B and C it is mixed bending and twist. Thus, **the wire of a spring subjected to bending at its ends is alternately in twist and in bending when proceeding along that wire in steps of 90 deg.**

Problems 139 *to* 141.

CHAPTER VIII

THE METHOD OF WORK

26. A Single Rigid Body. The "method of work" is a procedure for solving problems in statics that is different from what we have seen so far and that is of practical advantage in applications to complicated systems involving many elements. The method is known by several other names, for which the reader is referred to the historical note on page 151.

In describing the method we start by defining the term "work."

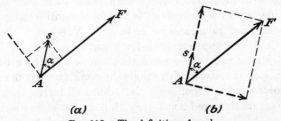

(a)   *(b)*

FIG. 118. The definition of work.

The work done by a force is the product of that force and the displacement of the point of action of that force in the direction of the force.

The meaning of this definition is illustrated in Fig. 118a, in which F is a force acting on the point of action A, which point is allowed to displace itself through a distance s in an arbitrary direction, including an angle α with the force. We resolve the displacement vector into components along F and perpendicular to F. The component along F, which is the "displacement of the point of action in the direction of the force," has the magnitude $s \cos \alpha$, and the work done by the force is, by definition,

$$W = Fs \cos \alpha$$

This expression can also be written as $s(F \cos \alpha)$, and $F \cos \alpha$ car interpreted as the rectangular component of the force along th of displacement. Thus **the work done by a force is also product of the displacement of its point of action and the (re**

133

component of the force along the line of displacement. This property is illustrated in Fig. 118*b*. As a particular case it follows that a force does no work when its point of action is displaced in a direction perpendicular to that force. The weight force does no work when a body is displaced horizontally. If a body is displaced in an arbitrary direction, partially horizontally and partially vertically, the work done by the weight force equals the weight multiplied by the vertical displacment, independent of the horizontal shift.

The work is considered positive if the displacement is in the same direction as the force and negative if these two directions are in opposition. Thus the weight force does positive work on a descending body, and it does negative work on a body that is being hoisted up, while the rope force that does the hoisting performs positive work.

Work is measured in inch-pounds or foot-pounds, which is the same measure as the moment of a force. Work has certain properties in common with moment: whereas work is the force multiplied by a length in line with itself, moment is the force multiplied by a length across itself. For moments we have seen Varignon's theorem on page 14. A similar theorem holds for work:

The algebraic sum of the amounts of work done by a number of forces acting on the same point equals the work done by the resultant of those forces. The proof is simple. Since all forces, including the resultant, act on the same point, they all have the same displacement.

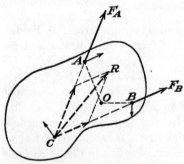

Let us resolve all forces, including the resultant, into their components along the displacement and perpendicular to it. The work of any force is the displacement multiplied by the in-line component of the force. Thus the work done by all forces (exclusive of the resultant) is the displacement multiplied by the algebraic sum of all in-line components. But by Fig. 5 (page 10) we know that the sum of all in-line components is the in-line component of the resultant, which completes the proof.

Fig. 119. Varignon's work theorem for forces acting on a rigid body.

Next we will try to generalize Varignon's work theorem from forces acting through a common *point* to the case of forces acting on a rigid *body*, of which the different points may have different displacements. For example, in Fig. 119, two forces F_A and F_B are acting on the rigid body at points A and B. We give to the body a *small* displacement by rotating it through a small angle about point O. The small displacement vectors are shown in the figure, and it is seen that F_A does positive work and F_B does negative work. It is by no means obvious (but true nevertheless) that the resultant R of F_A and F_B does an amount of work equal to the sum of the work done by F_A and F_B separately. To see this we must first prove that the axiom of transmissibility of forces (page 4) applies to work as well as to static equilibrium. Granted that it does, then we can shift F_A along its line of action to C, and at that location the force F_A does the same amount of work as it does at A. The same holds true for B, and then we have reduced our problem to that of forces acting through a common point C, for which Varignon's theorem was just proved.

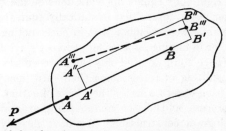

Fig. 120. The work done by a force on a small displacement of a rigid body is the same for all points of application of the force along its line of action.

To see that the axiom of transmissibility holds, consider Fig. 120, showing a rigid body on which a force P is acting at point A. We propose to shift the force along itself to make it act at another point B, and we propose to show that when the body is given an arbitrary *small* displacement, the work done by P at point A is the same as the work done by P at point B. Let the displacement of the body be broken up into several stages: first a parallel displacement of line AB (and with it the entire body) to the position $A'B'$; another parallel displacement to $A''B''$, perpendicular to the first displacement; and finally a small rotation about the mid-point between A'' and B'' to the final position $A'''B'''$. During the second and third stages of this displacement, the force P does no work, whether located at A or at B, because the displacements are perpendicular to P. Only during the first stage does the force perform work, to the amount $P \times AA'$ or $P \times BB'$. But, if the body is rigid, the length AB remains constant, or $AB = A'B'$, and consequently $AA' = BB'$. Thus we have proved that **a force acting on a rigid body which is given an arbitrary small** displacement performs an amount of work that is the same for all locations of the point of application of that force along its own line of action.

We note that this statement is true only for *small* displacements, because for a large rotation from $A''B''$ to $A'''B'''$ in Fig. 120, the displacement $A''A'''$ does have a component in the direction of P. We also note that it is true only for *rigid* bodies, because if the body were deformable, the length $A'B'$ would not have to be the same as AB.

By the explanations around Fig. 119 we have proved Varignon's work theorem for *concurrent* forces acting on a rigid body. We have not yet proved it for the more general case of non-concurrent forces, for which it is also true.

The work done by any set of forces on an arbitrary small displacement of a rigid body equals the work done on the same displacement by another set of forces that is statically equivalent to the first set. To prove this, we remember that in constructing the resultant of many forces, or in finding another set of forces statically equivalent to the first set (page 106), we used only three procedures over and over again: (a) the shifting of forces along their own lines of action, (b) the parallel shifting of forces across their lines of action, introducing a couple in the process, and (c) the compounding of intersecting forces by the parallelogram construction. We have already proved that Varignon's work theorem holds for operations (a) and (c), so that we have to prove it only for operation (b). That operation is accomplished by adding "nothing" to the force F in question in the form of two equal and opposite forces (Fig. 14, page 18). The algebraic sum of the work done by these two added forces is zero for any arbitrary displacement, because the forces are equal and opposite and have the same point of action and hence the same displacement. Therefore, Varignon's work theorem is true for the most general case.

The moment theorem of page 14 can be similarly generalized, and the proof is along the same lines as that for the work theorem just mentioned. The statement of **Varignon's generalized moment theorem** is

The moment of a set of space forces about an arbitrary line in space is equal to the moment about that same line of any other set of forces that is statically equivalent to the first set.

We shall have occasion to use this theorem later (page 318).

Next we investigate the *work done by a couple* on a given displacement. In Fig. 121 let the couple FF' act on the points 1 and 1' of a rigid body. Give the body a *small* displacement so that point 1 moves to 4 and point 1′ moves to 4′. This movement can be split up into three phases: first, a shift of 1-1′ along itself to 2-2′; second, a parallel shift perpendicular to itself to position 3-3′; and finally a rotation of 3-3′ about its center to 4-4′. During the first phase neither force does work. During the second phase one force does positive work and the other an equal amount of negative work, with a total of zero. During the third phase both forces do positive work. Let the angle of rotation be $d\varphi$, measured in radians, and let the moment arm 1-1′ be a. Then the displacements 3-4 and 3′-4′ are $a\,d\varphi/2$, and the work done by each force is $Fa\,d\varphi/2$. The work done by both forces, *i.e.*, by the couple, is $(Fa)\,d\varphi = M\,d\varphi$. Thus

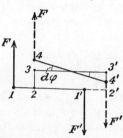

Fig. 121. The work done by a couple is $M\,d\varphi$.

A couple in a plane does no work if the body on which it acts is displaced parallel to itself in any direction; it does work only when the body is rotated (in the plane of the couple), and the amount of work done is the product of the couple and the angle of rotation in radians. As before, the work is considered positive if the directions of the couple and of the (rotational) displacement coincide, and negative if those directions are opposite to each other.

Generalizing this to three dimensions, we see in Fig. 121 that a displacement of the body in the third direction, perpendicular to the paper, again involves zero work by either of the forces, so that **a couple acting on a rigid body does no work on a parallel displacement of that body in any direction in space.**

Now we consider rotations of the body about three perpendicular axes in succession. First we investigate a small rotation about an

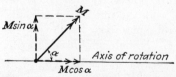

Fig. 122. A couple does work only with its component along the axis of rotation.

axis parallel to the force F, lying in the plane of the paper (Fig. 121). Such a rotation causes displacements at the points of actions of the forces in directions perpendicular to the forces; hence no work is done. The same is true for a small rotation about an axis in the plane of the paper parallel to line 1-2-1′-2′. The only work is done when the body rotates about an axis perpendicular to the paper, the work being $M\,d\varphi$ as before in the two-dimensional case. The most general case is shown in Fig. 122, where a couple M is acting on a rigid body, which is allowed to rotate through the small angle $d\varphi$ about an axis, oblique in space with respect to the couple vector and including an angle α with it. Then, by virtue of Fig. 96 (page 105), the moment vector can be resolved into a component $M \cos \alpha$ along the axis of rotation, and a component $M \sin \alpha$ perpendicular to it. We have seen that the latter component does no work on the rotation, and the first component does work to the amount $M \cos \alpha \, d\varphi$. Thus **the work done by a couple in space is the product of the angle of rotation and the (rectangular) component of the moment vector along the axis of rotation.** This statement is very similar to that for forces (page 133).

By the definition of work it is obvious that a force or a couple does no work on any displacement if that force or couple happens to be zero. Now, the condition of equilibrium of a rigid body (page 109) is that the sum of all forces acting on it shall have a zero resultant force and also a zero resultant couple. Then by Varignon's theorem we conclude that

The sum of the work done by all external forces (including support reactions) acting on a rigid body in equilibrium is zero for any small displacement that may be given to the body.

This gives us a means of writing down equations of equilibrium. For example, let the displacement of the body consist of a small shift dx parallel to the x axis. Then all x components of the forces do work; the y and z components do no work, and we must have, for equilibrium,

$$\Sigma X \, dx = 0$$

in which dx is a common multiplying factor for all terms. The equation therefore becomes

$$(\Sigma X) \, dx = 0 \qquad \text{or} \qquad \Sigma X = 0$$

which is the same as Eq. (4) of page 110. The other Eqs. (4) can be obtained similarly by computing the work on small displacements in the y and z directions and on small rotations about the three axes.

Nobody, except possibly the Duke of Marlborough,[1] likes to write dx's behind a number of terms for the purpose of erasing them again later, so that the method of work, as applied to a single rigid body is never used, since it only introduces an unnecessary complication into the component method of Eq. (4). However, the new method, when

---

[1] The Duke of Marlborough
   Had twenty thousand men.
   He marched them up a hill
   And marched them down again.

applied to systems consisting of a number of bodies connected together by hinges or ropes, will result in a considerable saving of work in many cases.

Problems 142 *to* 144.

27. Systems of Bodies. Consider in Fig. 123 a system consisting of two bodies, connected by a frictionless hinge, loaded by a number of known forces $F_1, F_2, \ldots, F_n$, and supported adequately, or in a statically determinate manner. Imagine the system broken loose from one of its supports, say R_2, and given a small displacement. The second support R_1, as well as the connecting hinge H, is left undisturbed; the bearing R_1 does not move, but the hinge H does. Apply

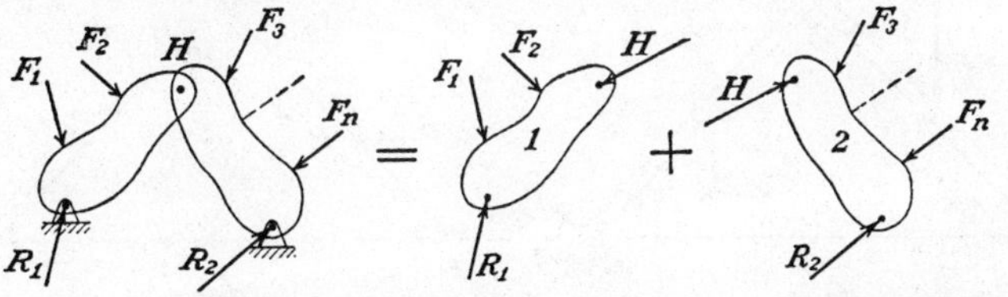

FIG. 123. The work done by a set of forces on a system of two bodies is the sum of the amounts of work done on each body separately.

the method of work to each one of the two constituent bodies, 1 and 2, separately. The hinge force connecting the two bodies is unknown, but by the axiom of action and reaction, the force H from 2 on 1 is equal and opposite to the $-H$ from 1 on 2.

Applying the method of work to the first body, which we observe to be in equilibrium, we conclude that the work done by R_1, F_1, F_2 and H is zero. Similarly, for the second body, the work done by $F_3, \ldots, F_n, -H$, and R_2 is zero. But the hinge, being a common point of the two bodies, has only one displacement, so that the work done by H on the first body is equal and opposite to the work done by $-H$ on the second one. Adding the amounts of work for both bodies, the H terms cancel, and we find that for the system of two bodies the work done by $F_1, F_2, \ldots, F_n, R_1$, and R_2 is zero. It is seen that the hinge force has dropped out of this equation. This argument can be extended to n bodies, held together by $n - 1$ hinges, with the same conclusion: the $n - 1$ hinge forces do not appear in the work equation. The forces $F_1, \ldots, F_n$ we will call *external forces* of the system to distinguish them from the hinge forces H, which we will call *internal forces* of the system. In this example the work done by the internal forces is zero. We then come to the general conclusion:

The work done by all external forces and support reactions on an arbitrary small displacement of a multibodied system in equilibrium is zero, provided the internal forces of the system do not do work.

Before proceeding with the theory, the advantage of the new method will be shown in an example, the two-bar system of Fig. 124a, loaded by a vertical load P in the middle of one of the bars. This load tends to push the two supports apart, and we want to know the consequent horizontal component of the support reactions. We break loose support 2 and allow the end of the bar 2-3 to slide to the right by the *small* amount ϵ (Fig. 124b). As a consequence the bar 1-3 turns about 1, and the point of application of P moves at 45 deg

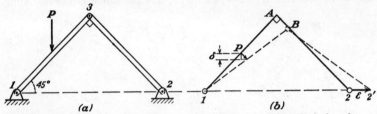

FIG. 124.   Determination of the bearing reaction 2 by the method of work.

toward the lower right; the vertical component of this displacement we call δ. The work done by the external forces (P only) and reactions (at 1 and 2) is

$$P\delta - X\epsilon$$

where X is the horizontal component of the force from the support 2 on the bar directed to the left. The vertical component of the reaction at 2 exists but does no work, since the displacement is purely horizontal. The reaction force 1 does no work, because there is no displacement there. The internal hinge force at 3 does no work, assuming no friction in that hinge.

For equilibrium the above expression must be zero, and

$$X = P\,\frac{\delta}{\epsilon}$$

The solution of the problem is thus reduced to finding the ratio of the small displacements δ and ϵ.

In Fig. 124b the new location B of the top hinge is on the right bar, because a small rotation about 1 moves A perper   ·larly to the radius 1A. Now, by symmetry of the two bars, ·   the horizontal distance between A and B. But that hor·   ⌐e between A

and *B* is again twice the horizontal displacement at *P*, since it is twice as far from the hinge 1. Again, by the 45-deg relation at *P*, the horizontal and vertical displacements at that point are the same; hence we conclude that $\epsilon = 4\delta$ and consequently that $X = P/4$.

By the old method we would have to set up equilibrium conditions for the two bars separately, and the internal hinge reaction at point 3 would appear in the analysis. In this case we would remark that the bar 2-3 is without intermediate loads, so that it can be in tension or compression only (page 23). Hence the force in it is $X \sqrt{2}$, in which *X* is the horizontal component at 2, as before. Now, writing the moment equation for bar 1-3 about 1 as a center, we have

$$Xa \sqrt{2} = Pa/2 \sqrt{2}$$

with the same result. The two analyses lead to the answer with about the same amount of work in this case; however, they are entirely different, and the advantage of the method of work is that the internal hinge force does not appear in the equations. In this problem we have *one* internal hinge only; the desirability of the work method over the resultant method increases with the number of internal reactions of the system.

Now we return to the theory and examine once more the statement of the method on page 140. That statement contains the word "provided," which weakens it considerably. It is in order, therefore, to investigate now under which circumstances internal forces do or do not perform work.

Forces exerted by one body of a system on another body are of various types, such as:

a. Normal forces at contact surfaces
b. Tangential forces at contact surfaces (friction)
c. Forces transmitted by (inextensible) ropes or struts
d. Forces transmitted by extensible springs
e. Forces, not classed under *a* to *d*

We will investigate the cases *a* to *d* one after the other, and leave out case *e*, which comprises forces that hardly ever occur in practice, such as magnetic, electric, or gravitational attraction forces *between* the various parts of a system.

Case *a* occurs in a frictionless hinge. In general two surfaces in contact must displace through the same distance in the normal direction, because if they did not, they would either draw apart or dig into each other. By the axiom of action and reaction the forces on the two surfaces in contact are equal and opposite, and with the same displacement, they perform equal and opposite amounts of work, totaling zero. The same conclusion is reached for case c; the two ends of an inextensible rope or strut must have the same displacement along the strut or rope, because if not, it would become longer or shorter (Fig. 120).

Now we turn to case b, which is different. Let a block rest on a horizontal plane, and let there be friction between the two: a force F from the plane on the block and $-F$ from the block on the plane. Now let the block slide along the plane through distance δ. Then the force F on the block performs work to an amount $-F\delta$ (negative because the friction force is always directed opposite to the motion), and the force $-F$ on the plane does zero work because the plane does not move. Hence **the internal friction forces perform work to an amount equal to the negative product of the force and the relative displacement between the two sliding surfaces.** The amount of work is proportional to the *relative* displacement, and it is independent of the actual or total displacements. Suppose, for example, that the block moves through distance $\epsilon + \delta$, and the plane through distance ϵ. Then the work done by the force F on the block is $-F(\epsilon + \delta)$, and the work by the force $-F$ on the plane is $F\epsilon$, the total being $-F\delta$, independent of the common displacement ϵ and proportional to the relative displacement δ only.

Now let us repeat the analysis of page 139 on the subject of Fig. 123, in which the internal force H consists partially of a normal force N and partially of a friction force F, and let the point of contact of body 1 move tangentially through $\epsilon + \delta$, while the point of contact of body 2 moves through ϵ. Then the work done by the forces on body 1 is the work done by F_1, F_2, R_1, and N plus the amount $-F(\epsilon + \delta)$, which sum is zero (page 138). On body 2 the work is that done by $F_3, \ldots, F_n$, R_2, and $-N$ plus the amount $F\epsilon$, which sum again is zero. Adding the two gives again zero, which is equal to the work of all external forces and reactions plus the amount $-F\delta$, or in other words, the work by the external forces equals the positive amount $F\delta$. This positive amount of work is usually called the "work dissipated by friction." Then, the general rule of page 140 becomes for this case

The work done by all external forces and support reactions on an arbitrary small displacement of a multibodied system in equilibrium, having friction in its joints, is equal to the work dissipated by these internal friction forces.

In order to calculate the friction work in a given case we have to know not only the relative displacements, but also the friction forces, which usually means that we have to compute the normal forces. Thus these internal forces have to be calculated anyhow, which takes away most of the direct simplicity of the method of work. Applications of this relation will be given in the next article.

Finally we investigate case d, where the internal forces between bodies are transmitted through extensible springs, as shown in Fig. 125, which is a modification of Fig. 123. The spring is shown in compression with force S. As before, we apply the method of work to the two bodies, 1 and 2, separately and add the amounts of work. The argument should be carried out by the reader, and the result is that the work by all external forces and reactions plus $S\delta$ is zero, or

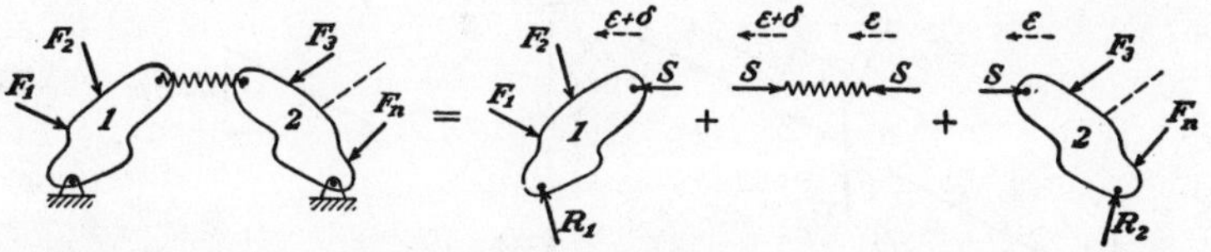

Fig. 125.   The work done on a system equals the sum of the amounts of work done on all constituent parts.

that the work of all forces and reactions equals $-S\delta$. Now δ is the extension of the spring, and consequently $-\delta$ is the shortening of the spring, and $-S\delta$ is the product of the compressive spring force and the shortening, which is the work done by the internal forces on the spring. Later (page 256) we will see that such work is stored in the spring in the form of elastic energy and that the spring is capable of giving back this stored work. Again, the argument with Fig. 125, like that of Fig. 123, can be extended to more than two bodies, and we can generalize the conclusion as follows:

The work done by all external forces and reactions on an arbitrary small displacement of a multibodied system in equilibrium, having extensible springs and friction between its parts, is equal to the sum of the work stored in the springs and the work dissipated by friction due to that small displacement, provided any other internal forces (category e, page 141) do not perform work on that displacement.

Unfortunately, we still have the word "provided" with us, but now it is hardly of practical significance to engineers, who seldom deal with magnetic or *internal* gravitational forces in equilibrium systems.

28. Applications. The foregoing theory will now be applied to a number of examples:

   a. The smooth inclined plane
   b. The multiple pulley
   c. A simple truss
   d. A balance
   e. Lazy tongs
   f. The rough inclined plane
   g. The buggy wheel
   h. A system with springs

   a. The Smooth Inclined Plane. In Fig. 126 the pulley as well as the plane and block are supposed to be without friction. For what ratio of weights W_2/W_1 is there equilibrium? Let W_2 go down through a small distance δ, and consequently let W_1 slide up the plane through

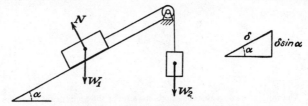

Fig. 126. The frictionless inclined plane.

the same distance δ. The internal connections are frictionless, and the rope is inextensible. Hence the statement of page 140 applies. There are three external forces and reactions: W_1, W_2, and N. (The rope tension is an internal force, which we should not even mention.) The force N does no work; the displacement is perpendicular to it. The displacement of W_1 in its own direction is $-\delta \sin \alpha$, negative because it is upward, while W_1 points downward. Then

$$W_2\delta - W_1\delta \sin \alpha = 0$$

or

$$W_2 = W_1 \sin \alpha$$

   b. The Multiple Pulley. Figure 127 shows a tackle consisting of a fixed pulley, supported from the ceiling at O, and a floating pulley carrying the load W. Assuming no friction in the pulley axles and an inextensible rope, what is the ratio P/W for equilibrium? Pull down at P through a small distance δ. This causes the upper pulley to rotate and the left branch of rope to rise through distance δ, including the point A of that rope. On the other hand, point B of the rope does not rise, because it is directly connected to the ceiling at O'. The rope does not slip on the lower pulley; thus the points A' and B' of the pulley, lying just opposite the points A and B of the rope, rise just like A and B. The center of the floating pulley, being midway between A' and B', therefore rises through a distance $\delta/2$. The external forces and reactions of the systems are P, W, and the supporting forces at O and O', the latter two doing no work because there is no displacement.

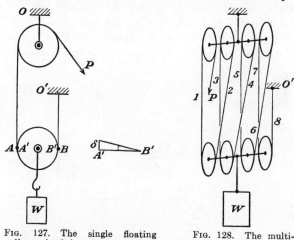

Fig. 127. The single floating pulley solved by the method of work.

Fig. 128. The multiple pulley.

P does positive work $P\delta$, and W does negative work: $-W\delta/2$. The rope tension is an internal force. Then

$$P\delta - \frac{W\delta}{2} = 0 \quad \text{or} \quad P = \frac{W}{2}$$

Figure 128 shows a similar tackle in which the upper or fixed block and the lower or floating block carry four pulley wheels each. The wheels are drawn apart to show how the rope passes over them; actually, of course, the pulleys are close together. Again we ask for the ratio P/W at equilibrium, with no friction in the pulleys and an inextensible rope. As in Fig. 127, the only forces doing work are P and W, and if P is pulled down through distance δ and as a consequence W is raised through distance ϵ, then $P\delta - W\epsilon = 0$ or

$$\frac{P}{W} = \frac{\epsilon}{\delta}.$$

To calculate ϵ/δ we could trace the displacement through all the branches of the rope from P to O', but it is easier to remark that if W is raised through distance ϵ, the eight pieces of rope holding up W

are shortened by 8ϵ, and since the rope is inextensible, this must come out at P in the form of δ. Thus $\epsilon/\delta = \frac{1}{8} = P/W$.

c. *A Truss.* Figure 129 is a truss consisting of five hinged bars, forming a square with one diagonal, loaded with forces P along the other diagonal. We want to know the force in the diagonal bar under this loading. To solve this we remove the diagonal bar and replace it by forces X and $-X$, which then are considered as external forces on the system of the four remaining bars. The small displacement we give to the system consists of allowing the forces P to spread apart

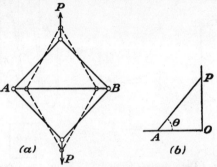

FIG. 129. To find the force in the diagonal bar.

by a distance 2δ (each force P is allowed to move through distance δ while the center of the square remains in place). As a result of this the square becomes a diamond, and the diagonal of the forces X becomes shorter, say by distance 2ϵ.

If the forces X are considered positive if they pull the joints A and B closer together, then the X forces do positive work on the shortening distance and

$$2P\delta + 2X\epsilon = 0 \qquad \text{or} \qquad \frac{X}{P} = -\frac{\delta}{\epsilon}$$

The forces X come out negative, *i.e.*, opposite in sign to what we considered positive: they push the joints A and B apart, and the bar AB is in compression. The ratio δ/ϵ has to be found by geometry, which can be done as follows (Fig. 129b): Let the length of a side of the square be l; then

$$OP = y = l \sin \theta, \qquad OA = x = l \cos \theta$$
$$\delta = \Delta y = l \cos \theta \, \Delta\theta, \qquad \epsilon = \Delta x = -l \sin \theta \, \Delta\theta$$
$$-\frac{\delta}{\epsilon} = \frac{l \cos \theta \, \Delta\theta}{l \sin \theta \, \Delta\theta} = \cot \theta = \cot 45° = 1 = \frac{X}{P}$$

Hence the compressive force in the diagonal bar of Fig. 129a is equal to force P.

d. A Balance.   Figure 130 shows a system consisting of four rigid bodies: two straight bars and two T-shaped pieces, hinged together at four points.   The system is supported at A by a hinge, and at B a pin on the lower straight bar can ride up and down between vertical guides without friction.   This gives the plane system three supports, two at A and one at B, which makes it statically determinate (page 110).   The loading consists of a force P at distance $2a$ from the center, held in equilibrium by a load X at distance $3a$ from the center.   We want to determine X when P is given.

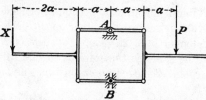

FIG. 130.   In Roberval's balance the magnitude of the force X is independent of its horizontal location.

To solve this we give the system a small displacement by inclining the upper bar through a small angle, so that its right-hand hinge goes down δ and its left-hand hinge goes up δ.   The lower bar follows this movement and remains parallel to the upper one; the two T pieces move parallel to themselves, the right one down and the left one up by the same amount δ.   The external forces are P, X, and the reactions at A and B.   These reactions do not perform work because the points A and B do not move.   Hence

$$P\delta - X\delta = 0 \qquad \text{or} \qquad X = P$$

This result is surprising, and the reader is advised to derive it by the ordinary method of statics, involving four hinge forces with horizontal and vertical components.   This will entail an amount of labor greatly in excess of that by which the answer was found here.   However, after the analysis is completed, we not only have the result that $X = P$ but also we know all the hinge forces and the bending moments in the various bars.

e. Lazy Tongs.   The system of Fig. 131 consists of six bars, each of length $2a$, hinged together at their ends and at their mid-points, with two extra bars on top of length a.   It is loaded with force P on top, and we ask for the forces X at the bottom that are required to keep P in equilibrium. To solve, we give the system a small displacement by allowing the forces X to come nearer each other by distance δ, which has the result that the top P moves up through distance ϵ with respect to the bottom. Then

$$X\delta - P\epsilon = 0 \qquad \text{or} \qquad X = P\frac{\epsilon}{\delta}$$

and the problem is reduced to the geometric one of finding the ratio ϵ/δ. Now in the triangle ABC the length AC is shortened by $\delta/2$, and

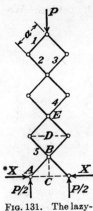

consequently BC is lengthened by a quantity called x, while AB, being a bar, does not change its length. This is just the same case as Fig. 129b, where we found that the elongation of one side is equal to the shortening of the other, or BC in Fig. 131 becomes longer by $\delta/2$. The geometry of BD, DE, etc., is the same as that of BC, so that ultimately the top point P moves up seven times as much as B, or

$$\frac{\delta}{\epsilon} = \frac{7}{2} \qquad \text{and} \qquad X = 3\frac{1}{2}P$$

Fig. 131. The lazy-tongs mechanism

On account of the large number of bodies in this system, the derivation of this result by the ordinary method of repeated equilibrium equations for the consecutive bodies is very cumbersome indeed. The reader should do this, first by noticing that the top bar 1 has no loads along its length and is therefore in pure compression (of $P/\sqrt{2}$). This establishes the load (by action equals reaction) on the top hinge of bar 2. This bar has unknown forces at the lower end and in the middle. Of the force in the middle we can reason that it must be horizontal, because if it had a vertical component, say an upward one, there should be an equal downward component on the middle of bar 3, by action equals reaction, but there should be an equal upward component there, by symmetry, because 2 and 3 are situated exactly alike. A force that is upward and downward at the same time must be zero. Now the three unknown force components on bar 2 (one in the middle, two at the lower end) can be solved for, and we are ready to proceed to bar 4 and, after it, to bar 5. This is a lot of work, which not only consumes time but also involves the likelihood of computing errors. This example vividly illustrates the fact that **for systems of many bodies without friction or springs the method of work is usually preferable to the method of components.**

f. The Rough Inclined Plane. Consider the inclined plane of Fig. 126, but this time with friction under the block, as in Fig. 86 (page 87). We apply the statement of page 142 to a small downward displacement δ of W_2:

$$W_2\delta - W_1\delta \sin \alpha = fN\delta \quad \text{(just going up the plane)}$$

On the other hand, if W_2 is given a small displacement δ upward, and W_1 goes down the plane, we write

$$-W_2\delta + W_1\delta \sin \alpha = fN\delta \quad \text{(just going down the plane)}$$

When these equations are divided by δ, they are no different from those of page 88. Besides, in order to finish the solution we must calculate N, for which we need an equilibrium equation again, and so we lose the prime advantage of the method of work, which consists of the non-appearance of internal forces in the equations. Suppose we should make the problem more complicated and ask for W_2 for the case of friction both in the block and in the pulley axle, then the above equations would acquire one more term on the right-hand side, expressing the work dissipated in friction in the pulley. But the pulley friction force is proportional to the normal pressure, which again is proportional to the rope tension T. Therefore, we would have to find T, and every advantage of the method of work would be lost.

g. The Buggy Wheel. We return to the question of Fig. 89 (page 92), in which we want to find the ratio P/W of the horse's pull to the buggy weight in the presence of friction in the journals. We give the system a small displacement by allowing the axle to go forward through distance δ and the wheel to roll over the ground without slipping, while the wheel turns through a small angle about the axle. The external forces are the pull P, the weight W, the reactions from the ground vertically up and horizontally. We do not care how large these reactions are because neither of them performs work during the displacement. (The ground friction force does no work because there is no slipping or relative displacement between wheel and ground horizontally.) Neither does the buggy weight W perform work. Thus,

$$P\delta = \text{work dissipated in friction.}$$

The friction force in the journal is fN, where N is approximately equal to W. Now we need to know the distance through which the bearing slips over the surface of the journal. The wheel turns through an angle $\varphi = \delta/R$, where R is the wheel radius. Then the distance of slip at the axle is $\varphi r = \delta r / R$, where r is the journal radius. Thus the work dissipated in friction is $fW\delta r/R$, and upon equating this to the external work $P\delta$, we obtain the result of page 93.

h. *A System with Springs.* Figure 132 represents a system of three bars of equal length a hinged together and cross-braced by two springs, of such lengths that they are unstressed when the bars form a square as shown. The springs have stiffnesses k and $2k$ lb./in., and the system

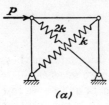

(a)

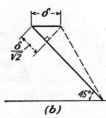

(b)

FIG. 132. The method of work applied to a system involving springs.

is loaded by a sidewise force P. We want to know the deflection under P and the forces in the two springs. Before proceeding with the solution, the reader should turn to page 256 for an understanding of the meaning of a spring stiffness k and for the fact that if an unstressed spring is either lengthened or shortened by an amount δ by a force, the work necessary to accomplish this is $\frac{1}{2}k\delta^2$.

Now we give the system (Fig. 132) a small displacement, by allowing the top bar to shift to the right by an amount δ, which transforms the square into a parallelogram. As a consequence, the spring $2k$ is shortened by $\delta/\sqrt{2}$, and the spring k is lengthened by $\delta/\sqrt{2}$, as becomes clear from the sketch of Fig. 132b.

Now in Fig. 132a, the external forces are P and the reactions at the bottom, but only P does work. The spring forces are internal. Due to the displacements the springs store an amount of work equal to

$$\frac{1}{2} k \left(\frac{\delta}{\sqrt{2}}\right)^2 + \frac{1}{2} 2k \left(\frac{\delta}{\sqrt{2}}\right)^2 = \frac{3}{4} k\delta^2$$

The work done by the force P on the displacement δ is not $P\delta$, but $1/2P\delta$, because when δ is zero the force P for equilibrium also must be zero. The force P grows proportionally with δ for equilibrium, so that the work done is $1/2P\delta$.

Applying the statement of page 143 to this case we have

$$\frac{1}{2} P\delta = \frac{3}{2} k\delta^2 \text{ or } \delta = \frac{2}{3} P/k$$

The force in a spring is k times its displacement which is $\delta/\sqrt{2}$, so that the flexible spring k sustains a tensile force of $(\sqrt{2}/3)P$ and the stiff spring $2k$ a compressive force of $(2\sqrt{2}/3)P$.

Historical Note. The method of work, as explained in pages 138 to 143, was perfected more than two centuries ago and was then called the **principle of virtual velocities.** There was nothing strange about such a name at a time when one of the most eminent writers used the word "movement" for force or velocity indiscriminately (page 217) and when kinetic energy (page 251) was called "living force" or "vis viva."

Most of those names died a natural death in the course of time, and by about 1850 the nomenclature had boiled down to the simple one now in general use, with the curious exception of "virtual velocity." By "virtual velocity" was meant "small displacement," and the work done by a force was called the "virtual moment" of that force. In the book "Statics" by Todhunter,[1] which was a classic in its day, we read:

"The word *virtual* is used to intimate that the displacements are not really made, but only *supposed*. We retain the established phraseology, but it is evident from these explanations that the words 'virtual velocity' might be conveniently replaced by 'hypothetical displacement' . . . .

"The virtual moment of a force is the product of its intensity by the virtual velocity of its point of application estimated in the direction of the force. . . . If a system of particles is in equilibrium, the sum of the virtual moments of all the forces is zero, whatever be the virtual velocity."

After 1900 the "virtual moments" died and became "work," and the "virtual velocities" were toned down to "virtual displacements," while the principle became that of "virtual work," but the word "virtual" itself has persisted to this day in most books. The term "method of work" as employed in this book is not in general use.

Problems 145 *to* 153.

29. Stability of Equilibrium. Everybody (except Columbus) knows that a body standing upright on a point support will fall over, although by our theories, if the center of gravity is vertically above the point support, the body is in equilibrium. The equilibrium of the body is said to be "unstable," and in this article we propose to investigate under which conditions equilibrium is stable or unstable.

Figure 133*a* shows a rod hinged at one end and subjected at the other end to a force F in line with the rod. We assume absence of gravity forces, *i.e.*, a weightless rod, for the purpose of this discussion. The rod is in equilibrium, because the hinge reaction is a force F equal, opposite, and in line with the force on top of the rod. Is this equilibrium stable or unstable? The question, as stated, is meaningless, because its answer may be one way or the other depending on circumstances we have not yet specified. So far we have talked about

[1] Fourth ed., 1874, Cambridge, England.

"stability" only in the usual sense of the English language without a precise scientific definition. Such a definition we now propose to give.

The equilibrium of a body is stable if the forces or moments acting on it cause a small deviation from the position of equilibrium to be decreased; the equilibrium is unstable if the forces and moments tend to increase a small deviation from that position; the equilibrium is indifferent if the body remains in equilibrium in positions deviating on either side from the position of equilibrium.

From this definition we see that stability cannot be judged in the equilibrium position itself and that the behavior of the system in a deviated position must be studied before conclusions can be drawn. Now we return to Fig. 133*a* and deviate the bar by turning it through

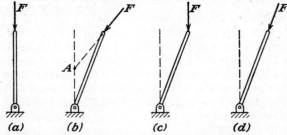

(a)   (b)   (c)   (d)

FIG. 133.   Stable (*b*), unstable (*c*), and indifferent (*d*) equilibrium.

a small angle about the hinge. What does the force F tend to do? The question is unanswerable because we have specified F only in the equilibrium position and have not stated yet how the force will act in the deviated position. Suppose that the force F is always passing through point A, as in Fig. 133*b*, for any small angle of deviation. Then, clearly, F tends to push the bar back to its vertical position, and, by definition, the equilibrium is *stable*. In Fig. 133*c* the force F is vertical, and in 133*d* it is in line with the rod for all deviations, so that, by the definition, the equilibrium of Fig. 133*a* for the behavior of 133*c* is *unstable*, and for the behavior of 133*d*, it is *indifferent*.

Other examples of the application of this definition occur if in Fig. 133*a* we omit the end force F and replace it by the weight force of the bar, passing through its center of gravity. Then we have a situation akin to that of 133*c*, which is unstable. Or, if we turn 133*a* upside down and have a hanging pendulous bar under the influence of gravity only, the equilibrium is stable. The rule or definition given is sufficient to investigate the stability or instability of any system, however complicated. But there is another way of judging stability, by the method of work.

A force does positive work if its point of application moves in its own direction, and it does negative work if the point of application moves against it. Then, if the external forces tend to increase a deviation, they evidently push in the direction of that deviation, and if the deviation is allowed to increase, the external forces do positive work. In the same manner we can reason that if the external forces tend to decrease a deviation, they will do positive work if the deviation is allowed to decrease, or they will do negative work on an increasing deviation. Thus the definition of page 152 is equivalent to the following:

The equilibrium of a body is stable if the external forces do negative work on a small deviation; it is unstable if these forces do positive work on a small deviation, and it is indifferent if they perform zero work on such a small deviation from the equilibrium position.

Consider Fig. 133c, for example. In moving the bar from the vertical or zero position to the deviated position shown, the top of the bar describes a circular arc about the hinge as center. Thus the force F goes down slightly between those two positions and does positive work. In the case of Fig. 133d, the force F is always perpendicular to its (circular) path and does no work. In Fig. 133b, if the deviation is increased somewhat, the force does negative work, because its tangential component is directed against the increase in displacement.

But on page 140, we saw that the work done on a small displacement in a system in equilibrium is zero, neither positive nor negative, which seems to contradict the statement just made. The difficulty lies in the order of magnitude of the work done: as a first approximation, the work done in each case, 133b, c, or d, is zero for a sufficiently small displacement, because in each case the force is substantially vertical and the displacement substantially horizontal. It is only in a further, second, approximation that we see that the work done is not quite zero; it is said to be "zero in quantities small of the first order and positive (or negative) in quantities small of the second order."

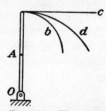

Fig. 134. Illustrates the stability of the equilibrium of Fig. 133.

Let us look at the situation from another angle, with the help of Fig. 134. In that figure we see three curves: b is a circle about A as center, d is a circle about the hinge O as center, c is a horizontal line. The letters b, c, and d refer to the corresponding ones in Fig. 133, and the curves are those on which the point of application of the force would have to move in order to do zero work. For instance, in case c the force is always vertical and will do no work if moved along the horizontal line c of Fig. 134. In case 133b, the force will do no work if moved on a circle about A, because then it is always perpendicular to its path. Now, returning to Fig. 134, the *actual* path of the point of application of the force is the path of the top of the bar, curve d, so that only for case d the work done is zero for displacements that are finite. In case c the force does positive work, because *if* moved along c, it would do no work; in reality it moves along d, and thus the force descends. In case b the force does negative work. But in all three cases, the work done for *very* small displacements from the center position is zero (to first approximation) because all three curves have the same horizontal tangent at the top.

Now let us investigate this analytically and take case c as an example. Let the small angle be θ, and let the length of the bar be l. Then the descent of the force or of the end of the bar is

$$l - l \cos \theta = l(1 - \cos \theta).$$

We can expand $\cos \theta$ into a power series (Maclaurin or Taylor series)

$$\cos \theta = 1 - \frac{\theta^2}{2} + \frac{\theta^4}{4!} - \cdots$$

Neglecting terms with θ^4 and higher powers, the descent of F is $l\theta^2/2$, an expression in which the term with θ to the first power is absent, and the term proportional to the second power of θ is positive. The work is said to be "zero in quantities small of the first order and positive in quantities small of the second order." Summarizing, we can state that

The work done by all external forces and reactions of a system on an arbitrary small displacement is zero in terms of quantities small of the first order if the system is in equilibrium; the equilibrium is stable if the work is negative in terms of quantities small of the second order; the equilibrium is unstable if the work is positive in terms of quantities small of the second order; the equilibrium is indifferent if the work is zero also in terms of quantities of the second order; all of this provided that the internal forces do no work on the displacement.

The combination "stable = negative work, unstable = positive work" can be remembered more easily if we look at it as follows: The scientific definition of "work" coincides in many respects with the ordinary conception of that word, but in one respect the scientific "work" differs radically from the "work" of common speech. A force, unless restrained by bearings or supports, will push a body in its own direction. A force, therefore, will do positive (scientific) work unless forcibly restrained from doing it and will continue to do positive (scientific) work until it can do no more, which is completely opposite to what any man or animal will do with (common-speech) work.

As long as a small displacement is associated with positive work, the force will do it and continue to do it; the system is unstable. When a small displacement means negative work, the force refuses to do it (water refuses to flow uphill), and the system is stable. If the external forces of the system consist of gravity forces only, the above statement means that for a small displacement the center of gravity G moves horizontally, or rather on a curve with a horizontal tangent; the equilibrium is stable if G moves on a curve with an upward curvature, as in the bottom of a trough; unstable if G moves on top of a hill. This is evident if we think of a small heavy particle in a smooth bowl or on top of a smooth sphere, but it is equally true for more complicated cases, such as Fig. 135, for example. That figure represents a half sphere to which is attached a cone; the whole piece can roll without sliding on a horizontal plane. Is it stable or unstable in the upright position? To investigate this, let the body roll through a small angle, so that the center line is no longer vertical. Instead of redrawing the figure in the deviated position, we draw a new, dotted, ground line and turn our book until the dotted line is horizontal. Suppose that the center of gravity of the combined body were at C, the center of the

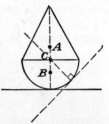

Fig. 135. This is stable in the upright position when the center of gravity is below C; unstable when above C.

circle; then C would be just as far from the dotted ground line as from the solid one; the equilibrium is indifferent. In case the center of gravity were at B, the equilibrium would be stable, because B is farther from the dotted ground line than from the solid one. A location of the center of gravity at A would mean instability because A is closer to the dotted line. When we sketch the paths of points A, B, and C with respect to the solid ground line for different positions of the body, we recognize that B is at the bottom of a trough, A is on top of a hill, and C moves on a horizontal line.

Problems 154 *to* 160.

CHAPTER IX

KINEMATICS OF A POINT

30. Rectilinear and Angular Motion. In this chapter we propose to study the motion of a point in space without considering the forces that cause the motion. In the next chapter the relations between force and motion will be considered, and then the "point" will be made a heavy point, or a mass-endowed point, which usually is called a "particle." Now, however, we do not care whether our particle is heavy or not; we only study the geometry of the motion.

The first motion we consider is the simplest possible one, *i.e.*, the motion of a point P along a straight line. The position of point P at any time t is determined by its distance x from a fixed origin O on the straight line. Thus x depends on the time or

$$x = f(t)$$

Consider two instants of time t and $t + \Delta t$ with the corresponding displacements x and $x + \Delta x$. During the time interval Δt the displacement increases by Δx, and the quotient $\Delta x/\Delta t$ is called the "average velocity during the interval Δt." If, in the manner of the calculus, we let Δt become smaller indefinitely, the average velocity converges to a limit,

$$\lim_{\Delta t \to 0} \frac{\Delta x}{\Delta t} = \frac{dx}{dt} = \dot{x} = v$$

which is called the *instantaneous velocity*, or shorter, the *velocity* of point P at time t. The notation $\dot{x}$, which is shorthand for dx/dt and is pronounced "x dot" is the original notation employed by Newton (1642–1726) for the differential quotient. It is in use at present only in the subject of mechanics and **the dot always represents differentiation with respect to time,** and never with respect to any other variable.

The velocity $\dot{x}$ or v of the point P may be different at different instants of time. Let the velocity at time t be v and at time $t + \Delta t$, let it be $v + \Delta v$. Then during the time interval Δt the velocity increases by Δv and the quotient $\Delta v/\Delta t$ is called the "average acceleration during the interval Δt." If Δt shrinks to zero we have

$$\lim_{\Delta t \to 0} \frac{\Delta v}{\Delta t} = \frac{dv}{dt} = \frac{d^2x}{dt^2} = \dot{v} = \ddot{x}$$

which is called the *instantaneous acceleration*, or shorter, the *acceleration* of the point at time t. The Newtonian notation $\ddot{x}$ (pronounced "x double dot") again refers to two differentiations with respect to *time*. The differential quotient dy/dx, representing the slope of the curve $y = f(x)$ is sometimes written y' (y dash), but never $\dot{y}$. In some books $\ddot{x}$ is denoted by a; that notation will be avoided in this text, as a usually is reserved for a constant quantity, mostly a length.

The displacement, velocity, and acceleration, all being time functions, can be plotted graphically. This has been done for a hypothetical case in Fig. 136, and the usual relations between ordinates, slopes, curvatures, and areas, familiar from the differential calculus, are illustrated in Figs. 136a, b, and c. Figure 136d is a cross-plot between Figs. 136a and b, obtained by taking the ordinates of those two figures for the same time t and plotting them against each other in Fig. 136d. The reader is advised to study Fig. 136 carefully, and as an exercise before proceeding, to sketch the diagrams $\ddot{x} = f(x)$, and $\ddot{x} = f(\dot{x})$.

A simple and important example of the foregoing is the freely falling stone. Let the point (the stone) at time $t = 0$ be at a certain height above the ground, where we place our origin $x = 0$, counting x positive downward from that position. We specify that the downward acceleration of the point is constant with respect to time, and we denote this constant acceleration by g. What are the various diagrams? The analytic solution involves integration:

$$\ddot{x} = \dot{v} = g = \text{constant}$$

$$\dot{x} = v = \int \ddot{x}\, dt = \int g\, dt = gt + C_1$$

$$x = \int \dot{x}\, dt = \int (gt + C_1)\, dt = \frac{gt^2}{2} + C_1 t + C_2$$

The constants of integration C_1 and C_2 are specified by the initial conditions. We have stated that at $t = 0$, $x = 0$, and substituting that into the above result for x leads to $C_2 = 0$. In addition, we now specify that at $t = 0$, $\dot{x} = 0$, which means that the stone has no initial velocity and starts falling from rest. Substituting this into the expression for $\dot{x}$ leads to $C_1 = 0$. This enables us to construct the diagrams 137a to c. To find $v = f(x)$, we eliminate the time t from $v = gt$ and $x = \frac{1}{2}gt^2$ with the result $v^2 = 2gx$, plotted in Fig. 137d.

The first man who studied and partially understood these relations was Leonardo da Vinci (1452–1519), almost two centuries before the invention of the calculus. He describes, in his famous notebooks, the apparatus of Fig.

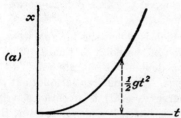

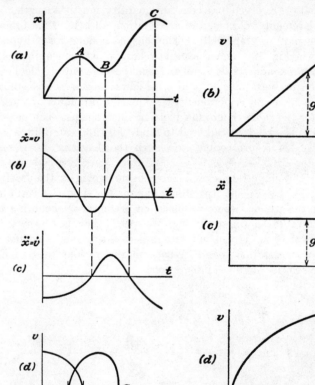

FIG. 136. Displacement, velocity, and acceleration diagrams.

FIG. 137. The freely falling body.

138, consisting of two vertical boards, hinged together on one side and covered with blotting paper on the inside faces. A leaking water tap lets drops fall down between the boards at presumably equal intervals of time. When a string is suddenly pulled, the boards are clapped together and the positions of the drops on the blotters can be inspected. Leonardo observed that the distances between consecutive drops increased in a "continuous arithmetic proportion," which is a way of saying that $v = gt$.*

Another simple and important example is the rectilinear motion of a vibrating point, known as the *simple harmonic motion*, and described by the equation

$$x = a \sin \omega t$$

where a is a constant length, called

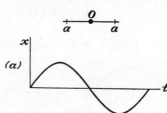

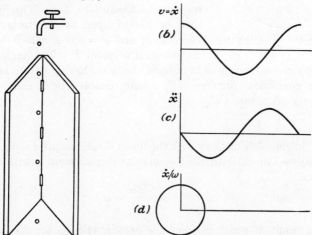

FIG. 138. Leonardo da Vinci's experiment.   FIG. 139. The simple harmonic motion.

the "amplitude" of the vibration, and ω is a constant of dimension 1/time called the "circular frequency" of the motion. Differentiating this equation twice gives

$$v = \dot{x} = a\omega \cos \omega t$$
$$\dot{v} = \ddot{x} = -a\omega^2 \sin \omega t = -\omega^2 x$$

plotted in Figs. 139a to c. They show the vibrating motion along a straight line between two points at distance a on either side of the

* Leonardo da Vinci, "Del Moto e Misura dell'Acqua," reprinted in Bologna, 1923, p. 188.

center origin O. Positive ordinates on the diagrams mean displacements, velocities, or accelerations to the right; negative ordinates mean that these quantities are to the left. The relation between velocity and displacement (Fig. 139d) is interesting. From the equations we can write

$$\left(\frac{x}{a}\right)^2 + \left(\frac{v}{a\omega}\right)^2 = \sin^2 \omega t + \cos^2 \omega t = 1$$

or $x^2 + (v/\omega)^2 = a^2$, which shows that if we plot the ratio v/ω against
x, the curve is a circle of radius a.

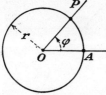

Fig. 140. Illustrates angular motion.

Now we turn our attention to another subject, *angular motion*. In Fig. 140 let a point P move on a circle of radius r and let the radius OP of the moving point P be at angle φ with respect to a fixed radius OA. This line OA can then be looked upon as the "origin" of the variable φ; and $\varphi = f(t)$ is called the *angular displacement* of point P. The angle φ can be measured in degrees or in radians, the latter having the advantage that the curvilinear distance AP, usually denoted by s, can be directly expressed in terms of φ

$$s = r\varphi$$

In complete analogy with the linear displacement x along a straight line, we can differentiate the angular displacement φ with respect to time.

$$\dot{\varphi} = \frac{d\varphi}{dt} = \omega$$

the result of which is called the *angular velocity*, usually denoted by ω, and measured in radians per second, revolutions per second, revolutions per minute, or other similar units. Since one revolution equals 2π radians, an angular velocity of 600 rpm = 10 rps = 20π radians/sec.

The angular velocity or angular speed ω can be differentiated once more:

$$\ddot{\varphi} = \dot{\omega} = \frac{d^2\varphi}{dt^2} = \frac{d\omega}{dt}$$

which is named the *angular acceleration*. In some books this quantity is written α, a notation which will not be used in this text, because α by common usage means a constant angle. The angular acceleration is the rate of change of angular speed of a wheel, and is zero for a wheel rotating at uniform angular speed. It is measured in radians per second squared, or in similar units, such as revolutions per minute squared, or even in mixed units like revolutions per minute per second. The latter units are not recommended because they easily lead to numerical errors. If at all possible, one should always work with only one unit of length and one unit of time in each calculation.

The question might be asked why we stop here, and why we don't keep on differentiating to $\dddot{x}$ and higher derivatives. The answer to this, as we shall see in the following chapters, is that the second derivative of the displacement, linear or angular, is of great importance, whereas the third derivative hardly ever occurs in practice. The quantity $\dddot{x}$, the time rate of change of acceleration, has some physical meaning and has even been given a name: it is called "jerk" (Problem 309, page 432) but it is of no particular importance.

Problems 161 *to* 168.

31. Motion in Space.

The location of a point P in space is described by three numbers, usually the x, y, and z coordinates of a rectangular coordinate system. When the point moves through space its location becomes a function of time and the coordinates x, y, and z are all functions of time. The three equations

$$x = f_1(t), \qquad y = f_2(t), \qquad z = f_3(t)$$

then describe a curve in space; they are known as the three "parametric" equations of a curve with the time t as "parameter."

By eliminating t from between the three equations, we obtain two equations in x, y, and z, not containing t. Each of these equations represents a surface in space and the pair of equations determines the curve of intersection of the two surfaces.

In Fig. 141 the space curve along which the point moves is drawn, and at time $t = 0$ the point is at a certain location on the curve. This point we call O, the origin, and lay our coordinate axes through it. At some other time $t = t$ the point is located at P, and a little later, at $t = t + \Delta t$, the point is located at P'. The vector OP is often denoted by $\mathbf{s}$, the first letter of the Latin *spatium*, meaning space or distance. The small vector PP' then is written $\Delta\mathbf{s}$, and OP' is $\mathbf{s} + \Delta\mathbf{s}$, where the $+$ sign has to be understood vectorially; *i.e.*, a parallelogram construction sum, and not an algebraic sum. The vector $OP = \mathbf{s}$ has the three components x, y, z; the vector $OP' = \mathbf{s} + \Delta\mathbf{s}$ has the components $x + \Delta x$, $y + \Delta y$, $z + \Delta z$, while the small vector $\Delta\mathbf{s}$ has the components Δx, Δy, Δz, as can be seen in Fig. 141. Then the quotients

$$\frac{\Delta x}{\Delta t}, \frac{\Delta y}{\Delta t}, \text{ and } \frac{\Delta z}{\Delta t}$$

are called the "average velocities" during the interval Δt in the x, y, and z directions, and by letting Δt become small indefinitely, in the

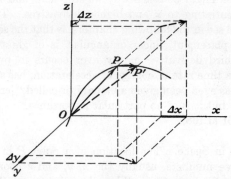

FIG. 141.   The displacement and velocity of a point moving in space.

manner of the calculus, these average velocities go to the limiting values

$$\dot{x}, \dot{y}, \dot{z} \qquad \text{or} \qquad \frac{dx}{dt}, \frac{dy}{dt}, \frac{dz}{dt}$$

which are the (instantaneous) velocities in the x, y, and z directions. The numerical value of the length of the vector $\Delta \mathbf{s}$ is designated as Δs, and the quotient $\Delta s/\Delta t$ is called the "average velocity during the interval Δt," and its limiting value for small Δt is

$$\dot{s} \text{ or } \frac{ds}{dt}$$

the velocity of point P. In Fig. 141 it is seen that $\Delta \mathbf{s}$ is related to Δx, Δy, and Δz as the diagonal of a small parallelepiped with sides Δx, Δy, and Δz. Division by Δt and going to the limit does not alter this relation, and therefore

The velocity of a point P in space is a vector of value $\dot{s}$, directed tangent to the path of P, and is the vector sum of three component velocities, $\dot{x}$, $\dot{y}$, and $\dot{z}$.

Accelerations are deduced from velocities by a process of differentiation, just as velocities were deduced from displacements by differentiation, **and the** properties of accelerations are very similar to those of velocities. In order to simplify the figure, we consider motion in a plane instead of in three-dimensional space, and Fig. 142a is the two-dimensional equivalent of Fig. 141. The curve of Fig. 142a is the path of the point in the plane, showing the point in two positions, at time t and at time $t + \Delta t$. The increment Δs and its two components Δx and Δy are almost equal to $\mathbf{v}\,\Delta t$, $\dot{x}\,\Delta t$ and $\dot{y}\,\Delta t$ (but not exactly equal, because Δt is not exactly zero); this is indicated by the symbol $\approx$, meaning "approximately equal to." The velocity vectors $\mathbf{v}$, $\dot{x}$, $\dot{y}$ in Fig. 142a are in the direction of the increment vectors Δs, Δx, Δy.

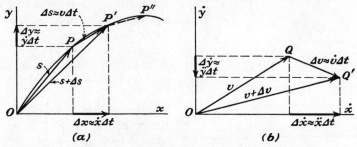

FIG. 142. Velocity and acceleration of a point moving along a plane curve.

Now we turn to Fig. 142b, where $OQ = \mathbf{v}$ is (approximately) parallel to $PP' = \Delta s$ of the previous figure, and the velocity vector $\mathbf{v} + \Delta \mathbf{v}$ is approximately parallel to $P'P''$ in 142a. Again the velocity increment Δv, when divided by Δt, gives the average acceleration during the interval Δt, and when Δt becomes very small, this ratio converges to $d\mathbf{v}/dt = \dot{\mathbf{v}} = \ddot{\mathbf{s}}$, the acceleration.

The acceleration of a point P in space is a vector of value $\dot{\mathbf{v}}$ or $\ddot{\mathbf{s}}$, directed tangent to the curve Q, formed by the end points of the velocity vectors, **and is the vector sum of the component accelerations, $\ddot{x}$, $\ddot{y}$, and $\ddot{z}$.**

The part of this sentence that is of practical importance is printed in heavy type; the fact that the direction of the acceleration is tangent to the Q curve (Fig. 142b) is hardly ever used, but it *is* important to note that the acceleration vector is *not* tangent to the path, or P curve of Fig. 142a.

A useful way of visualizing and memorizing the direction of the acceleration is by resolving it into normal and tangential components. The curve of Fig. 143 is the path of point P, and two positions of that point are shown: P at time t and P' at time $t + \Delta t$. The velocity vectors at those two instances are drawn in heavy line, tangent to

164   *KINEMATICS OF A POINT*

the path, and at point P' the velocity $\mathbf{v}$ of point P is drawn again in a thin line. The vector difference between the velocity at P and the velocity at P' is $\Delta\mathbf{v}$ as shown. In the previous analysis (of Fig. 142) this $\Delta\mathbf{v}$ was resolved into dx and dy components. Here we do something else and resolve $\Delta\mathbf{v}$ into components along and across the velocity vector $\mathbf{v} + \Delta\mathbf{v}$, as shown. We now proceed to calculate these components for the case that P and P' are very close together. First we draw lines at P and P' normal to the path or, which is the same thing, normal to the velocity vectors. These lines intersect at C, and if

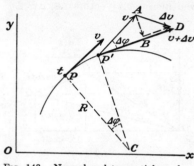

the distance PP' is made smaller indefinitely, the angle $\Delta\varphi$ at C becomes smaller also and C *converges to* a point called *the center of curvature of the path at point* P. The angle between the two velocity vectors is also $\Delta\varphi$. The line AB is normal to $P'B$ and for very small $\Delta\varphi$ it is practically normal to $P'A$ as well. Then, for $\Delta\varphi$ measured in radians, we have $AB = v\,\Delta\varphi$.

Fig. 143. Normal and tangential acceleration of a point.

But in triangle CPP' we have $PP' = \Delta s = R\,\Delta\varphi$, when R is the radius of curvature at P. Eliminating $\Delta\varphi$ from these equations, we find for the increment in speed in a direction normal to the velocity

$$AB = (\Delta v)_n = \frac{v}{R}\Delta s$$

Dividing both sides by the time interval Δt,

$$\left(\frac{\Delta v}{\Delta t}\right)_n = \frac{v}{R}\frac{\Delta s}{\Delta t}$$

For the limit as Δt goes to zero this becomes

$$(\dot v)_n = \frac{v}{R}v = \frac{v^2}{R}$$

This expression can be written in another form, by noting that in Fig. 143

$$\Delta s = R\,\Delta\varphi, \qquad \frac{\Delta s}{\Delta t} = R\frac{\Delta\varphi}{\Delta t}, \qquad \dot s = R\dot\varphi$$

$$v = R\omega = R\dot\varphi \tag{5}$$

Substituting this into the above we can write for the *normal accelera-tion*

$$(\dot{v})_n = \frac{v^2}{R} = \omega^2 R = \dot{\varphi}^2 R = V\omega \qquad (6a)$$

This acceleration is directed toward the center of curvature, C, and therefore is often called the *centripetal acceleration* (which means the center-seeking acceleration).

Now, in Fig. 143, we turn our attention to the component BD of the velocity increment, which, for smaller and smaller $\Delta\varphi$, comes nearer and nearer to $(v + \Delta v) - v = \Delta v$, and, dividing by Δt and going to the limit $\Delta t = 0$, we find for the *tangential acceleration*

$$(\dot{\mathbf{v}})_t = \dot{v} = \ddot{s} \qquad (6b)$$

The acceleration vector in space can be resolved into all sorts of components, of which practically the most important ones are **the Cartesian components, $\ddot{x}$, $\ddot{y}$, $\ddot{z}$, or the normal (v^2/R), and tangential $(\dot{v})$, components.**

It is significant that there are *three* Cartesian components, while in the other resolution there are only *two* components, normal and tangential. The derivation of the latter components was done on the two-dimensional curve Fig. 143. In case of a skew space curve we take three points on that curve P, P', P'', as in Fig. 142a, and pass a plane through these three points. When the three points get closer and closer together the plane becomes the *osculation* (kissing) *plane* of the curve at point P. The analysis of Fig. 143 can then be applied in the osculation plane of the curve at the point P, with the result that the normal component of acceleration in the plane of osculation is v^2/R, while the normal component of acceleration perpendicular to that plane is zero.

It depends on the problem at hand, which of the two methods of resolution happens to be most convenient or useful. As an example of the equivalence of the two ways of resolving the acceleration, consider the uniform motion of a point P around a circle (Fig. 144) with a constant speed $v = \omega R$. The angle varies with time: $\varphi = \omega t$. The velocity v is resolved into Cartesian components

$$\dot{x} = -v \sin \omega t \qquad \text{and} \qquad \dot{y} = v \cos \omega t$$

of which $\dot{x}$ is negative because it points to the left, the negative x direction. Differentiation leads to

$$\ddot{x} = -v\omega \cos \omega t \qquad \text{and} \qquad \ddot{y} = -v\omega \sin \omega t$$

with both components of the acceleration negative or pointing toward the origin O, while the ratio of their two magnitudes is as $\cos \omega t : \sin \omega t$, which makes the resultant point toward the origin, as a simple sketch quickly shows. The magnitude of the resultant acceleration is (the two components are perpendicular):

$$\sqrt{(-v\omega \cos \omega t)^2 + (-v\omega \sin \omega t)^2}$$
$$= v\omega \sqrt{\cos^2 \omega t + \sin^2 \omega t} = v\omega = \frac{v^2}{R} = \omega^2 R$$

Approaching the same problem from the other direction and resolving by Eqs. (6), we realize that the tangential component is zero

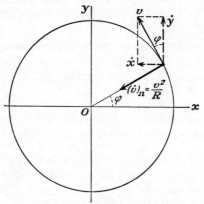

Fig. 144.   Uniform motion of a point along a circle.

because the tangential speed is constant, so that $\dot{v} = 0$. The normal component is v^2/R, giving the same answer as before with much less work. For this example, therefore, the Cartesian method of resolution is awkward and unsuited, whereas the other method is naturally adapted to it. In the next example the converse will be true.

Consider the motion of a point P, such that the horizontal acceleration $\ddot{x} = 0$ and the vertical acceleration $\ddot{y} = -g$ (Fig. 145), which, as we will see later, represents the motion of a bullet or projectile. Using the rectangular-component method, we integrate the two equations specifying the motion.

$$\ddot{x} = 0, \qquad\qquad \ddot{y} = -g$$
$$\dot{x} = C_1, \qquad\qquad \dot{y} = -gt + C_3$$
$$x = C_1 t + C_2, \qquad y = -\frac{gt^2}{2} + C_3 t + C_4$$

The four integration constants are determined by four initial conditions. Suppose we specify that at time $t = 0$, both x and y are zero, which is another way of saying that we lay the origin of coordinates at the location of point P at the instant $t = 0$. From these two conditions we see that $C_2 = C_4 = 0$, and only C_1 and C_3 are left, which, as we see from the equations, represent the x and y velocities at time

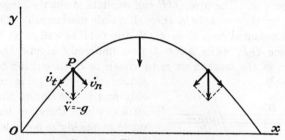

FIG. 145. The motion of a projectile.

$t = 0$, which we then might as well call v_{0x} and v_{0y}. Thus

$$x = v_{0x}t \quad \text{and} \quad y = -\frac{gt^2}{2} + v_{0y}t$$

are the parametric equations of the path. The ordinary equation of the path is obtained by eliminating the parameter t from between the two above equations, giving

$$y = -\frac{g}{2}\left(\frac{x}{v_{0x}}\right)^2 + v_{0y}\left(\frac{x}{v_{0x}}\right)$$

which, being a quadratic equation between x and y, represents a conic section, in this case the parabola of Fig. 145. At each point P of this parabola the acceleration is directed downward and is equal to g, which is very simple. To put this result in the other form requires first finding the value of the velocity at P and the value of the radius of curvature, R, of the parabola at that point. The normal component of acceleration is then v_P^2/R and the tangential is $\dot{v}_P$. This, however, represents a large amount of algebra, and obviously this is a problem to which the Cartesian-coordinate approach is much more suited than the normal-tangential-component method.

Problems 169 *to* 171.

32. Applications. The foregoing theories will now be applied to three cases:

    a. The quick-return mechanism of shaping tools

    b. The crank mechanism

    c. The rolling wheel, or cycloidal motion

    a. The Quick-return Mechanism. This device is shown schematically in Fig. 146 and consists of a crank CA rotating at uniform angular speed $\dot{\varphi}$. The arm OAP can oscillate about the hinge O, and is pushed by the crankpin A through a slide mechanism. The point P is again mounted on a slide on the arm OAP as well as on the horizontal guide O_1P, while point A runs uniformly around the circle. The object of the mechanism is to attach to P the cutting tool of a

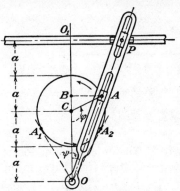

shaper that cuts going one way only and is idle during the return stroke, which therefore has to be made as quickly as possible. Now P moves from its extreme right to its extreme left position while A moves from A_2 to A_1, via the top of the circle, while the return stroke takes place while A moves from A_1 to A_2 via the bottom of the circle in considerably less time. We want to calculate and plot the velocity and acceleration of point P. In order to do this we first choose as origin the point O_1 and designate

Fig. 146.   The quick-return mechanism.

$O_1P = x$. Then we have to express x in terms of φ by geometry. To that end, draw the perpendicular AB and note that

$$AB = a \sin \varphi$$
$$OB = OC + CB = 2a + (-a \cos \varphi) = a(2 - \cos \varphi)$$

By the similarity of triangles OAB and OPO_1, we have

$$\frac{O_1P}{O_1O} = \frac{AB}{OB} \quad \text{or} \quad \frac{x}{4a} = \frac{a \sin \varphi}{a(2 - \cos \varphi)}$$

so that

$$x = 4a \frac{\sin \varphi}{2 - \cos \varphi}$$

This formula enables us to calculate the position x of P for every position φ of point A, and the relation is plotted in Fig. 147. To find the velocity of P, we differentiate the displacement.

$$v_P = \dot{x} = \frac{dx}{dt} = \frac{dx}{d\varphi}\frac{d\varphi}{dt} = \dot{\varphi}\frac{dx}{d\varphi}$$

$$= 4a\dot{\varphi}\frac{\cos\varphi(2 - \cos\varphi) - \sin\varphi\sin\varphi}{(2 - \cos\varphi)^2}$$

$$= 4a\dot{\varphi}\frac{2\cos\varphi - 1}{(2 - \cos\varphi)^2}$$

This result again has been plotted in Fig. 147. We note that $a\dot{\varphi}$ is the tangential velocity of point A along the circle and that the other factors in the expression are pure numbers.

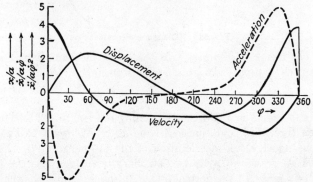

Fig. 147. Displacement, velocity, and acceleration of the rectilinear motion of point P of Fig. 146.

To find the acceleration of P, the velocity has to be differentiated, and in that expression the length a as well as the angular velocity $\dot{\varphi}$ are constant with time, while only φ is variable. Thus

$$\ddot{x}_P = \frac{dv_P}{dt} = \frac{dv_P}{d\varphi}\frac{d\varphi}{dt} = \dot{\varphi}\frac{dv_P}{d\varphi}$$

$$= 4a\dot{\varphi}^2\frac{-2\sin\varphi(2 - \cos\varphi)^2 - (2\cos\varphi - 1)2(2 - \cos\varphi)\sin\varphi}{(2 - \cos\varphi)^4}$$

$$= -8a\dot{\varphi}^2\frac{\sin\varphi(2 - \cos\varphi) + (2\cos\varphi - 1)\sin\varphi}{(2 - \cos\varphi)^3}$$

$$= -8a\dot{\varphi}^2\frac{\sin\varphi(1 + \cos\varphi)}{(2 - \cos\varphi)^3}$$

This is the third curve plotted in Fig. 147. The quantity $a\dot{\varphi}^2$ is the centripetal acceleration of the point A and the other factors in the expression are pure numbers. Thus, although the analysis and the differentiations are a little complicated, the result obtained is very general; we have the displacement, velocity, and acceleration for every position of angle φ. The displacement of P is expressed in units equal to the radius of the crank; the velocity of P is in units equal to the speed of the crankpin A; and the acceleration of P in units equal to the acceleration of A. Later on (page 208) we will see a graphical method for finding these results.

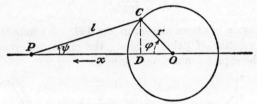

FIG. 148.   The crank mechanism.

b. The Crank Mechanism.   In the crank mechanism of Fig. 148 the crank OC is supposed to rotate at uniform angular velocity

$$\omega = \dot{\varphi} = \text{constant.}$$

The crank radius is r and the connecting-rod length PC is l, usually between three and four times as large as r. We want to know the position, velocity, and acceleration of the piston, point P. For the analysis we draw and write two auxiliary quantities into the figure: $\angle CPO = \psi$ and CD, the normal on OP. Choose point O as the origin and call $OP = x$, positive to the left. The angle φ is counted from line OP, and we have $\varphi = \omega t$. Then

$$x_P = OP = OD + DP = r \cos\varphi + l \cos\psi$$

For the purpose of eliminating the auxiliary angle ψ we write the length CD in two ways: as a side of triangle OCD and as a side of PCD.

$$CD = r \sin\varphi = l \sin\psi$$

Hence

$$\sin\psi = \frac{r}{l} \sin\varphi,$$

and

$$\cos\psi = \sqrt{1 - \sin^2\psi} = \sqrt{1 - \frac{r^2}{l^2}\sin^2\varphi}$$

Substitute this into the expression for x_P.

$$x_P = r \cos\omega t + l \sqrt{1 - \frac{r^2}{l^2}\sin^2\omega t}$$

which is the answer for the position of the piston expressed as a function of the time t. We should now differentiate this expression twice to obtain the other two required answers for $\dot{x}$ and $\ddot{x}$, and we should then plot them in a figure like 147. However, we do not do that, for the reason that the differentiation of a square root is more complicated than we like it to be and that we can make a simplification leading to an answer which is more clearly understandable and of sufficient accuracy. We note that r/l is about $\frac{1}{4}$, so that $r^2/l^2 = \frac{1}{16}$ and $r^2 \sin^2 \omega t/l^2$ is less than that.

An expression of the form $\sqrt{1 - \epsilon}$ can be developed into a power series,

$$\sqrt{1 - \epsilon} = 1 - \frac{\epsilon}{2} - \frac{\epsilon^2}{8} - \cdots$$

and for $\epsilon = \frac{1}{16}$ the third term is about $1/2,000$. If, therefore, the third and all following terms are neglected, an error of 1 part in 2,000 is involved, which is entirely acceptable. Thus

$$x_P \approx r \cos \omega t + l - \frac{r^2}{2l} \sin^2 \omega t$$

of which the first term is OD in Fig. 148, the second term is CP, and the third is almost the difference between CP and CD. Differentiate this expression

$$\dot{x}_P \approx v_P = -r\omega \left(\sin \omega t + \frac{r}{l} \sin \omega t \cos \omega t \right)$$

$$= -r\omega \left(\sin \omega t + \frac{r}{2l} \sin 2\omega t \right)$$

Differentiating once more, we obtain

$$\ddot{x}_P \approx -r\omega^2 \left(\cos \omega t + \frac{r}{l} \cos 2\omega t \right)$$

The parentheses in both expressions are pure numbers; the factor $r\omega$ in the expression for $\dot{x}_P$ is the (tangential) velocity of the crankpin C and $r\omega^2$ is the (centripetal) acceleration of the crankpin. Figure 149 shows these relations, being a plot of the two above formulas for the case $r/l = \frac{1}{4}$. It is seen that the maximum piston acceleration is 25 per cent greater than the centripetal acceleration of the crankpin and occurs at the dead-center position far away from the crank center. The thin dotted lines represent the first terms in the parentheses of the formulas, and they can be interpreted physically as the velocity

and acceleration in the case of a very long connecting rod $l/r = \infty$.
In that case the motion reduces to a simple harmonic one (page 159).
On page 199 we shall return to this problem with a graphical method.

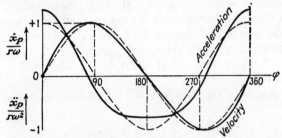

FIG. 149.   Velocity and acceleration of the piston P of Fig. 148 for a uniformly rotating
crank as a function of the crank angle.

c. *The Rolling Wheel.*   The third and last example of this article
will be the motion of a point on the periphery of a rolling wheel of
radius a.   Figure 150 shows the wheel in full line in the position it

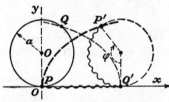

FIG. 150.   Construction of a point P'
of the path of a peripheral point of a
rolling wheel.

occupies at $t = 0$ with the point P,
to be studied, at the bottom.   The
dotted position occurs after the wheel
has rotated (at uniform speed)
through an angle $\varphi = \omega t$, without
slipping, so that the wavy straight
line PQ' and the wavy circular arc
$P'Q'$ have the same length and have
been rolling over each other.   Thus
we can graphically construct the

position of P for each angle φ, and analytically we can read from the
figure

$$x_P = \varphi a - a \sin \varphi, \qquad y_P = a(1 - \cos \varphi)$$

in which the angle $\varphi = \omega t$ is measured in radians.   The above pair of
equations determine the path: they are the parametric form of these
equations.   The ordinary form, being a single equation in x and y,
could be found by eliminating φ between the above two equations.
However, in this case that is very difficult algebraically, and therefore
we will not do it but will retain the parametric form.   The path of P
is known as a *cycloid*, and has the shape shown in Fig. 151.   The next
step is differentiation of the position equations.

$$\dot{x} = a\dot{\varphi}(1 - \cos \varphi), \qquad \dot{y} = a\dot{\varphi} \sin \varphi$$
$$\ddot{x} = a\dot{\varphi}^2 \sin \varphi, \qquad \ddot{y} = a\dot{\varphi}^2 \cos \varphi$$

In these expressions $\dot{\varphi} = \omega$ is the constant angular velocity of the the wheel. Again $a\dot{\varphi}$ can be interpreted as the speed the point P would have, if the wheel were not rolling but were only rotating with point O fixed; and $a\dot{\varphi}^2$ would be the centripetal acceleration of P in the same case. The plot of Fig. 151 shows these relations for four points, 90 deg apart in φ. The velocity vectors at these points are shown in thin lines; the acceleration vectors in heavy lines. In both

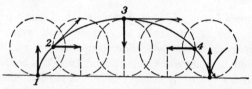

Fig. 151. Velocities and accelerations in various positions along the cycloidal path.

cases the radius of the circle is the scale for the unit velocity $a\dot{\varphi}$ and the unit acceleration $a\dot{\varphi}^2$. The reader should check the results shown for these four points against the analytical results and sketch in for himself some intermediate points $\varphi = 30, 45$ deg, etc. Also, the reader should deduce from the formulas that the total acceleration vector has the magnitude $a\omega^2$, independent of the location φ, and is always directed towards the center of the wheel. The velocity vector is tangent to the path (page 163); the acceleration vector is not, and is purely tangential at point 1, purely centripetal at point 3, and mixed at 2 and 4. To find the acceleration at point 2 from its tangential and normal components, involving a calculation of the radius of curvature of a cycloid, obviously would be a complicated procedure. In this example, therefore, the Cartesian-component method is greatly to be preferred. In the chapter on relative motion (page 298) we shall understand more clearly than we can now why the acceleration vector is always directed toward the center of the wheel.

Problems 172 *to* 182.

CHAPTER X

DYNAMICS OF A PARTICLE

33. Newton's Laws. In this chapter we shall deal with dynamics, which is the part of mechanics concerning itself with the relation between forces and motions. Sometimes the word "dynamics" is used to include kinematics as well and then the term "kinetics" applies to what we call "dynamics" here. A *particle*, also sometimes called "material particle" or "material point," is defined as a *massive point*, or as a body with a definite mass but of zero dimensions. This, of course, is a mathematical abstraction; no such "particle" in the strict sense of the definition can possibly exist in reality, but the laws of mechanics are commonly expressed in terms of it. The actual bodies encountered in practice under certain conditions will "act like particles." Later (page 213) we will see what these conditions are precisely; for the time being we will think of a particle as a small body with a certain mass or weight.

The laws of dynamics were entirely unknown to the ancient Greeks and Romans; their discovery was the first great scientific achievement of our present Western civilization. It started with *Leonardo da Vinci* (1452–1519), who made many experiments and wrote his findings down in a series of notebooks, which, however, were not published until several centuries later, so that Leonardo's influence on the development of the science was practically zero. The next great name is *Galileo* (1564–1642), who published his theories in a book entitled "Discorsi e dimostrazioni mathematiche," which contained the laws of dynamics in a primitive form, and on which Newton based his work. *Isaac Newton* (1642–1726) in 1687 published his famous "Principia" ("Philosophiae naturalis principia mathematica"), in which the laws of dynamics were not only set forth, but were treated with a new mathematical method, very much suited to the purpose, the calculus. Newton, however, was so far ahead of his time that only a very few people could understand his writings, and during practically the whole following eighteenth century a series of brilliant scientists worked out the consequences of Newton's great publication. The most important of those were Euler (1707–1748), John Bernoulli

(1667–1748), Daniel Bernoulli (1700–1782), d'Alembert (1717–1783), and Lagrange (1736–1818). Therefore, the discovery and formulation of the laws of mechanics, like most great discoveries, was not the work of one man alone, but Newton's contribution is so preponderant, that the laws are called by his name. They are

First Law. A particle on which no forces are acting has zero acceleration.

Second Law. A particle on which a force is acting experiences an acceleration in the direction of that force, proportional to the force and inversely proportional to the mass of the particle.

Third Law. Action equals reaction, or the forces acting between two particles are equal and oppositely directed.

In all of these statements, "force" is as defined on page 3, and includes only direct-contact pushes or pulls and the force of gravity, but excludes inertia forces or centrifugal forces, which will be introduced and defined on page 211.

The first law is no more than a special case of the second law. It is the basis of all statics, but it is more than that, since the term "zero acceleration" not only comprises the state of rest or equilibrium, but also the state of uniform velocity. Therefore, by the first law, a particle on which no force is acting retains its velocity indefinitely, in magnitude as well as in direction. An automobile coasting on a level road would retain its speed forever, if no friction force or other retarding forces were acting. A bullet in horizontal flight will retain its horizontal speed except for the retarding action of air resistance.

Newton's third law of action and reaction is of great importance in statics as well as in dynamics and has been discussed on page 4 and in the several subsequent applications.

We now turn to the second and most important of Newton's laws, and in order to express it in a formula, we choose a coordinate system with the x axis along the direction of the force F_x, and denote the mass of the particle by m. Then

$$\ddot{x} = \text{constant} \times \frac{F_x}{m}$$

In this equation the force is measured in pounds and the acceleration in feet per second squared or inches per second squared, but since this is the first time we encounter "mass," we have as yet no unit for it. In order to make the equation as simple as possible, we choose the unit of mass so as to make the constant equal to unity, and thus

The unit of mass equals the unit of force divided by the unit of acceleration, or 1 lb/1 ft/sec² or 1 lb ft⁻¹ sec².

In case it is not convenient to choose the x axis in the direction of the force and hence the force **F** is directed obliquely in space, we can resolve that force into its three Cartesian components F_x, F_y, and F_z, by page 102. Similarly by page 163 we can resolve the acceleration $\ddot{s}$ into three components $\ddot{x}$, $\ddot{y}$, and $\ddot{z}$. By Newton's second law the total acceleration $\ddot{s}$ is in the direction of the total force **F**, and by pages 102 and 163 the parallelepiped of acceleration is geometrically similar to the parallelepiped of forces, or

$$\frac{F_x}{\ddot{x}} = \frac{F_y}{\ddot{y}} = \frac{F_z}{\ddot{z}} = \frac{F}{\ddot{s}}$$

But, by the second law, the last ratio equals the mass m, so that we deduce

$$\left.\begin{array}{l} F_x = m\ddot{x} \\ F_y = m\ddot{y} \\ F_z = m\ddot{z} \end{array}\right\} \qquad (7)$$

a set of equations expressing Newton's second law in terms of Cartesian coordinates. More generally, if the force acting on a particle is resolved into components along any three arbitrary directions in space (not necessarily perpendicular or Cartesian), each component force equals the mass multiplied by the corresponding component of the acceleration vector. Applied to the normal (centripetal) and tangential directions, the equations become

$$\left.\begin{array}{l} F_t = m\ddot{s} \\ F_n = m\dfrac{\dot{s}^2}{R} \end{array}\right\} \qquad (7a)$$

A particular case of Eq. (7) is the freely falling stone. The only force acting on the particle is the attractive force of gravity, called the "weight," denoted by W. If we choose the coordinate system with the x axis vertical, pointing downward, Eqs. (7) become

$$\begin{array}{l} W = m\ddot{x} = mg \\ 0 = m\ddot{y} \\ 0 = m\ddot{z} \end{array}$$

The downward acceleration due to gravity has been measured carefully by many experiments; it is commonly denoted by the symbol g

and is found to be on the average

$$\frac{W}{m} = g = 32.2 \text{ ft/sec}^2 = 386 \text{ in./sec}^2 \qquad (8)$$

The value of g differs with the location on earth and with the height above ground; these variations are so small that they can be neglected in almost all engineering applications. Equation (8) expresses the relation between mass and weight, and enables us to visualize the size of the unit of mass. Consider for example a pound weight, which is $1/0.28$ or about 4 cu in. of steel. The earth pulls on this piece of steel with a force of 1 lb, and that force gives it an acceleration of 32.2 ft/sec² or 32.2 units. The same force of 1 lb applied to a larger piece of steel will accelerate it slower, by the second law; in particular, if the 1-lb force acts on a piece of steel weighing 32.2 lb, the acceleration will be 1 ft/sec² or 1 unit. Thus the unit of mass is a piece of steel weighing 32.2 lb with a volume of $32.2 \times 1/0.28 = 115$ cu in. This unit will always be referred to in this book as lb ft⁻¹ sec²; in other books it is sometimes called "slug."

The pound, foot, and second are the units in common use among engineers in the English-speaking world. Other systems of units exist, differing from the above in two respects: (*a*) by the adoption of mass as the fundamental unit, with force derived from it instead of the other way around as above, or (*b*) by the adoption of metric units instead of English units. The four possibilities are shown in the table below, the fundamental units being printed in heavy type.

	No. 1	No. 2	No. 3	No. 4
Length.....	**foot (ft)**	**ft**	**meter**	**centimeter (cm)**
Time.......	**second (sec)**	**sec**	**sec**	**sec**
Force......	**pound (lb)**	lb ft sec⁻² (poundal)	**kilogram (kg)**	g cm sec⁻²(dyne)
Mass.......	lb ft⁻¹sec²(slug)	**pound (lb)**	kg m⁻¹ sec²	**gram (g)**
Name......	English engineering	. . . .	Metric engineering	The cgs system
Used by....	Engineers in U.S. & U.K.	Nobody	Engineers outside U.S. & U.K.	Physicists everywhere

In this book we will always use the English engineering system of units, No. 1 in the table, but the word "slug" will be avoided. The cgs system, No. 4 in the table, is universally used by physicists, who prefer to make mass fundamental and force derived, because the subject of statics is of no importance to them. Also physicists now generally use metric measure even in English-speaking countries. The system No. 2 has had some champions in the recent past, but is now almost abandoned, certainly by engineers, who do not like to use the weight of about ½ oz of steel (the poundal) for the unit of force in all their statical calculations. But strictly speaking, the expression "nobody" in the table should be interpreted as "hardly anybody," as in Gilbert and Sullivan's "Pinafore."

34. Rectilinear Motion of a Particle. In this article the foregoing theory will be applied to the following cases:

   a. The ball with two strings
   b. The smooth inclined plane
   c. The rough inclined plane
   d. The simple vibrating system

   a. The Ball with Two Strings. Figure 152 shows a heavy cast-iron ball, several inches in diameter, of weight W, suspended from above

FIG. 152. The classical experiment of the heavy ball with the two thin cotton threads.

by a thin cotton thread, and having an identical thread hanging down from it. When we start pulling down on the lower string, which one of the two strings will break first? Considered as a problem in statics the answer is obvious. The bottom string sustains the force F, and the upper string the force $F + W$, which is larger, so that the upper string will break first. This will actually happen if we pull down slowly. But when we give a sudden, sharp pull to the lower thread, it will break and the ball remain suspended. The explanation of this curious behavior is that the threads are elastic and have a certain elongation associated with the tensile force sustained by them. Before the upper thread can carry more than the weight W, the ball must be allowed to go down somewhat. By giving a quick pull to the lower thread the force in the lower thread can be made quite large and this large force will accelerate the ball downward. But this takes some time and before there is any appreciable downward displacement the lower string has snapped. A variation of this experiment consists in placing the ball with just a single thread attached to it on a flat table. By a gentle pull, the ball can be dragged along the table at uniform speed, the force in the thread being equal to the friction force between ball and table. A quick pull, however, can easily bring the thread force to $10W$, which will break the thread in, say, 0.001 sec. The ball then is subjected to an acceleration $10g$ during that short time, and will hardly move. It will have a very small velocity at the end of the 0.001 sec, and that velocity will soon be destroyed by the retarding action of the friction force, and if the pull has been quite sudden, the ensuing displacement will be hardly visible.

b. The Smooth Inclined Plane. The block of Fig. 153 lies on a smooth inclined plane, and is shown with all forces acting on it. These forces are not in equilibrium, and hence the block or "particle" will not remain at rest. We choose a coordinate system with the x axis along the incline pointing downward, and with the y axis perpendicular to the incline pointing upward, and assume that the block does not leave the

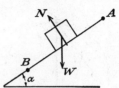

Fig. 153. The frictionless inclined plane.

plane, so that $y = 0$, and consequently $\dot{y} = \ddot{y} = 0$. Then the Newton equations (7) are

In the x direction: $\qquad W \sin \alpha = m\ddot{x}$
In the y direction: $\qquad N - W \cos \alpha = m\ddot{y} = 0$

The second of these gives $N = W \cos \alpha$ as in the static case. Remembering that by Eq. (8) $m = W/g$, the other equation leads to

$$\ddot{x} = g \sin \alpha$$

or the acceleration down the plane is a fraction of g, and the motion down the plane can be described as a retarded free fall. For $\alpha = 0$ there is no acceleration, and for $\alpha = 90$ deg there is a free fall with acceleration g. This result was known to Galileo, who put forward the following question. Suppose the inclined plane is hinged at A (Fig. 153), enabling us to change the inclination at will, and suppose we start the block from A at rest at time $t = 0$. How far will it slide during the first t sec for a given constant value of α, and if we plot the end point of travel on the plane as point B, what will be the locus of all points B for various angles of inclination α? Obviously for $\alpha = 0$ the block will not move, and B coincides with A. For another angle α the distance $AB = x$ is found by integrating twice the expression $\ddot{x} = g \sin \alpha$ with the result

$$x = (g \sin \alpha)\frac{t^2}{2}$$

Then, to answer Galileo's question, consider t constant and α variable, so that x = constant $\times$ sin α. In Fig. 154 the distance AB is plotted, and the line BC is drawn perpendicular to AB, leading to point C. It is seen that $AB = AC$ sin α, so that if we keep AC constant, we can vary α, and find point B' for a different α, by making $\angle CB'A = 90°$. Therefore, the locus of B is the locus of the apex of a right-angled triangle on AC as hypothenuse, and this locus is a circle with AC as diameter. Galileo used this result for an experimental test setup to check the law $s = \frac{1}{2}gt^2$ of free-falling bodies. He could not do it for the real case, because the bodies fell faster than his primitive time-measuring apparatus could handle. In order to slow down the free-falling motion, he used an inclined plane, and in order to avoid friction he used rolling cylinders instead of sliding blocks. The "particle" theory does not apply to rolling bodies, as we will see on page 242, but Galileo did not know that and was satisfied with the results.

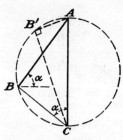

Fig. 154. Galileo's experiment. The locus of the end points B of travel of a block sliding down a smooth inclined plane from a given starting point A, during a given time t, for various angles of inclination α, is a circle.

c. The Rough Inclined Plane. Figure 155 shows a (block behaving like a) particle on a rough inclined plane, with the forces acting on the particle. The friction force has been given the magnitude and direction corresponding to a downward sliding motion. Choose a coordinate system along and across the incline, as shown, and write the Newton equations [Eqs. (7)], assuming that the particle does not jump off the incline but slides down with an unknown acceleration $\ddot{x}$.

$$W \sin \alpha - fN = m\ddot{x}$$
$$-W \cos \alpha + N = m\ddot{y} = 0$$

From the second equation we solve for the normal force N and substitute the result into the first equation, remembering that by Eq. (8) $m = W/g$. This gives

$$\sin \alpha - f \cos \alpha = \frac{\ddot{x}}{g}$$

or

$$\ddot{x} = g(\sin \alpha - f \cos \alpha)$$

The parenthesis is a pure number, less than 1, so that the acceleration down the plane is seen to be a certain fraction of g. It is seen that for a sufficiently large value of the friction coefficient, the second term in the bracket becomes larger than the first one, so that $\ddot{x}$ becomes negative and the block slides uphill, which is contrary to the assumption under which this result was obtained. If the block would actually go uphill, the force fN would be reversed, and by Fig. 155 clearly the block could not go uphill. In that case, *i.e.*, in the case that

$$f \cos \alpha > \sin \alpha$$

or $f > \tan \alpha$, the assumption of downward sliding does not allow us to satisfy Newton's equations, and therefore is untenable. We must make another assumption and try to assume no motion at all. Then $\ddot{x} = 0$ is known, while previously it was unknown. But the friction force F is now unknown while previously it was known to be fN. Then we write two equations, which are equations of statics,

$$W \sin \alpha - F = m\ddot{x} = 0$$
$$-W \cos \alpha + N = m\ddot{y} = 0$$

and find that $F = N \tan \alpha$. For the assumed case that $f > \tan \alpha$, this becomes $F < fN$, which is required for no slipping. On the other hand, if we should try to apply the non-slip analysis to the case of little friction $f < \tan \alpha$, this would give $F > fN$, which is physically impossible.

This analysis is typical of all dynamical problems involving friction. At the start we do not know whether the system slips or not, or when it *does* move, we often do not know which way it tends to move. We start by assuming one of these possibilities and carry out the analysis; at the end we check whether the answer agrees with the assumption. If it does not

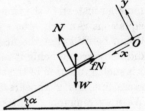

Fig. 155. The rough inclined plane.

agree, we start again with another assumption, and if the problem is physically possible at all, we will find a satisfactory answer to it.

d. The Simple Vibrating System. Figure 156 shows a weight W hanging on a spring of stiffness k lb/in. By this we mean that it takes k lb of force to make the spring 1 in. longer or shorter, and that the extension x is proportional to the force F, or that

$$F = kx$$

We want to know what motions the weight is capable of in an up-and-down direction. First we choose an origin of coordinates, taking point O at the location where the particle is when in static equilibrium. If the weight has some size, as it always has, we replace it by a particle at the center of gravity of the weight (page 213). In

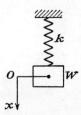

the position of static equilibrium there are two forces acting on the particle, its weight W downward and a spring tension W upward, the spring therefore is somewhat elongated already. Count the displacement x downward from this origin. Now consider the particle in position x. On it act two forces: the weight W downward and a spring force upward that is now $W + kx$. Therefore there is a resultant upward force kx acting on it, or a force $-kx$ in the positive x or downward direction. Newton's law then can be written

Fig. 156. A mass on a spring is the simplest vibrating system.

$$m\ddot{x} = \frac{W}{g}\ddot{x} = -kx$$

or

$$\ddot{x} = -\frac{k}{m}x \qquad (9a)$$

This is *the differential equation of the simple vibrating system.* It cannot be integrated directly because x appears in the right-hand member instead of the time t. Readers familiar with linear differential equations can write the solution immediately; for those who are not, it can be said that the first person who solved the equation probably did it by pronouncing it in words thus: "In Eq. (9a) the displacement x is such a function of the time, that, when it is differentiated twice, the same function x appears again, multiplied by a negative constant." We may remember that sines and cosines behave just like that, and after some trials find that

$$x = \sin\left(\sqrt{\frac{k}{m}}\,t\right) \qquad \text{or} \qquad x = \cos\left(\sqrt{\frac{k}{m}}\,t\right)$$

satisfies Eq. (9a), and therefore both are solutions. Further it is noticed that Eq. (9a) contains two differentiations, that its solution therefore is tantamount to two integrations, and that the general solution ought to contain two integration constants, C_1 and C_2. After some more trials we find that

$$x = C_1 \cos\left(\sqrt{\frac{k}{m}}\,t\right) + C_2 \sin\left(\sqrt{\frac{k}{m}}\,t\right)$$

is a solution of Eq. (9a) for any value of C_1 and C_2, which can be verified by substitution. The constants C_1 and C_2 are to be determined by the initial conditions of the problem. Suppose for example that we specify that at time $t = 0$ the weight is at position x_0 and has a speed

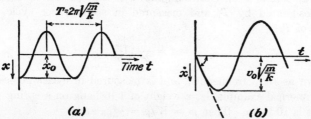

FIG. 157. Motions of the weight of Fig. 156, (a) when starting from rest from a position x_0 below the equilibrium position, and (b) when starting from the equilibrium position with an initial downward speed v_0.

$\dot{x} = v_0$. Substituting these into the solution gives

$$x_0 = C_1 \cos 0 + C_2 \sin 0 = C_1$$
$$v_0 = -C_1 \sqrt{\frac{k}{m}} \sin 0 + C_2 \sqrt{\frac{k}{m}} \cos 0 = C_2 \sqrt{\frac{k}{m}}$$

or, solved for C_1 and C_2,

$$C_1 = x_0, \qquad C_2 = \sqrt{\frac{m}{k}} \, v_0$$

Substitute this into the solution.

$$x = x_0 \cos \left(\sqrt{\frac{k}{m}} \, t \right) + v_0 \sqrt{\frac{m}{k}} \sin \left(\sqrt{\frac{k}{m}} \, t \right) \tag{9b}$$

This equation can be considered as the *general solution of Eq. (9a), in which the initial displacement x_0 and the initial velocity v_0 can be looked upon as arbitrary constants.*

Figure 157a shows the solution for the case of an initial displacement only, v_0 being zero, which means that in Fig. 156 the weight is pulled down a distance x_0 and then released from rest. Figure 157b shows the solution for the case of an initial speed only, the displacement being zero.

The most important point of Eq. (9b) is the time that elapses during a full up-and-down vibration of the weight. The equation shows that this occurs while the angle ($\sqrt{k/m}t$) of the sine or cosine increases by 360 deg or 2π radians. Thus

$$\sqrt{\frac{k}{m}}\, t = 2\pi \qquad \text{or} \qquad t = 2\pi \sqrt{\frac{m}{k}}$$

is the duration of a full vibration, known as the "period" of the vibration, designated by T, and measured in seconds per cycle. The inverse of T

$$\frac{1}{T} = f = \frac{1}{2\pi}\sqrt{\frac{k}{m}} \tag{9c}$$

is known as the "frequency" and is measured in cycles per second. As a numerical example, let a weight of 1 lb hang on a spring whose stiffness is 10 lb/in. Then $m = W/g = \frac{1}{386}$ and

$$f = \frac{1}{2\pi}\sqrt{10 \times 386} = \frac{62}{2\pi} = 9.9 \text{ cycles/sec}$$

Note that $g = 386$, and not 32.2, because k was taken as pounds per inch and not as pounds per foot.

Problems 183 to 192.

35. Curvilinear Motion of a Particle. This article contains four more applications of Newton's laws to simple particles:

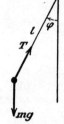

    a. The simple pendulum
    b. The spherical pendulum
    c. The path of a projectile
    d. A particle rolling off a sphere

    a. The Simple Pendulum. A simple pendulum is a particle of mass m suspended from a weightless string of length l, which is supposed to move in a vertical plane. Let the position of the mass be determined by the angle φ, measured in radians, as in Fig. 158. Two forces act on the particle, its weight mg and the unknown string tension T. The path of the particle is a circle with O as center, and this problem seems suited for treatment with tangential and normal components rather than with rectangular ones. The force mg has components $mg \cos \varphi$ radially or normal to the path, and $mg \sin \varphi$ tangentially. Remembering Eqs. (6) (page 165), we write the Newton equations

Fig. 158. The simple pendulum.

radially: $T - mg \cos \varphi = m\dot{\varphi}^2 l$
tangentially: $-mg \sin \varphi = ml\ddot{\varphi}$

In the tangential equation the force is written with a − sign, because the acceleration on the right-hand side is positive for *increasing* angle φ, which in Fig. 158 is seen to be associated with a motion of the particle to the left, whereas the force component is to the right. Of these two equations the radial one enables us to calculate the string tension T after we know the motion, while the tangential equation determines the motion φ without containing the unknown T. In the latter equation we see that the mass m can be canceled: *the motion of the pendulum is independent of its mass,* and

$$\ddot{\varphi} = -\frac{g}{l}\sin\varphi$$

This differential equation cannot be solved by elementary means (the solution involves elliptic functions, which are known only in the form of infinite series). However, if we limit ourselves to investigating motions with small angles φ, the $\sin\varphi$ can be developed into a power series

$$\sin\varphi = \varphi - \frac{\varphi^3}{6} + \cdots \approx \varphi$$

and if all terms except the first one are neglected, the error involved is small for small angles. For example, if $\varphi = 5.7$ deg or 0.1 radian, the first term is 0.1 and the next one $1/6,000$, so that the error is 1 part in 600, which is acceptable. Then the differential equation reduces to

$$\ddot{\varphi} = -\frac{g}{l}\varphi$$

which is of the same form as Eq. (9a) (page 182), and consequently has the same solution. In particular, the frequency of the pendulum is [Eq. (9c), page 184]

$$f = \frac{1}{2\pi}\sqrt{\frac{g}{l}} \tag{9d}$$

and, if the pendulum is started ($t = 0$) from the position φ_0, the solution is [Eq. (9b)]

$$\varphi = \varphi_0 \cos\left(\sqrt{\frac{g}{l}}\,t\right)$$

The angular velocity is, by differentiation,

$$\dot{\varphi} = -\varphi_0\sqrt{\frac{g}{l}}\sin\left(\sqrt{\frac{g}{l}}\,t\right)$$

This completes the solution for the motion.   From the radial equation
of Newton we can now solve for the string tension.

$$T = mg \cos \varphi + m\dot{\varphi}^2 l$$

At the ends of the stroke where $\varphi = \pm \varphi_0$ and $\dot{\varphi} = 0$, this string tension
is smaller than at any other position:

$$T_{\min} = mg \cos \varphi_0 \approx mg \left(1 - \frac{\varphi_0^2}{2}\right)$$

whereas in the center, for $\varphi = 0$, it is as large as it ever becomes:

$$T_{\max} = mg + ml\dot{\varphi}_{\max}^2 = mg + ml\varphi_0^2 \frac{g}{l}$$

$$T_{\max} = mg(1 + \varphi_0^2)$$

Thus the string tension varies from a maximum value in the mid-
position to a minimum at the extremes.

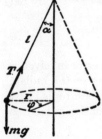

FIG. 159.   The sim-
ple spherical pendu-
lum.

   b. The Spherical Pendulum.   The spherical
pendulum is the same device as the simple pendulum, except that instead of moving in a vertical
plane, the particle moves at constant speed along
a circular path in a horizontal plane, as indicated
in Fig. 159.   We want to know under what circumstances such a motion is possible and what
the frequency is.   In this case the most convenient coordinates are the vertical one, z, and the
normal and tangential ones in the horizontal plane
of motion.   The displacement z is constant in
time, hence $\ddot{z}$ is zero, and the corresponding Newton equation is

$$mg - T \cos \alpha = m\ddot{z} = 0$$

or

$$T = \frac{mg}{\cos \alpha} = \text{constant}$$

Then the resultant of the forces T and mg is a force $T \sin \alpha = mg \tan \alpha$
in the horizontal plane, directed radially inward.   In particular, the
tangential component of force is zero and hence the tangential acceleration $r\ddot{\varphi}$ is zero, or $\dot{\varphi}$ is constant, as assumed.
   The radial Newton equation is

$$mg \tan \alpha = m\dot{\varphi}^2 r$$

As in the simple pendulum, the mass cancels out, or *the motion is independent of the mass of the particle*. Solving for the angular speed, we obtain

$$\dot{\varphi} = \sqrt{\frac{g \tan \alpha}{r}} = \sqrt{\frac{g}{l \cos \alpha}}$$

The angle φ equals $\dot{\varphi} t$, and one full revolution occurs when φ increases by 360 deg or 2π radians, or

$$T \sqrt{\frac{g}{l \cos \alpha}} = 2\pi$$

and

$$f = \frac{1}{T} = \frac{1}{2\pi} \sqrt{\frac{g}{l \cos \alpha}}$$

The frequency depends on the apex angle of the cone described by the pendulum. For a small angle α the frequency is the same as that of the "simple" pendulum (page 185), while for α approaching 90 deg the frequency becomes very large.

c. The Path of a Projectile. Consider a bullet of mass m in its flight and assume the air friction to be negligible, so that the only force acting on the bullet is its weight. If we choose a coordinate system with x measured horizontally to the right and y vertically upward, the Newton equations are

horizontally: $\qquad m\ddot{x} = 0$
vertically: $\qquad m\ddot{y} = -W = -mg$

With these equations the accelerations of the particle are known and are independent of the mass or size of the particle. The rest of the problem is one of kinematics and was discussed on page 167.

A question that may be asked is which region in space is safe from being hit, if the gun can be pointed at any angle of elevation and is limited only by a definite muzzle velocity v_0. Referring to the results of page 167 and calling the angle of elevation of the gun α, we have (Fig. 160).

$$v_{0x} = v_0 \cos \alpha, \qquad v_{0y} = v_0 \sin \alpha$$

The equations of the path of flight, with this notation, are

$$x = v_0 \cos \alpha \, t, \qquad y = -\frac{gt^2}{2} + v_0 \sin \alpha \, t$$

Suppose we pick at random a point x_0, y_0 (Fig. 160) and ask whether that point can be hit or not by firing the gun at a suitable elevation α. Obviously when we choose x_0 and y_0 too large, the gun cannot reach it. Let the point x_0, y_0 be hit after t_0 sec. Then, in the equations

$$x_0 = v_0 \cos \alpha\, t_0, \qquad y_0 = -\frac{gt_0^2}{2} + v_0 \sin \alpha\, t_0$$

there are two unknown quantities, t_0 and α, while x_0 and y_0 are known. We are not interested in t_0 but do want to know α. Eliminate t_0.

$$y_0 = -\frac{g}{2} \left(\frac{x_0}{v_0 \cos \alpha}\right)^2 + v_0 \sin \alpha \left(\frac{x_0}{v_0 \cos \alpha}\right)$$

From this equation we must attempt to solve for α. If we find an answer it

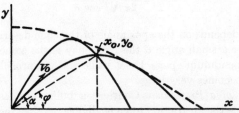

Fig. 160. A given point in space can be hit with two different angles of elevation of a gun with prescribed muzzle velocity, provided the point does not lie outside the parabola of safety.

means that x_0, y_0 can be hit by shooting with the α so found; if the answer for α comes out imaginary, the point x_0, y_0 cannot be hit.

$$y_0 \cos^2 \alpha = -\frac{gx_0^2}{2v_0^2} + x_0 \sin \alpha \cos \alpha$$

$$y_0 \frac{1 + \cos 2\alpha}{2} = -\frac{gx_0^2}{2v_0^2} + \frac{x_0}{2} \sin 2\alpha$$

$$x_0 \sin 2\alpha - y_0 \cos 2\alpha = y_0 + \frac{gx_0^2}{v_0^2}$$

In Fig. 160 we see the auxiliary angle φ, the angle of sight of the target, and

$$\sin \varphi = \frac{y_0}{\sqrt{x_0^2 + y_0^2}}, \qquad \cos \varphi = \frac{x_0}{\sqrt{x_0^2 + y_0^2}}$$

Substitute this into the above.

$$\cos \varphi \sin 2\alpha - \sin \varphi \cos 2\alpha = \frac{y_0 + gx_0^2/v_0^2}{\sqrt{x_0^2 + y_0^2}}$$

The left-hand side equals sin $(2\alpha - \varphi)$, and the right-hand side of this expression is known; so that we can look up $2\alpha - \varphi$ in a table of sines, and since φ is known, we find α. But, in general, $\sin \beta = \sin (180° - \beta)$; so that if we find a numerical answer for $2\alpha - \varphi$, the value $180 - (2\alpha - \varphi)$ will do as well. Thus we find two answers for the elevation α; two possible parabolic paths can hit a given point, as shown in Fig. 160. When do we fail to find an answer for α? Obviously this is the case when the right-hand side of the above equation becomes larger than 1, because the sine of any angle is always less than 1. Therefore, we are on the border line of safety if

$$\frac{y_0 + (gx_0^2/v_0^2)}{\sqrt{x_0^2 + y_0^2}} = 1$$

or, worked out, if

$$x_0^2 + \frac{2v_0^2}{g} y_0 = \frac{v_0^4}{g^2}$$

If x_0 and y_0 satisfy this relation, the point P can just be hit. Now consider x_0 and y_0 to be variables; then the above equation represents a curve, which is recognized to be a parabola again: the parabola of safety, sketched in Fig. 160.

d. Particle Rolling Off a Sphere. Let a particle rest on top of a sphere of radius r, and be permitted to slide down without friction (Fig. 161). At what angle φ will the particle leave the sphere? Consider the general position φ, and note that there are two forces acting on the particle, the weight mg and the normal pressure N of the sphere. Set up Newton's equa-

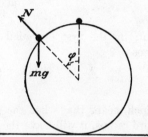

Fig. 161. The particle sliding down a sphere will leave the sphere at a certain point φ.

tions in the tangential and normal directions with the help of Eqs. (6) (page 165).

radially: $\qquad mg \cos \varphi - N = m(\dot{\varphi}^2 r)$

tangentially: $\qquad mg \sin \varphi = m(\ddot{\varphi} r)$

The second equation will enable us to solve for the motion φ, and then the first one can be solved for N. It is obvious that for small values of φ the normal pressure N is positive; for large φ, certainly for $\varphi = 90°$, N should come out negative, which means that the sphere must pull on the particle. Somewhere in between, N becomes zero, and if the sphere cannot pull on the particle, there the particle will leave it.

Therefore, we set $N = 0$ in the first equation, finding

$$g \cos \varphi = \dot{\varphi}^2 r$$
$$g \sin \varphi = \ddot{\varphi} r$$

for the point where the particle leaves the sphere. Integrate the second equation by the standard procedure when the independent variable (t in this case) is absent. That procedure is

$$\ddot{\varphi} = \frac{d\dot{\varphi}}{dt} = \frac{d\dot{\varphi}}{dt} = \frac{d\dot{\varphi}}{d\varphi}\frac{d\varphi}{dt} = \frac{d\dot{\varphi}}{d\varphi}\dot{\varphi} = \frac{\dot{\varphi}\,d\dot{\varphi}}{d\varphi}$$

and the second equation becomes

$$g \sin \varphi \, d\varphi = r\dot{\varphi} \, d\dot{\varphi}$$

Integrated:

$$-g \cos \varphi \,\Big|_0^\varphi = r \frac{\dot{\varphi}^2}{2} \,\Big|_0^{\dot{\varphi}}$$
$$g(1 - \cos \varphi) = \frac{r\dot{\varphi}^2}{2}$$

But, by the first or radial Newton equation this is also equal to $g \cos \varphi/2$ or

$$g(1 - \cos \varphi) = \tfrac{1}{2}g \cos \varphi$$
$$\cos \varphi = \tfrac{2}{3}$$

which means that when the particle has vertically descended by one-third radius, it will leave the sphere. Since this result is independent of g, it is even true for a particle and sphere placed on the moon. *Problems 193 to 202.*

36. Systems of Two Particles. We will now discuss four examples in which two particles tied together by massless connecting members form a system. They are

   a. Atwood's machine
   b. An incline and pulley combination
   c. A chain slipping off a table
   d. A flyball engine governor

   a. Atwood's Machine. Atwood's machine, illustrated in Fig. 162, consists of two nearly equal weights W and $W + w$, connected together by a string and slung over a pulley, which for this analysis will be assumed to have no mass and no friction. What motion will take place? Let the unknown tensile force in the string be T. This force is constant all along the length of the string, because neither the string nor the pulley have mass. If T were not constant, we could isolate a piece of string with a resultant force different from zero, which for zero mass would lead to an infinite acceleration by Newton's second law. This is absurd and we conclude in general that *the forces on a massless body have a zero resultant even if the body is being accelerated.* (See also Fig. 211 on page 234.)

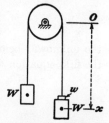

Fig. 162. The apparatus of George Atwood (1746–1807), a tutor in Trinity College, Cambridge, England, who was "considered particularly happy in the clearness of his explanations," according to the "Encyclopaedia Britannica."

Returning to Atwood's machine, we choose an origin and a coordinate x as indicated in the figure. We isolate the weight $W + w$ by cutting the string and write

$$W + w - T = \frac{W + w}{g} \ddot{x}$$

For the other weight we have a similar equation, and remembering that if one weight goes down the other must go up by the same amount, we write

$$T - W = \frac{W}{g} \ddot{x}$$

This pair of equations has two unknowns, T and $\ddot{x}$. Eliminating T by adding the two and solving for $\ddot{x}$ leads to

$$\ddot{x} = \frac{w}{2W + w} g$$

The system has a constant acceleration very much smaller than g.

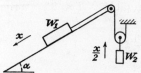

Fig. 163. A smooth incline and pulley combination.

The apparatus was designed by Atwood (1746–1807) for the purpose of demonstrating a slowed-down free-falling motion. Its intent therefore was the same as that of Galileo's rolling cylinders (page 180).

b. Incline and Pulley. Next, consider the system of Fig. 163, again assuming a massless string and pulley and no friction anywhere. From the geometry we reason that if W_1 goes down a distance x along the incline, the weight W_2 goes up $x/2$. Let the string force again be T, constant all along its length.   Then we write

$$W_1 \sin \alpha - T = \frac{W_1}{g} \ddot{x}$$

$$2T - W_2 = \frac{W_2}{g} \frac{\ddot{x}}{2},$$

the unknowns being T and $\ddot{x}$. We can eliminate T by adding twice the first equation to the second one and find

$$\ddot{x} = g \, \frac{4W_1 \sin \alpha - 2W_2}{4W_1 + W_2}.$$

In case W_2 is large, this expression is negative, which means that the system accelerates in a direction opposite to that first assumed. The particular case $\ddot{x} = 0$ expresses the relation between W_1 and W_2 for static equilibrium.

   c. The Chain Sliding Off the Table.   Figure 164 shows a flexible chain of total length l and total weight w_1l, lying partly on a table without friction.   How fast does it slide down?   To solve this we consider the chain as consisting of two particles: the piece x hanging down and the piece $l - x$ lying on the table, directly connected to each other at the corner.   This time we are farther away than ever from our exact definition of particle (page 174), and the reader should

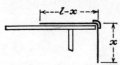

FIG. 164. The flexible, frictionless chain slipping off the table.

protest.   But on page 213 it will be shown that we are indeed justified in considering these two pieces of chain as particles, because every point of each body moves in exactly the same way and all forces acting on the body pass through the center of gravity.   Let the tensile force in the chain at the corner be T, and this force is guided around the corner by a short piece of elbow tubing attached to the table through which the chain passes.   (Explain that the force between the elbow and the table is $T \sqrt{2}$.)   Write Newton's equations

$$T = \frac{w_1(l - x)}{g} \ddot{x}$$

$$w_1 x - T = \frac{w_1 x}{g} \ddot{x}$$

Eliminate T by adding and solve for $\ddot{x}$,

$$\ddot{x} = g \, \frac{x}{l}$$

This is a differential equation, almost like Eq. (9a); only the factor before x is here positive, while it was negative in the previous case. The solution is found by the same process as used there, which leads to

$$x = C_1 e^{\sqrt{\frac{g}{l}}\,t} + C_2 e^{-\sqrt{\frac{g}{l}}\,t}$$

Suppose that at time $t = 0$ we have $x = a$ and $\dot{x} = 0$, which means that we start from rest with an overhang of length a. Substituting

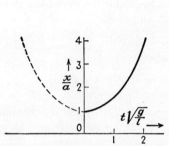

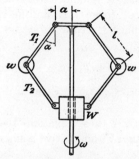

Fig. 165. Shows the relation of the slipped-off length x of the chain of Fig. 164 and the time, when starting from rest with $x = a$.

Fig. 166. The position α of the flyball governor depends on the speed ω.

these conditions into the general solution and solving for the integration constants C_1 and C_2 gives the result

$$x = a\,\frac{e^{\sqrt{\frac{g}{l}}\,t} + e^{-\sqrt{\frac{g}{l}}\,t}}{2}$$

or

$$x = a \cosh\left(\sqrt{\frac{g}{l}}\,t\right)$$

with the hyperbolical cosine, which function was encountered previously in connection with the catenary on page 66. The relation is plotted (from a table in Peirce's "Short Table of Integrals") in Fig. 165, and it is noted that both ordinate and abscissa are pure numbers. This one plot, therefore, gives the time of fall of a chain of any length l, starting from any overhang a, of any unit weight w_1.

  d. The Flyball Governor. The flyball engine governor is illustrated in Fig. 166. It consists of two equal particles of weight w and a third, heavier, particle of weight W, linked together by weightless hinged bars of length l as shown. The weight W can slide freely up and down the central rod and the entire figure rotates at uniform angular speed ω. We want the relation between the angle α and the angular speed ω in terms of the various weights and lengths. Let the tensile force in the two upper bars be T_1 and the tensile force in the two lower bars be T_2. In the state of steady rotation the acceleration of W is zero; the acceleration of each of the w's is directed radially inward and is equal to $\omega^2(a + l \sin \alpha)$. The forces acting on the various particles are the bar forces T_1 and T_2 and the weight. The reader is reminded that he never even heard the word "centrifugal force" (page 211), and he is advised to reread the definition of force on page 3. Now we set up Newton's equations

$W\downarrow$: $W - 2T_2 \cos \alpha = 0$

$w\downarrow$: $T_1 \cos \alpha - T_2 \cos \alpha - w = 0$

$w \rightarrow$: $T_1 \sin \alpha + T_2 \sin \alpha = \dfrac{w}{g} \omega^2(a + l \sin \alpha).$

In these three equations the unknowns are T_1, T_2 and α. Eliminate T_1 and T_2 and derive a single equation, containing only α.

$$\frac{w}{W + w} \frac{l}{g} \omega^2 = \frac{\tan \alpha}{\sin \alpha + a/l}$$

This equation is written in a dimensionless form; both sides are pure numbers. In case the numerical values for the weights, lengths, and speed ω are given, the above equation is not fit to calculate α directly. That, therefore, has to be done by trial and error. Note that for zero ω, the angle α is zero, and for large ω the angle α goes to 90 deg.

Problems 203 *to* 211.

CHAPTER XI

KINEMATICS OF PLANE MOTION

37. Velocities. In this chapter we will study the relations between the time and the displacements of the various points of a rigid body moving in a plane. If the body itself is plane, the paths of all of its points lie in the plane of motion; in general however the body is three-dimensional, and then the paths of the various points lie in parallel planes, and the velocities of two points of the body, located in a line perpendicular to the plane of motion, are equal and parallel. The various points of the body lying in one of the planes of motion, how-ever, all have different velocities in general, so that the problem is one of considerable complication.

Consider a rigid body in a plane (Fig. 167) and draw on it a line 1-2. After a certain time, the body has displaced itself to the new position 1'-2'. The displacements 1-1' and 2-2' must be so related that the distance 1-2 equals 1'-2', because the body has

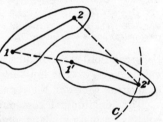

FIG. 167. Two different positions of a rigid body in a plane.

been presumed rigid. If we are entirely free to move the body where we like in the plane, we can choose point 1' at will, but after having chosen 1', our choice for the new position 2' is limited to the circle C shown in the figure.

Our next step is to redraw Fig. 167 for the case that the two con-secutive positions of the body are close together

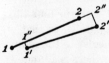

FIG. 168. The longitu-dinal components of the *small* displacements of two points of a rigid body are equal.

and the time interval Δt very small (Fig. 168). In Fig. 167 the curved outlines of the body are shown, but they are not essential because the position of the body is completely determined by the position of the two points 1 and 2; hence in Fig. 168 we draw only the lines 1-2 and 1'-2', omitting the curved outlines. The small displacement 1-1' is considered to consist of two components 1-1'', the longitudinal displacement, and 1''-1', the transverse dis-

195

placement. In the figure the length 1-2 is exactly equal to 1'-2', and the longitudinal displacements 1-1'' and 2-2'' are approximately equal; they would be exactly equal if the bar 1'-2' were parallel to 1-2.

Call

$$1\text{-}2 = 1'\text{-}2' = l, \qquad 1\text{-}1'' = \Delta s_{l_1}, \qquad 1''\text{-}1' = \Delta s_{t_1}$$
$$2\text{-}2'' = \Delta s_{l_2}, \qquad 2''\text{-}2' = \Delta s_{t_2}$$

Calculate Δs_{l_2} in terms of the other quantities, neglecting powers of the Δ's higher than the second, and find:

$$\Delta s_{l_2} = \Delta s_{l_1} - \frac{(\Delta s_{t_2} - \Delta s_{t_1})^2}{2l}$$

It can be said that if in Fig. 168 the displacements are small of the first order, then the difference between 1-1'' and 2-2'' is small of

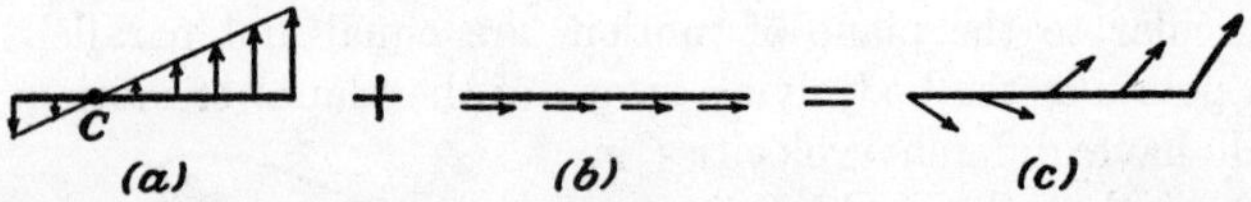

FIG. 169. The velocity pattern of a rigid body (c) can be considered as the sum of a rotation (a) and a translation (b).

the second order. Dividing all these displacements by the time interval Δt, going to the limit $\Delta t = 0$, and hence neglecting second-order quantities, we reach the conclusion that

The longitudinal components of the velocities of two arbitrary points of a body in plane motion are equal.

The transverse velocity components of the two points 1 and 2 can be arbitrarily chosen, but as soon as that is done, the transverse velocities of all other points of the line 1-2 follow, as is shown in Fig. 169a. The total motion of any point on the line 1-2 (and of any point of the entire body of which that line forms a part) is then looked upon (Fig. 169c) as the sum of a rotation about a point C (169a) and a longitudinal translation (169b). In Fig. 170 we see how the velocity of an arbitrary point D, not on the connecting line 1-2 can be found by compounding the translational speed v_l (equal and parallel to that of points 1 or 2) with the rotational speed ωr (perpendicular to the radius CD). When this is understood, we can ask whether there is a point D somewhere in the body, for which the velocity is zero. If such a point exists, obviously the rotational velocity component ωr must cancel the translational velocity component v_l. We draw a line CE perpendicular to the line 1-2. The rotational velocity of any point on this line is directed to the left, opposite to v_l, and by adjusting the distance CE, we can make the value of the rotational velocity anything we please. We then choose point P_v so that the two velocity components cancel. This construction is possible in one way only, and thus we recognize that there is one point P_v in the body for which the velocity is zero. Then the total motion of the rigid body cannot be anything else but a rotation of that body about the point P_v, and we see in Fig. 170 that the total velocities of the points 1, 2, and D are directed perpendicular to the respective radii P_v1, P_v2, and P_vD.

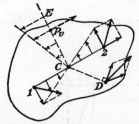 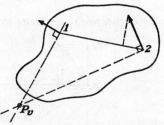

Fig. 170. Construction of the velocity of an arbitrary point D from the longitudinal speed of line 1,2 and the rotational speed about C.

Fig. 171. Construction of the velocity pole P_v from the speeds at two points 1 and 2.

The point P_v is called the **velocity pole** or the **instantaneous center of rotation.** The last-mentioned property of the velocity pole can be used as a means of finding that point when the velocities v_1 and v_2 of two points are given. This is shown in Figure 171, where we draw lines through points 1 and 2 directed perpendicular to their respective velocities. The intersection of these two lines is the pole P_v.

The velocity pole is not necessarily located in the body we are studying; in Fig. 171 it falls just outside that body. The point P_v is called "*instantaneous* center of rotation," because in general it remains fixed in the body only for very small displacements, or for a small time increment Δt. After this displacement has occurred, the body finds itself in a different geometrical position and the pole may or may not be in the same position. In general it is not, as will be seen in the following examples.

We have seen two methods for finding the velocity pole, (a) by the canceling of a longitudinal and rotational speed as in Fig. 170 and (b) by the intersection of two normal lines as in Fig. 171. It will now be proved that these two methods lead to identical results. In Fig. 172, let A_1 and A_2 be the two points, whose velocities A_1V_1 and A_2V_2 are given. The pole P_v has been constructed by intersecting A_1P_v, which is $\perp A_1V_1$, with A_2P_v, which is $\perp A_2V_2$. Drop $P_vN \perp A_1A_2$. From this we want to prove (a) that $A_1L_1 = A_2L_2$

and (*b*) that point N is the point C of Fig. 170, where the transverse velocity of A_1A_2 is zero, and finally (*c*) that the rotational velocity at P_v cancels the translational speed there.   The proof is as follows:

$$\angle T_1A_1V_1 = 90° - \angle V_1A_1L_1 = \angle L_1A_1P_v \quad \text{(marked } \cdot\text{)}$$

Therefore

$$\triangle A_1T_1V_1 \approx \triangle A_1P_vN \quad \text{(equal angles)}$$

Therefore

$$\frac{A_1V_1}{A_1P_v} = \frac{A_1L_1}{P_vN} = \frac{A_1T_1}{A_1N}$$

From the other side of the figure we have the same argument, with the same

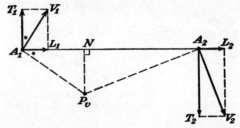

Fig. 172.   Proof of a property of the velocity pole P_v.

result, except that the subscripts 1 are replaced by subscripts 2:

$$\frac{A_2V_2}{A_2P_v} = \frac{A_2L_2}{P_vN} = \frac{A_2T_2}{A_2N}$$

Now, if P_v is the pole, the velocities A_1V_1 and A_2V_2 must be proportional to their distances from the pole A_1P_v and A_2P_v.   Hence all six ratios above are equal to each other.   Equating the two middle ones, we have $A_1L_1 = A_2L_2$, which proves (*a*).   Equating the last two shows that N lies on the line connecting points T_1 and T_2, and hence is the center of transverse motion, point C of Fig. 170, which proves (*b*).   The rotational velocity of P_v, about N as center, is NP_v/NA_1 times the transverse velocity A_1T_1, or by the above equalities that is A_1L_1, which proves (*c*).

The foregoing theory will now be applied to four examples: the crank mechanism, the rolling wheel, a three-bar linkage, and Watt's parallelogram.

a. The Crank Mechanism.   The crank mechanism is shown in an arbitrary angular position in Fig. 173, with the crankpin speed ωr known.   The first plane body we consider is the crank OC.   Its motion is simple, and obviously O is the velocity pole, which, moreover, remains at O for all positions of the crank.

The second plane body we consider is the connecting rod. The velocity of its end C is known to be ωr. The velocity of its other end P is only partially known: we know it must be directed along PO, but we do not know the value as yet. There are two graphical means of determining it. First, we resolve ωr at C into its longitudinal and transverse components; we then slide the longitudinal speed v_l unchanged along the rod to P and then add enough transverse speed to it there to make the result come out along OP.

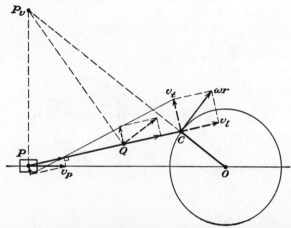

Fig. 173.   Velocities of the connecting rod of a crank mechanism.

The other method is by means of the velocity P_v, which we can construct immediately, by erecting normals to the speeds at the points C and P. The velocity v_P then is equal to ωr multiplied by the ratio $P_v P / P_v C$.

The velocity of an arbitrary point Q of the rod again can be found in two ways, both indicated in Fig. 173. The easiest one is to draw $P_v Q$, and to lay off v_Q perpendicular to it, giving it a value corresponding to the distance $P_v Q$.

The instantaneous motion of the rod can be visualized by covering the drawing with a sheet of transparent paper, tracing PC on that top sheet, inserting a thumbtack through both sheets at point P_v and then rotating the top sheet over the bottom one through a *small angle*. It is obvious that for larger angles of rotation about the thumbtack at P_v, the motion is no longer correct. After a small displacement of the transparent sheet we have to construct a new velocity pole and transfer the thumbtack. When we do this for a large number of positions, for 360-deg rotation of the crank, we find many pinpricks in both sheets of paper, forming a curve on each.   The locus of P_v on the lower sheet is shown in Fig. 174, and the various points on it should be checked by the reader.   The curve is known as the *pole curve of the steady plane;* the first man who constructed it, Poinsot (1777–1859), knew classic Greek and called the curve the *polhodie* (from the Greek *hodos*, path). The pin pricks in the upper transparent sheet form a curve, which the reader should attempt to sketch for himself; it has four branches to

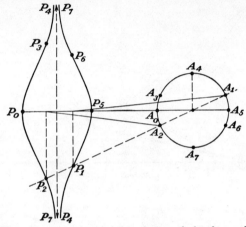

FIG. 174.   The pole curve or locus of the velocity pole for the crank mechanism.

infinity.   This curve is the *pole curve on the moving plane* and was called by Poinsot the *herpolhodie.*   At any instant one point of one pole curve coincides with a point of the other pole curve at the thumb-tack.   Any plane motion can be regarded as a rolling of one pole curve on the other one.

Comparing the graphical method with the analytical one of page 170, leading to the complete result, Fig. 149, we see that the analytical approach is much more direct and suitable in this case.   In general, it can be said that when an analytical treatment is possible, it is preferable over the graphical method.   But in many cases of greater complication, the graphical method just described is the only feasible one.

b. The Rolling Wheel.   The next example to be discussed is the rolling wheel (Fig. 175).   Let the center C of the wheel move horizontally to the right with velocity v; imagine the wheel motion to consist of two components, a sliding parallel to itself to the right with speed v, and a clockwise rotation about C. The latter has to be such that the peripheral speed of the wheel is also v, because at the point of road contact there is no horizontal slip of the wheel, so that its velocity there must be zero. This contact point then is the velocity pole P_v.

The speed at any other point D, for instance, can be found in either one of two ways: first, by laying it off perpendicular to the radius P_vD of the pole, or second, by adding the sliding speed v to the peripheral speed v, as indicated. We can again cover the drawing with a transparent sheet, trace the wheel outline on it, push a thumbtack through

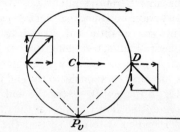

Fig. 175. Velocities of the various points of a rolling wheel.

both sheets at P_v, and turn the upper sheet through a small angle. The pole curve on the bottom sheet, or the succession of pinpricks, is the horizontal road line; the pole curve on the upper transparent sheet is the wheel circle. Thus the rolling of one pole curve on the other one is immediately visible here.

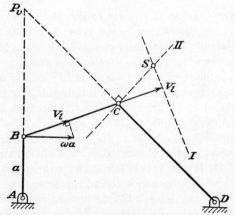

Fig. 176. Construction of velocities in a three-bar linkage.

c. Three-bar Linkage. The third example is the three-bar linkage of Fig. 176, consisting of three bars and four hinges A,B,C,D. The bar AB rotates at constant angular velocity ω about A; we want to know the velocity of all points; in particular, that of hinge C. There are three bodies in the system. The bars AB and CD have very simple motions: they rotate about their fixed hinges, which are their velocity poles. The bar BC however has a more complicated motion. We know the velocity of point $B = \omega a$, and of the velocity of point C we know only the direction, which must be perpendicular to CD, because C rotates about D. As in the previous examples, there are two ways in which the problem can be handled. First resolve $v_B = \omega a$ into its longitudinal v_l and transverse components, and slide the longitudinal speed along the bar to C. The total speed at C consists of v_l plus an unknown amount of transverse speed. Adding this transverse component to the end point of the v_l arrow, in the manner of Fig. 5 (page 10), places the end point of the v_c vector somewhere on the dotted line I, which is thus the first locus of v_c. But v_c must be perpendicular to CD, so that the dotted line II is the second locus of v_c. The two loci intersect at S, so that CS is the desired velocity vector of point C.

The second method is to construct the velocity pole of bar BC, which is point P_v, and then construct CS perpendicular to P_vC and of magnitude

$$\omega a \frac{P_vC}{P_vB}$$

This, by the proof of Fig. 172, ought to give the same answer.

Before leaving this example we note that three-bar linkages occur very frequently under many guises. For example the crank of Fig. 173 can be considered one, in which two of the three bars (OC and CP) are immediately visible, and the third bar is imaginary; it starts at P, is directed perpendicular to OP, and is infinitely long. In Fig. 176 the paths of B and C are circles about the centers A and D. In Fig. 173 the path of P is a straight line, or a "circle of infinite radius."

   d. The Parallelogram of Watt. The last example of this article is the parallelogram of Watt, illustrated in Fig. 177. It is a link mechanism, invented by James Watt (1736–1819) for the purpose of coupling the rectilinear motion of a piston rod to the circular arc path A of a point of the "walking beam" OA of his engine. The principle is shown in Fig. 177a, where OC, CD, and DF are the bars of a three-bar linkage, with fixed hinges at O and F. Point C can move on a circular arc about O; point D can move in a similar arc about F. The midpoint E of bar CD describes a complicated curve, which is approximately a straight vertical line even for reasonably large motions of C and D, as a graphical construction shows. The linkage is shown in a deflected position $OC'E'D'F$, and for that position P_v is the velocity pole of bar *CD*. It is seen that the P_vE' is almost parallel to *OC* or *DF*, so that the velocity of E' is almost perpendicular to *OC*, independent of the deviation. If, in Fig. 177a, *OC* is made the right half of the "walking beam," the top of the piston rod could be attached to

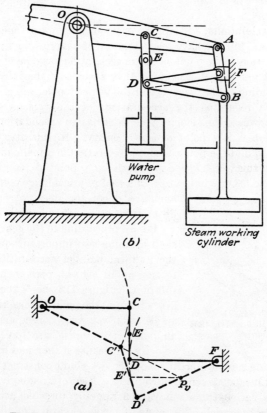

(b)

(a)

Fig. 177. In Watt's steam engine the midpoint *E* of the three-bar linkage *OCDF* describes approximately a straight line; by the four-bar parallelogram *CDBA*, the motion of *E* is doubled at *B*.

point *E*. This construction has the practical disadvantage that the anchor *F* comes to lie far from the engine. Therefore (Fig. 177b) Watt placed the link *CED* in the middle of the right half of his walking beam instead of at the end of it, and used point *E* to drive, not the main piston, but the auxiliary water pump. He then added the links *AB* and *DB*, forming a parallelogram *ABCD*, and chose the dimensions such that *OC* = *AC* and that *CE* = *DE*. Then, from geometry, the points O, E, and B are always in a straight line and $OB = 2OE$. The point B then describes a path similar to that of E (and twice as large), which therefore also approximates the desired straight line.

Problems 212 to 220.

38. Accelerations. In many previous examples (such as Figs. 136, 139, 147, 149) we have seen that a point of zero velocity may have an acceleration, or a point of zero acceleration may have a velocity. We should not expect, therefore, that a velocity analysis of a plane motion, leading to the discovery of the velocity pole, will give us much information regarding the accelerations of the various points. In almost every case the velocity pole itself has an acceleration, although by definition it is a point of zero velocity. In this article we shall learn how to find the point of zero acceleration in a plane motion, and, in general, this acceleration pole will be a point different from the velocity pole. Only in the case of a real physical hinge point, which is anchored down and does not move at all, do the two poles coincide with the hinge.

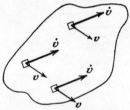

Fig. 178. A plane body in uniform parallel acceleration and consequently with uniform parallel velocities.

Before proceeding to the general case we shall examine two important special cases: the uniform parallel acceleration and the motion about a fixed, permanent center of rotation. Figure 178 shows the first case; every point of the body has exactly the same acceleration $\dot{v}$. As a consequence of this, all points then must also have the same velocity v, because if they had not and two points should have different velocities, we could construct a velocity pole or instantaneous center of rotation. In a rotating body with some speed, the various points have centripetally directed accelerations, which are not parallel. Therefore, if the accelerations of all points are parallel and alike, so must be the velocities. However, the velocities are not necessarily in the same direction as the accelerations. For example, imagine a bullet or shell of some size flying through the top point of a parabolic trajectory (Fig. 145, page 167) without rotation. The velocities are horizontal and the same for all points of the shell; the accelerations are vertical, and again the same for all points, equal to g.

The other important special case, that of rotation about a fixed center, is illustrated in Figs. 179a, b, and c. In the first of these the body rotates about C at uniform angular speed ω (ω = constant; $\dot{\omega} = 0$), and we know by page 165 that the accelerations $\omega^2 r$ are directed radially inward and are proportional to the distance r. In the second picture (179b) the angular velocity is zero ($\omega = 0$; $\dot{\omega} \neq 0$), and the body is accelerated angularly from rest. The accelerations are $\dot{\omega} r$, directed tangentially and again proportional to the distance r. Finally in Fig. 179c we see the mixed case; both ω and $\dot{\omega}$ exist. The acceleration at each point has two components; the angle between the acceleration vector and the radius is the same for all points; and the acceleration is proportional to the radius r.

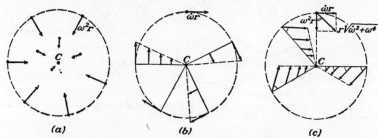

(a)　　　　　(b)　　　　　(c)

FIG. 179. Accelerations of a body rotating about a fixed center C.

Returning to Fig. 178 we might say that the parallel motion of Fig. 178 is a special case of the more general motion of Fig. 179c, in which the center of rotation C is very far away (at infinity). To understand this, consider in Fig. 179c a small square area of 0.01-in. side located 2 in. from C. In the various points of the small square, the radii from C are almost parallel, and the accelerations and velocities are almost parallel and equal. Magnify the picture of the small square and let the center C go farther and farther away; we then approach Fig. 178 more and more.

In now turning to the general case, we consider two points 1 and 2 of the body (Fig. 180), and ask whether there is any relation between the accelerations $\dot{v}_1$ and $\dot{v}_2$ of those two points. We remember that for velocities (Fig. 170 and page 196) the longitudinal components had to be equal. For accelerations this is not so, as can be seen from Fig. 179c by connecting the center C with any other point.

In order to find the relation, we reduce Fig. 180 to the state of Fig. 179c by superposing or adding to all points of the figure a speed $-v_1$ and an acceleration $-\dot{v}_1$, which puts point 1 to rest. The acceleration of point 2 then is the vector sum of $\dot{v}_2$ and $-\dot{v}_1$ (by the theorem of page 165), and this acceleration, called the acceleration of point 2

relative to point 1, cannot have a longitudinal component directed
away from point 1, by Fig. 179.   Thus the vector $\dot{v}_2 - \dot{v}_1$ at point 2
cannot have a longitudinal component directed away from point 1,
or the vector $\dot{v}_2$ cannot have a radial component in the direction away

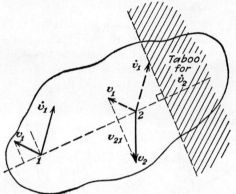

Fig. 180.   If the acceleration $\dot{v}_1$ of point 1 is given, then the acceleration $\dot{v}_2$ of point 2
cannot have its end point in the shaded region.

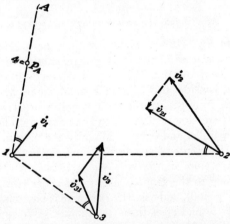

Fig. 181.   Given the accelerations of two points 1 and 2 of a rigid body, to construct
the accleration of a third point 3, and to find the acceleration pole 4.

from point 1 that is greater than the similar component of $\dot{v}_1$.   In
Fig. 180 this is shown by the shaded region: if $\dot{v}_1$ is given, the end
point of the $\dot{v}_2$ vector, plotted from point 2 as a start, cannot lie in that
shaded region, but there is no other restriction on the $\dot{v}_2$ vector.

Now supposing that in Fig. 181 the accelerations $\dot{v}_1$ and $\dot{v}_2$ of the
points 1 and 2 of a rigid body are given in accordance with the limitation just found, how can we construct from it the acceleration of an

arbitrary point 3 of the body? This is done by adding the acceleration vector $-\dot{v}_1$ to $\dot{v}_1$ and to $\dot{v}_2$. This makes the new $\dot{v}_1$ equal to zero and the new $\dot{v}_2$ equal to $\dot{v}_{21} = \dot{v}_2 - \dot{v}_1$, as shown. The acceleration $\dot{v}_{21}$ is pronounced "acceleration of 2 relative to 1." The pattern of the $\dot{v}_{21}$ vector, relative to the now-steady point 1, is as in Fig. 179c. Taking an arbitrary point 3 then, we construct $\dot{v}_{31}$ by making angle $13\dot{v}_{31}$ equal to angle $12\dot{v}_{21}$ and by adjusting the length of $\dot{v}_{31}$ proportional to the radius 1-3. To find the actual acceleration of point 3, we have to undo the previous addition of $-\dot{v}_1$, i.e., we have to add $\dot{v}_1$, giving $\dot{v}_3$ as shown. Now let us try to find a point 4 in the plane where $\dot{v}_4$ becomes zero; in other words let us try to find the pole of acceleration. Obviously then $\dot{v}_{41}$ must be equal and opposite to $\dot{v}_1$. Turn the radius 1-3 about point 1 to the position $1A$, so that angle $\dot{v}_1 1A$ equals angle $12\dot{v}_{21}$. Then the direction of $\dot{v}_{A1}$ for any point on line $1A$ is opposite to $\dot{v}_1$. Choose a point 4 on $1A$ so far from point 1 that the magnitude of $\dot{v}_{41}$ is equal to that of $\dot{v}_1$. This point is the acceleration pole we are seeking. As a check on our construction, we must find that angle $P_a2\dot{v}_2$ equals angle $P_a1\dot{v}_1$ equals angle $P_a3\dot{v}_3$. The verification of this last statement is left to the reader in Problem 225.

Before applying the theory to a few examples, we return to Fig. 180 for the purpose of deriving the magnitude of the in-line component of acceleration from the velocities. In Fig. 179 we see that the radial or in-line component is determined by the velocity itself (without dot) whereas the tangential or across component has a dotted velocity or an acceleration in its expression. In Fig. 180 let v_1 and v_2 be the velocities of the two points, and superpose $-v_1$ on the figure, setting point 1 at rest, and making the velocity of point 2 equal to $v_2 - v_1 = v_{21}$ (pronounced "v of 2 relative to 1"). This v_{21} must be perpendicular to line 1-2 by the proposition of page 196, and it is equal to $v_{2t} - v_{1t}$, where the transverse components are measured in the same direction. (In the figure it appears as the sum, but one of the velocity components is negative.) Letting the distance 1-2 be l, the acceleration of 2 relative to 1 due to this is

$$\dot{v}_{21l} = \frac{(v_{2t} - v_{1t})^2}{l}$$

by Eq. (6a) (page 165). The transverse component $\dot{v}_{21t}$ of the vector $\dot{v}_{21}$ cannot be found from the velocities directly, since it involves a differentiation [Eq. (6b)] that cannot be carried out graphically.

Now the theory will be applied to some of the examples of the preceding article.

a. The Crank Mechanism. For the crank mechanism with uniformly rotating crank (ω = constant), the velocities were derived in Fig. 173, and are reproduced in Fig. 182 as thin dotted lines. The acceleration of the crankpin $\dot{v}_1 = \omega^2 r$ is drawn as a heavy line having the length of a radius; all further accelerations will be drawn to the same scale. To find the acceleration $\dot{v}_2$ of the piston, point 2, we

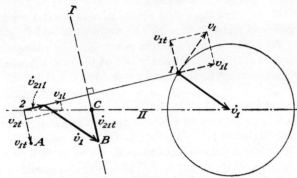

Fig. 182.   Given the acceleration $\dot{v}_1$ of the crankpin; to construct the acceleration of the piston.

state that $\dot{v}_2 = \dot{v}_1 + \dot{v}_{21}$, while the latter can be resolved into two components

$$\dot{v}_2 = \dot{v}_1 + \dot{v}_{21l} + \dot{v}_{21t}$$

By the result just found, $\dot{v}_{21l}$ equals $(v_{2t} - v_{1t})^2/l$, or $(2A)^2/2l$. We read graphically from the figure that $2A$ or $v_{2t} - v_{1t}$ is $0.70v_1$, and $l = 3r$, so that $\dot{v}_{21l} = \dot{v}_1 \dfrac{(0.70)^2}{3} = 0.16\dot{v}_1$, which is laid off in Fig. 182 to scale. To this is added $\dot{v}_1$, which brings us to point B. Then we have to add a vector perpendicular to 12 of unknown amount $\dot{v}_{21t}$, which places the end point of $\dot{v}_2$ on the dashed line I, the first locus for $\dot{v}_2$. A second locus is simply II, because the piston moves in a straight line. Therefore the stretch 2C is the required piston acceleration, drawn to the same scale as $\dot{v}_1 = \omega^2 r$.

b. Three-bar Linkage. In the three-bar linkage of Fig. 176 the bar AB has a constant angular speed ω; what is the acceleration of point C? The construction is shown in Fig. 183. We start by laying off $\dot{v}_1 = \omega^2 a$ to a convenient scale. Then $\dot{v}_2$ consists of three components, as before. The component $\dot{v}_{21l}$ is $(AB)^2/12$, which comes out, graphically, to be $0.29\dot{v}_1$. To this is added $\dot{v}_1$, which brings us to point C, and now we have to add a transverse component of unknown amount, giving the locus I for the end point of $\dot{v}_2$. But point 2 also moves on a circle as a point of the third bar $2D$, and hence its acceleration consists of a radial component $v_2^2/2D$ and a transverse component of unknown length. The radial component is found to be $0.49\dot{v}_1$ and is laid off; with the unknown transverse component this gives the locus II. The two loci intersect at E, and $2E$ is the acceleration of point 2; it is seen to be 0.96 times as large as $\dot{v}_1$.

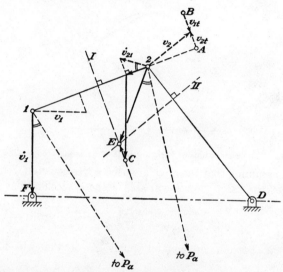

Fig. 183. In the three-bar linkage the acceleration $\dot{v}_1$ of point 1 is given; to construct the acceleration of point 2.

Now we ask for the location of the acceleration pole of bar 12. We lay off $\dot{v}_{21}$ at point 2 and note the angle $12\dot{v}_{21}$. This is laid off at point 1, angle $F1P_a$, and the pole P_a found such that

$$\frac{1P_a}{\dot{v}_1} = \frac{12}{\dot{v}_{21}}$$

The reader should check all of this, and also verify that angle P_a1F equals angle P_a2E, as it should.

c. The Rolling Wheel. The last example to be discussed in this article is the rolling wheel, for which Fig. 175 (page 201) shows the velocity diagram. If the wheel rolls at uniform speed the velocity of the center C is constant in direction as well as magnitude; thus C is the acceleration pole, and the pattern of accelerations is as in Fig. 179a. This is a result that was found analytically on page 172 (Fig. 151).

The rolling-wheel problem is more complicated when the wheel is accelerated; then the motion of the center C is determined not only by the horizontal speed $v_C = a\dot{\varphi}$, but also by the horizontal acceleration $\dot{v}_C = a\ddot{\varphi}$. To find the acceleration of a second point, we return to the equations on page 172 for a point on the periphery. The differentiations from the velocities $\dot{x}$ and $\dot{y}$ to the accelerations $\ddot{x}$ and $\ddot{y}$ were there

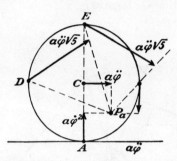

Fig. 184.   The accelerations of various points of an accelerated, rolling wheel.

performed assuming that the angular speed $\dot{\varphi}$ was constant, which now is no longer the case. For a variable $\dot{\varphi}$ these accelerations become

$$\ddot{x} = a\ddot{\varphi}(1 - \cos\varphi) + a\dot{\varphi}^2 \sin\varphi$$
$$\ddot{y} = a\ddot{\varphi} \sin\varphi + a\dot{\varphi}^2 \cos\varphi$$

and for the bottom or contact point A of the wheel $\varphi = 0$, so that

$$\ddot{x}_A = 0 \qquad \text{and} \qquad \ddot{y}_A = a\dot{\varphi}^2$$

which means that the acceleration of the contact point A is the same whether the wheel is accelerated or not; it depends on the velocity only. Figure 184 shows the case where the acceleration of the center $a\ddot{\varphi}$ is numerically equal to the acceleration of the contact point $a\dot{\varphi}^2$. The reader should verify that the acceleration pole P_a is located as indicated in the figure and that, as a consequence, the accelerations of the points $D, E,$ and F are as shown.

Problems 221 to 228.

CHAPTER XII

MOMENTS OF INERTIA

39. The Principle of d'Alembert. Up to this point we have carefully avoided the familiar terms "inertia force" and "centrifugal force," but now the time has come to introduce these concepts. Consider a particle of mass m on which a force F acts, so that it experiences an acceleration $\ddot{x} = F/m$. Now imagine that an additional force $m\ddot{x}$ is applied to the particle in a direction opposite to F (or a force $-m\ddot{x}$ in the direction of F). Obviously the particle is then in equilibrium. In any physical case the force $-m\ddot{x}$ does not exist (there is no push or pull from another body or a pull of gravity): the force $-m\ddot{x}$ is fictitious, and it is called the *inertia force*. Algebraically Newton's Law can be written

$$F = m\ddot{x}$$

or

$$F - m\ddot{x} = F + (-m\ddot{x}) = 0$$

and it can be stated that

The sum of all forces, including the "inertia force," acting on a particle is zero, or

A particle is in equilibrium under the influence of the forces acting on it, provided the "inertia force" is included among the forces.

For example, a freely falling stone has only one force acting on it, the weight W, and as a result it experiences a downward acceleration g; however if we would imagine two forces to be acting on it, the weight W downward and the inertia force $-mg$ downward or mg upward, the stone would be in equilibrium.

In case the acceleration of the particle is a centripetal acceleration, directed toward the center of curvature of the path, the corresponding outward inertia force is called the *centrifugal force*. Consider for example a stone whirling in a horizontal plane at the end of a string that forms the radius of the circular path of the stone, and for simplicity, let gravity be negligible. There is only one force acting on the stone, the string tension T, under the influence of which the stone experiences a centripetal acceleration T/m. In the new manner of talking, however, we would say that the stone is in equilibrium (has

zero acceleration) under the influence of the inward force T and the
outward centrifugal force mv^2/R. The balls of the governor of Fig.
166 (page 193) have a centripetal acceleration caused by the inward
horizontal components of tension of the bars, or in the new language,
the balls are in equilibrium under the influence of the inward horizontal
bar tensions and the outward centrifugal force.

All of this is of no particular usefulness on examples containing
only a single particle, but it does become important when applied to
larger bodies, which can be considered as built up of a large number
of particles rigidly tied together. In Fig. 185a let such a body be in a

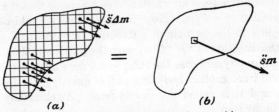

FIG. 185. A parallel acceleration of a rigid body can be caused by many small parallel
forces $\ddot{s}\,\Delta m$ or also by a single force acting on the center of gravity.

state of uniform acceleration $\ddot{s}$, which means that all points of the body
have equal and parallel accelerations, and let us ask what force we
have to exert on the body to bring that about. One way of doing
it, obviously, would be to attach a little string to each constituent
particle Δm, and pull on each such string with a force $\ddot{s}\,\Delta m$. If we did
it in that manner, no internal mechanical cohesion between the
particles would even be necessary: a swarm of loose grains would move
just as described, if each grain had the right pull exerted on its string.
But by the principles of statics (Chap. I) the many little string pulls
$\ddot{s}\,\Delta m$ are statically equivalent to their resultant, which is the single
force $\ddot{s}\Sigma\,\Delta m = \ddot{s}m$ acting through the center of gravity of the body.
Suppose then we replace the many small pulls of Fig. 185a by the single
large pull of 185b.

If we had a swarm of loose grains, the one grain in the center would
get a very large acceleration, and all the others would remain at rest.
But if the body hangs together, a state of uniform acceleration will
still be possible. On the center particle acts not only the large force
$\ddot{s}m$, but in addition to that, the pushes and pulls of the neighboring
particles in contact; these latter forces we call the "internal forces."
If there are N equal particles Δm and the total force is $N\ddot{s}\,\Delta m$, these
internal forces on the center particle are $-(N-1)\ddot{s}\,\Delta m$. On all

other particles we have only internal forces. The sum of all internal forces on *all* the particles is zero, because by action equals reaction they all cancel each other. In this manner the force $\ddot{s}m$ is distributed over all the constituent particles, and the complete body (185b) has the acceleration $\ddot{s}$ in all of its points.

With the new concept of "inertia force" we can repeat this story with different words, and with Fig. 186 we can say that the large body is in equilibrium under the influence of the sum of the externally acting forces ($\ddot{s}m$ in this case) and the inertia force ($-\ddot{s}\,\Delta m$). Or, to repeat it once more, the Newton equation of one constituent particle can be written

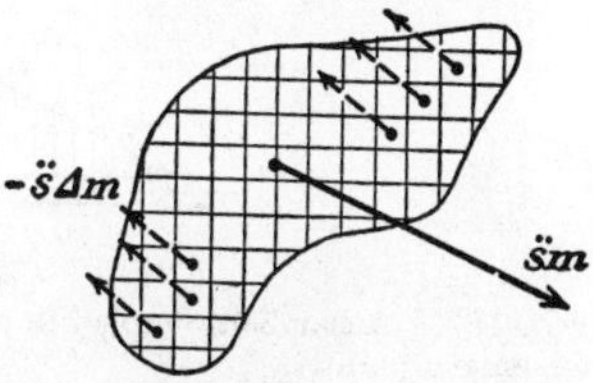

Fig. 186. A body is in equilibrium under the influence of its external force and the many (imaginary) inertia forces.

$$F_{\text{ext}} + F_{\text{int}} - \ddot{s}\,\Delta m = 0$$

The sum of all external forces on all the particles is R, the resultant of the external forces on the body. The sum of all internal forces on all the particles is zero, because they cancel by action equals reaction. The sum of all inertia forces is $-\ddot{s}\Sigma\,\Delta m = -\ddot{s}m$. Thus the body is acted upon and is in equilibrium by the two forces, R and the (imaginary) inertia force $-\ddot{s}m$. Thus the large body acts exactly like a single particle, and we come to the conclusion that

A rigid body of any size will behave as a particle if the resultant of its external forces passes through its center of gravity, or if all points of the body describe equal and parallel paths.

This is the justification for the procedure followed in the examples of Chap. X, in particular of that of Fig. 164 (page 192). We will return to this question once more on page 235.

The method just discussed, and illustrated by Figs. 185 and 186 applies not only when all particles of the body have the same acceleration, but also when these accelerations differ from point to point in the body. Then the method becomes of importance and gives us a means of analyzing the dynamics of such motions. It was originated by d'Alembert (1717–1783) and can be expressed as follows:

The principle of d'Alembert states that **the internal forces or stresses in a rigid body having accelerated motion can be calculated by the methods of statics on that body in a state of equilibrium under the influence of the external forces and of the inertia forces.**

As an example consider (Fig. 187) a uniform bar of total mass m being pulled in its own direction at a quarter-length point by a string in which the force is F.   The acceleration of the bar (which acts as a particle) is F/m, and the inertia force acting on an element Δl is $F\,\dfrac{\Delta l}{l}$ to the left.   The longitudinal force diagram, found by statics, is as sketched in the figure.   In particular, if we isolate a small piece Δl of the bar just at the point of attachment of the rope, the forward face

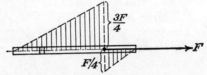

Fɪɢ. 187.   A bar being pulled in a longitudinal direction by a force F acting at a quarter-length point.

has a compressive force of almost $F/4$; the aft face a tensile force of almost $3F/4$; the sum of these being not quite equal to the rope pull F. The small difference is required to accelerate that short piece Δl. The front quarter bar is being pushed and the rear end is being pulled forward.   In connection with this example the reader is advised to return to the remarks on page 5, explaining the axiom of transmissibility of forces.

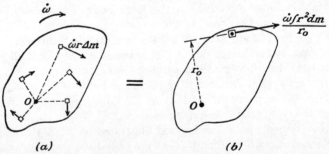

Fɪɢ. 188.   An accelerated angular motion about a fixed peg O (for zero speed) can be brought about by many small tangential forces $\dot{\omega}r\,\Delta m$, or also by a single force having the same moment about the peg O.

The next case to which we propose to apply d'Alembert's principle is that of a body which has an angular acceleration $\dot{\omega}$ and zero speed ω about a fixed axis.   In Fig. 188 that axis is perpendicular to the paper and thus appears as the *point* O.   Consider the body again as a swarm of unconnected particles Δm; then with the motion as prescribed, the tangential component of acceleration of a particle at radius r is $r\dot{\omega}$, and the tangential force required to move that particle is $r\dot{\omega}\,\Delta m$.   This time we do not care about the resultant of all these forces, but we do ask for the resulting moment about the axis O. The moment of one small force is $\dot{\omega}r\,\Delta mr = \dot{\omega}r^2\,\Delta m$, in which $\dot{\omega}$ is the same for all particles, while the (double-headed) arrow of the moment is directed perpendicular to the paper and is also the same for all particles. Therefore the total moment is simply the algebraic sum of all the small moments or

$$\text{Moment} = \Sigma\dot{\omega}r^2\,\Delta m = \dot{\omega}\Sigma r^2\,\Delta m$$

or in the limit when the Δm's become very small and very numerous

$$\text{Moment} = \dot{\omega}\textstyle\int r^2\,dm$$

This moment can be applied to the body in the form of many small tangential pulls $\dot{\omega}r\,\Delta m$, in which case the loose swarm of unconnected particles will move as a whole. If, however, we replace the moment of the many small pulls by an equal moment caused by one large force, as in Fig. 188b, the body will still accelerate with $\dot{\omega}$ as before, but now internal forces are required between the individual particles. Figure 188 can be said to be the rotational equivalent of Fig. 185.

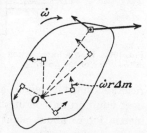

The same story in the language of d'Alembert's principle leads to Fig. 189, equivalent to Fig. 186. Here the body is in rotational equilibrium about the axis O under the influence of the clockwise moment of the single large force and the counterclockwise moment of the many small

Fig. 189. A body in rotational "equilibrium" about an axis O under the influence of one external force and many small "inertia forces."

inertia forces. The counterclockwise moment of these inertia forces is

$$\text{Moment} = \dot{\omega}\textstyle\int r^2\,dm = \dot{\omega}I \qquad (10a)$$

in which the integral, denoted by I, is called the *moment of inertia*. Thus the definition of that term is

The moment of inertia of a body about an axis is the moment of the inertia forces caused by a unit angular acceleration of the body about that axis, and is expressed by $\int r^2\,dm$.

The third and last application of d'Alembert's principle in this article is to the case of Fig. 190, where a flat plate is shown in the xy plane. This plate rotates at uniform angular speed ω about the x axis, so that the position shown is only an instantaneous one. We want to know the bearing reaction forces caused by the rotation ω only (and we are not interested just now in the static bearing forces due to the weight of the plate). We apply d'Alembert's principle directly by introducing the inertia forces, which are here centrifugal forces. They are directed outward, *i.e.*, opposite to the accelerations of the particles. Under the influence of these centrifugal forces and the bearing reaction forces the body must be in equilibrium. The sum of all the centrifugal forces is

$$\text{Total centrifugal force} = \int \omega^2 y \, dm = \omega^2 \int y \, dm$$

which is zero only if the center of gravity lies on the x axis (see page 33), and we will suppose this to be the case. Then the two bearing

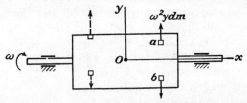

Fig. 190. Inertia forces acting on a rotating flat plate, with its instantaneous position in the plane of the drawing.

reactions must be equal and opposite for vertical equilibrium. The next equilibrium condition to be satisfied is that the moment of all inertia forces about point O in the xy plane cancels the moment of

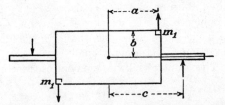

Fig. 191. If to a perfectly symmetrical rotor two extra masses m_1 are added, the rotor acquires a product of inertia about the bearing axis, and bearing reaction forces appear.

the two bearing forces. The moment arm about O of a small centrifugal force is x, and thus the moment of all the centrifugal forces about O is

$$\text{Moment} = \int \omega^2 xy \, dm = \omega^2 \int xy \, dm = \omega^2 I_{xy} \qquad (10b)$$

The integral of this expression is known as the "centrifugal moment" or as the "product of inertia." Thus we define

The product of inertia, also called the "centrifugal moment" of a plane body, referred to a set of perpendicular axes x, y, is the moment about an axis perpendicular to the xy plane of the centrifugal forces caused by unit angular velocity about the x axis, and is expressed by ∫xy dm.

The bearing forces in Fig. 190 thus are $\omega^2 I_{xy}$ divided by the distance between the bearings. We see from Fig. 190 that for a symmetrical plate the integral I_{xy} is zero because the moment of particle a cancels that of particle b, etc. In technical language, the bearing forces due to the rotation of a "balanced" rotor are zero. A rotor is "unbalanced," if as in Fig. 191, it has in addition to a symmetrical body two extra masses m_1 in opposite corners. Then

$$\int xy \, dm = m_1 ab + m_1(-a)(-b) = 2m_1 ab$$

and the bearing reaction forces are $2m_1\omega^2 ab/2c$.

Historical Note. The original statement of d'Alembert's principle as it appears in his "Traité de dynamique," page 74, 2d ed., Paris, 1758, is as follows:

"From the foregoing we deduce the following Principle for finding the movement of several bodies that act on one another. When the movements A, B, C, etc., that are imposed on each body are resolved each into two components, a, α; b, β; c, γ; etc., in such manner that if only the movements a, b, c had been imposed on the bodies, they could have conserved those movements without hindering each other reciprocally; and if only the movements α, β, γ etc. had been put on the bodies, the system would have remained at rest; then it is clear that A, B, C are the movements which these bodies take in virtue of their action."

The reader undoubtedly will have some difficulty understanding this, because the nomenclature of mechanics in 1758 was not as well established as it is now. D'Alembert uses the French word *mouvement* first in the sense of "motion," then in the sense of "force." If we interpret the "movements A, B, C" as "external forces acting on the various elements," the "movements α, β, γ" as "internal forces," and the "movements a,b,c" as "negative inertia forces $+\ddot{s} \, dm$," the meaning comes out a little better. The phrase that "if only the forces $+\ddot{s} \, dm$ had been imposed on the bodies, they could have conserved their motion without hindering [*i.e.*, pushing] each other reciprocally" refers to the swarm of unconnected particles discussed in connection with Figs. 185 and 188. It is interesting to note that d'Alembert in 1758 used the phrase "it is clear that . . . " in about the same sense as modern writers do.

Problems 229 to 234.

40. General Properties. Moments and products of inertia are of great importance in mechanics. Therefore we will devote this entire article to a study of the general properties of integrals of the form $\int x^2\, dm$ and $\int xy\, dm$.

Consider in Fig. 192 a particle dm, part of a larger body, situated at location x, y, z. The moments of inertia about the three coordinate

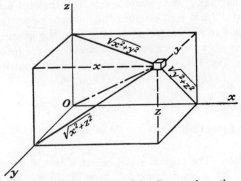

FIG. 192. A particle dm located at x,y,z and its distance from the various axes and planes.

axes, by the definition of page 215, are

$$\int (y^2 + z^2)\, dm, \qquad \int (x^2 + z^2)\, dm, \qquad \int (x^2 + y^2)\, dm$$

We can write all sorts of similar integrals, for instance:

$$\int x^2\, dm \qquad \int y^2\, dm \qquad \int z^2\, dm,$$

which are called *planar* moments of inertia, because x, y, and z are the distances to the coordinate *planes*, but a body cannot rotate about a plane and hence these integrals have no physical meaning. In analogy, the real moments of inertia are sometimes called *axial* moments of inertia, because their distances are measured to the coordinate *axes*. Another integral without much physical meaning is

$$\int (x^2 + y^2 + z^2)\, dm$$

called the *polar* moment of inertia, because the distances are to the origin of coordinates or to the *pole*. From these definitions we see that

The polar moment of inertia equals the sum of the three planar moments of inertia and also equals half the sum of the three axial moments of inertia.

Sometimes the planar or polar moments of inertia are easier to calculate than the axial ones, and then the above relations are useful for finding axial moments of inertia. In mechanics we are not interested in planar or polar moments; they serve only as auxiliary quantities in the calculation of the axial moments, which we need.

In case the body is flat, or two-dimensional, the coordinate z is zero, and the axial moments about the three axes become (Fig. 193)

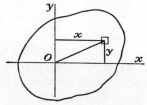

$$\int x^2 \, dm, \qquad \int y^2 \, dm, \qquad \int (x^2 + y^2) \, dm$$

The last expression is the axial moment about the z axis, but it is also the polar moment about the origin O. It is usually

FIG. 193. For a flat plate the polar moment of inertia about O (the z axis) is the sum of the moments of inertia about the x and y axes.

called "polar" moment, which in this case *does* have physical meaning. It is seen to be the sum of the moments about two perpendicular axes in the plane of the flat body.

An important property of moments of inertia is expressed by the *parallel-axis theorem*, to which we now turn our attention. Suppose we know the moment of inertia about a certain axis, say the z axis of

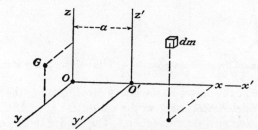

FIG. 194. Illustrating the proof of the parallel-axis theorem.

Fig. 194; how can we derive from that the moment of inertia about another axis parallel to the first one, say z'? Let the coordinates of a particle dm be x,y,z in the O coordinate system; then from Fig. 194 we see that these coordinates in the displaced O' system are $x' = x - a$, $y' = y$, and $z' = z$. We then write

$$
\begin{aligned}
I_{z'} &= \int (x'^2 + y'^2) \, dm = \int [(x - a)^2 + y^2] \, dm \\
&= \int (x^2 - 2ax + a^2 + y^2) \, dm \\
&= \int (x^2 + y^2) \, dm + \int a^2 \, dm - \int 2ax \, dm \\
&= I_z + a^2 m - 2a \int x \, dm
\end{aligned}
$$

The last integral we have seen before on page 34, and we recognize that this integral is zero if the x of the center of gravity is zero, or if G lies in the yz plane. This we will now assume to be the case and to indicate it we add a subscript G to I_z, which now becomes I_{z_G}. Then

$$I_{z'} = I_{z_G} + a^2 m \tag{11a}$$

or in words:

The moment of inertia of a body about any axis equals the moment of inertia about the parallel axis through the center of gravity plus the product of the mass of the body and the square of the distance between the two parallel axes.

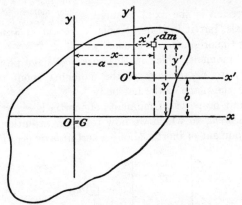

Fig. 195.   Toward the proof of the parallel-axis theorem for products of inertia.

The reader might object to the above proof, arguing that it holds only for a parallel displacement in the x direction (Fig. 194) and not for the y direction or a skew direction. The answer to this objection is that coordinate systems exist for the purpose of serving us and if the z axis and the direction of parallel displacement of that axis are prescribed, we can lay the x axis in the latter direction, and therefore the proof, as given, holds for a parallel displacement in any direction.

An important corollary of the parallel-axis theorem is that among the moments of inertia about many parallel axes, the smallest possible I belongs to the axis passing through the center of gravity.

There is a similar parallel-axis theorem for products of inertia. If the product of inertia of the flat plate of Fig. 195 about the center O is $I_{xy} = \int xy \, dm$, what then is $I_{x'y'}$ with respect to the displaced axes x', y'? The coordinates of any particle dm are $x' = x - a, y' = y - b$.

Then we can write

$$
\begin{aligned}
I_{x'y'} &= \int x'y'\, dm = \int (x - a)(y - b)\, dm \\
&= \int (xy - ay - bx + ab)\, dm \\
&= \int xy\, dm - a\int y\, dm - b\int x\, dm + ab\int dm \\
&= I_{xy} + abm - a\int y\, dm - b\int x\, dm
\end{aligned}
$$

The last two integrals, by page 34, are zero if the origin O of coordinates is the center of gravity, which we now suppose to be the case,

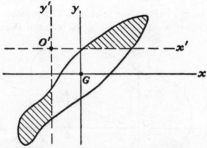

Fig. 196. The product of inertia about the x, y axes through the center of gravity G is a positive quantity; by shifting the axes to O' the product of inertia is made zero, by balancing the positive, shaded contributions against the negative, non-shaded part.

so that I_{xy} is written with subscript G. Then

$$
I_{x'y'} = I_{xyG} + abm \tag{11b}
$$

or in words: **The product of inertia with respect to two axes x',y' that are parallel to the axes x,y of the center of gravity is given by Eq. (11b), in which a and b are the x and y coordinates of the new origin, O', with respect to the center of gravity.**

In this case the I_{xyG} is *not* the smallest possible product of inertia, because the quantity abm to be added to it may be positive or negative, depending on the signs of a and b. In fact, Eq. (11b) indicates that it is always possible to find a new center O' for which the product $I_{x'y'}$ becomes zero, by making a and b of opposite sign. This is illustrated in Fig. 196, where a is negative, and b positive, such that the positive contributions to the integral about O' (which are shaded), are canceled by the unshaded negative contributions.

Besides a parallel shifting of axes a *rotation of the coordinates axes about the origin* leads to some interesting properties. In the Gxy system of Fig. 196 the product of inertia is seen to be positive, because most of the mass of the body lies in the quadrants where x and y are of equal sign. By turning the axes through approximately 45 deg about G, the axes become roughly symmetry axes of the body and the product of inertia is zero. For still further rotation the product becomes negative. We will now study this situation in detail. In Fig. 197 the origin O is an arbitrary point of a flat, two-dimensional body; in general O is *not* the center of gravity. We choose through O a set of rectangular coordinate axes xy so that the product of inertia I_{xy} is zero for those axes. We have just seen that this is possible for a body shaped like Fig. 196; we will verify later that that is always possible. Then we determine

$$I_x = \int y^2\, dm \qquad \text{and} \qquad I_y = \int x^2\, dm$$

and ask whether we can express the moments and product of inertia $I_{x'}$, $I_{y'}$, and $I_{x'y'}$ with respect to a set of axes x',y' in terms of the original I_x and I_y

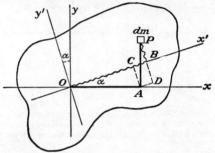

Fig. 197.   The coordinates of a point dm can be expressed as x,y or also as x', y'.

(I_{xy} is zero). In Fig. 197 the coordinates x,y of a particle dm are drawn in the heavy lines OA and AP; the coordinates x',y' in the wavy lines OB and BP. The first step is to express x',y' in terms of x,y, which is done by projecting the heavy L-shaped line OAP, first on the x' axis, then on the y' axis, with the result

$$x' = OB = OC + BC = OA \cos \alpha + AP \sin \alpha$$
$$= x \cos \alpha + y \sin \alpha$$
$$y' = PB = PD - BD = AP \cos \alpha - OA \sin \alpha$$
$$= y \cos \alpha - x \sin \alpha$$

Then we calculate the new moments of inertia.

$$I_{x'} = \int y'^2\, dm = \int (y \cos \alpha - x \sin \alpha)^2\, dm$$
$$= \cos^2 \alpha \int y^2\, dm + \sin^2 \alpha \int x^2\, dm - \sin \alpha \cos \alpha \int xy\, dm$$
$$= I_x \cos^2 \alpha + I_y \sin^2 \alpha - \text{zero}$$
$$I_{x'y'} = \int x'y'\, dm = \int (x \cos \alpha + y \sin \alpha)(y \cos \alpha - x \sin \alpha)\, dm$$
$$= \sin \alpha \cos \alpha (\int y^2\, dm - \int x^2\, dm) + (\cos^2 \alpha - \sin^2 \alpha) \int xy\, dm$$
$$= \sin \alpha \cos \alpha (I_x - I_y) + \text{zero}$$

Thus the problem is solved, and it is seen that for $\alpha = 0$ the new I values equal the old ones, as they should. There is a convenient graphical representation

of these results due to **Mohr** (1880).   To show it, we rewrite the equations in a
different form:

$$I_{x'} = \frac{I_x + I_y}{2} + \frac{I_x - I_y}{2} \cos 2\alpha$$

$$I_{x'y'} = \frac{I_x - I_y}{2} \sin 2\alpha$$

It is easy enough to verify that these expressions are the same as the previous
ones, and much harder to derive the new ones from the old ones.   But the
new expressions are capable of graphical interpretation, as shown in Fig. 198.
First we lay off the value of I_x as OA, and I_y as OB.   Then we construct a
circle on AB as diameter.   The radius of this circle is seen to be $(I_x - I_y)/2$
and the distance $OC = (I_x + I_y)/2$.
Then $OE = I_{x'}$ and $DE = I_{x'y'}$ if we
make $\angle ACD = 2\alpha$.   Comparing Fig.
197 with Fig. 198 we can say that the
abscissa and ordinate of point D in Fig.
198 represent the moment and product
of inertia of axis x' in Fig. 197, or
shorter that point D represents the
inertia properties of axis x'.   Similarly
point A represents axis x, and point
B represents axis y.   The points A

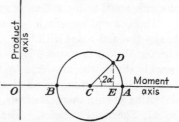

FIG. 198.   Mohr's circle.   The abscissa
OE of a point D on the circle represents
the moment of inertia about the x' axis in
Fig. 197.   Similarly the ordinate ED of D
represents the product of inertia about
the axes x',y' of Figs. 197.

and B in Fig. 198 are 180 deg apart;
their corresponding axes in Fig. 197
are 90 deg apart, as they should be.
We can read off Fig. 198 that the
maximum and minimum moments of inertia occur for those axes where the
product of inertia is zero.   These axes are called the "principal axes of
inertia."   Also we see that the product of inertia reaches a maximum at 45
deg between two principal axes of inertia.   In case the two principal moments
of inertia are equal then all moments of inertia about any axis are equal, because the whole Mohr's circle of Fig. 198 shrinks together to a point.

Mohr's circle is more important in strength of materials than it is in
dynamics.   In the strength of materials it affords a means of finding the
tensile and shear stresses in the material.

In the theory of bending of beams, expressions like

$$\int x^2 \, dA \qquad \text{or} \qquad \int y^2 \, dA$$

appear where dA is an element of *area*, instead of *mass* as before.
These quantities are measured in inches⁴ or feet⁴ and, of course, have
nothing to do with inertia, because there is no mass in them.   Nevertheless, they are often called *area moments of inertia*.   The theory for

"area" moments of inertia is identical with that for "mass" moments of inertia. If an area moment of inertia for a flat plate has been calculated, the mass moment of inertia $\int x^2\, dm$ follows by multiplication with the mass per unit area of that plate.

The units in which a moment of inertia is measured are length squared times mass (or area).

We can write

$$I = \int r^2\, dm = mk^2 \tag{12}$$

in which k is a length, called the "radius of gyration." **The moment of inertia of a body about an axis is not changed if the body is replaced by a concentrated particle of equal mass located at a distance from the axis equal to the radius of gyration.**

This concept is useful for visualization. The radius of gyration of a heavy-rimmed wheel with light spokes is almost R, while for a uniform circular disk it is about $0.7R$. After a numerical calculation we can quickly divide the answer I by the mass m, and compare the k thus found with the dimensions of the body, which gives us a means of detecting gross numerical errors.

41. Specific Examples. The two most important objects in engineering mechanics for which we want to know the moment of inertia are the *bar* and the *disk*, because a large number of practical shapes can be reduced to these two elements. First, we consider a uniform linear bar of length l, of negligible sidewise dimensions, and of total mass m (Fig. 199). We take the origin of coordinates in the center of gravity G and lay the x axis along the bar. The mass of an element dx is $m\, dx/l$, and the moment of inertia about the y or z axis is

$$I = \int_{-\frac{l}{2}}^{\frac{l}{2}} x^2 \left(\frac{m}{l}\, dx\right) = \frac{m}{l} \frac{x^3}{3}\Big|_{-\frac{l}{2}}^{+\frac{l}{2}} = \frac{m}{l}\left(\frac{l^3}{24} + \frac{l^3}{24}\right) = \frac{ml^2}{12}$$

and, by the parallel-axis theorem (page 220), the moment about a transverse axis at the end E is

$$\frac{ml^2}{12} + m\left(\frac{l}{2}\right)^2 = ml^2\left(\frac{1}{12} + \frac{1}{4}\right) = \frac{ml^2}{3}$$

Thus, summarizing, for a *linear bar:*

$$I_G = \frac{ml^2}{12}, \qquad I_{\text{end}} = \frac{ml^2}{3} \tag{13}$$

From this fundamental formula, which should be memorized, several others can be derived by inspection. Consider, for example, the flat plate of Fig. 200, and take first an axis in the plane through G. The distance of a mass element from that axis is not changed if the element is displaced parallel to the axis. Imagine, then, the whole plate

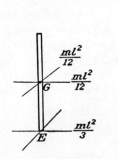

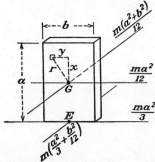

FIG. 199. Moments of inertia of a bar.

FIG. 200. Moments of inertia of a flat rectangular plate.

squashed horizontally into a heavy vertical line through G of length a, and of the same total mass m as the plate. By this squashing process no element has changed its distance from the axis, hence the value of I has not changed, and is $ma^2/12$ by Eq. (13). Exactly the same argument holds for the axis through the end E in the plane of the plate; $I = ma^2/3$. About an axis through G perpendicular to the plate, we have

$$I = \int r^2 \, dm = \int (x^2 + y^2) \, dm = \int x^2 \, dm + \int y^2 \, dm$$

The first of these integrals is one in which all mass elements are at distance x from G; *i.e.*, the plate is squashed into a vertical rod: $I = ma^2/12$. The second integral similarly means the plate squashed into a horizontal rod with $I = mb^2/12$, and the total moment is the sum of these two, as shown in Fig. 200. The axis through E perpendicular to the plate is at distance $a/2$ from and parallel to the previous axis, so that we have to add $m(a/2)^2$ to the previous answer, again shown in Fig. 200.

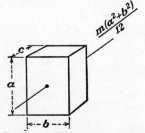

FIG. 201. The expression for the moment of inertia of a rectangular block is the same as that for a flat plate.

Figure 201 shows a solid rectangular block with its central axis parallel to the side c. This block can be squashed into a flat plate ab of zero thickness and the same mass as the block without changing the distance of any particle from the axis. Hence the moment of inertia is found by the same formula as the flat plate, and the dimension c appears only buried in the letter m, the value of which, of course, is proportional to c as well as to a and b.

The next case we consider is a flat uniform circular disk of radius R and total mass m (Fig. 202). We are to find the value of I about a central axis perpendicular to the disk. For an element we take a thin ring of thickness dr, of which all parts are at equal distance from the axis. The mass dm of this ring is

FIG. 202. A uniform flat circular disk.

$$dm = m\,\frac{dA}{A} = m\,\frac{2\pi r\,dr}{\pi R^2} = \frac{2m}{R^2}\,r\,dr$$

and hence the moment of inertia is

$$I = \int_0^R r^2\,\frac{2m}{R^2}\,r\,dr = \frac{2m}{R^2}\int_0^R r^3\,dr = \frac{mR^2}{2}$$

This result is indicated in Fig. 203. The moment of inertia about a diametral axis in the plane of the disk is obviously equal to that moment about any other diametral axis in the plane. On page 219 we saw that the polar moment of inertia equals the sum of two diametral-axis moments; hence that moment is half the polar moment, as indicated in Fig. 203. Summarizing, we have for a *uniform disk*

$$I_{\text{polar}} = \frac{mR^2}{2}, \qquad I_{\text{diametral}} = \frac{mR^2}{4} \quad (14)$$

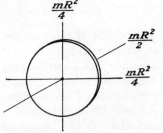

FIG. 203. Axial and diametral moments of inertia of a uniform circular disk.

which formulas should be memorized. The radius of gyration, k, for a rotating solid wheel (page 224) thus is $R/\sqrt{2}$ or $0.707R$, and the moment of inertia of such a uniform disk is the same as that of a heavy rim of 70 per cent radius of the same weight as the uniform disk.

Figure 204 shows a uniform cylinder of radius R and length l. The moment of inertia about the usual longitudinal center line is not changed by squashing the cylinder down to a thin disk of radius R and the same mass m. Next consider a diametral axis through the center G. To reach any mass element from G we first move a distance x along the center line and then a distance y vertically up. Hence

$$I = \int(x^2 + y^2) \, dm = \int x^2 \, dm + \int y^2 \, dm$$

The first term is the I of the cylinder when squashed radially into a heavy center line, while the second term is the I when the cylinder is squashed to a thin circular disk. Therefore

$$I = \frac{ml^2}{12} + \frac{mR^2}{4}$$

as indicated in the figure. Moments about other axes are found by the parallel-axis theorem.

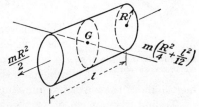

FIG. 204. Moments of inertia of a uniform solid cylinder.

Let us apply the foregoing to a composite example: the crank of Fig. 205 of crank radius R_c, shaft radius and length R_s and l_s, crankpin radius and length R_p and l_p, and crank cheek dimensions h, w, t, and a. First we calculate the weights of the various parts in pounds and find

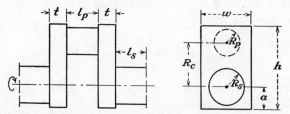

FIG. 205. A crank structure, made up of cylinders and parallelepipeds, for which the moment of inertia can be written immediately from Eqs. (13) and (14) and the parallel-axis theorem.

the answers W_p for the pin, W_c for one cheek and W_s for one piece of shafting. Then, the moment of inertia about the main shaft is

$$2\frac{W_s}{g}\frac{R_s^2}{2} + \frac{W_p}{g}\left(\frac{R_p^2}{2} + R_c^2\right) + \frac{2W_c}{g}\left[\frac{h^2 + w^2}{12} + \left(\frac{h}{2} - a\right)^2\right]$$

which should be carefully checked by the reader.

A problem arising in the operation of adjustable-pitch airplane propellers (page 238) consists in finding the product of inertia of a bar inclined at angle α with respect to the x axis (Fig. 206). Here, for an element $dm = m\, ds/l$, we have $x = s \cos \alpha$ and $y = s \sin \alpha$, so that

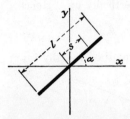

$$I_{xy} = \int xy\, dm = \frac{m}{l} \sin \alpha \cos \alpha \int_{-\frac{l}{2}}^{\frac{l}{2}} s^2\, ds$$

$$= \frac{ml^2}{12} \sin \alpha \cos \alpha = \frac{ml^2}{24} \sin 2\alpha$$

Fig. 206. To determine the product of inertia of a bar in a skew position α.

This answer can be read off immediately from the Mohr-circle diagram (page 223); it is seen to be zero for the positions $\alpha = 0$ and $\alpha = 90°$, which are the positions where the moments of inertia about the x axis are minimum and maximum, respectively.

To find the moment of inertia about a diameter of a uniform solid *sphere* of radius R, first solve the simpler problem of the polar moment of inertia. For the mass we take a spherical eggshell of thickness dr and radius r and of mass

$$m\, \frac{4\pi r^2\, dr}{\frac{4}{3}\pi R^3} = \frac{3m}{R^3}\, r^2\, dr$$

All points of this shell are at equal distance from the pole or center. The polar moment of inertia is

$$I_{\text{polar}} = \frac{3m}{R^3} \int_0^R r^4\, dr = \frac{3}{5}\, mR^2$$

By page 218 this is half the sum of the three diametral or axial moments of inertia, which, by symmetry, are all alike. Therefore the axial moment of inertia we are seeking is

$$I_{\text{diametral}} = \tfrac{2}{5} mR^2 \quad \text{(solid sphere)}$$

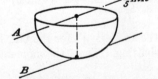

If the sphere is cut in half through the center, the moment of inertia of the half sphere about a diametral axis, of course, is half as large as that for the whole sphere. Nevertheless in Fig. 207 the

Fig. 207. The moment of inertia of a solid half sphere.

same formula is written as for the whole sphere, and the factor $\frac{1}{2}$ is neatly buried in the letter m, which always means the mass of the object we are dealing with. To find the value of I about a tangent line B at the bottom of the half sphere, we might be tempted to use the parallel-axis theorem and add mR^2 to the $\frac{2}{5}mR^2$ of the diametral axis A. This is wrong because axis A does not pass through the center of gravity of our object. The correct procedure for finding the moment of inertia about axis B is first to find the location of the center of gravity, then to shift from axis A to the gravity axis and from there to axis B, by the parallel-axis theorem. This is left as an exercise to the diligent reader.

The determination of the moments of inertia of other bodies with shapes expressible by mathematical formula, such as segments of spheres, cones, paraboloids of revolution, and the like, is a question of formal integration. For many practical objects in engineering no

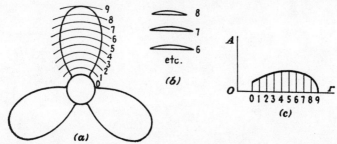

Fig. 208. The moment of inertia of a propeller is best determined by numerical integration in tabular form.

such expression can be found, as for example in the ship's propeller of Fig. 208. In such a case we proceed arithmetically in tabular form. First we draw a number of concentric circles on the propeller drawing; the greater the number n of circles, the greater the amount of work and the better the accuracy of the result. The blueprint drawing usually shows a number of cross sections as in Fig. 208b. These are planimetered and the cross-sectional areas A plotted against the radius in Fig. 208c and the curve smoothed out. Then in a table of n horizontal lines we write in the area A, the distance dr between circles, the volume $A\,dr$, and the product $r^2 A\,dr$. The n items in this last column are added and the result gives approximately

$$\int r^2 d \text{ volume}$$

or the "volume" moment of inertia, expressed in inches^5. This figure is multiplied by the weight per cubic inch (0.33 lb/cu in. for bronze; 0.28 lb/cu in. for cast steel) and the answer then, expressed in lb-in^2 or lb-ft^2 is the "weight" moment of inertia, usually denoted as WR^2 (and pronounced as written). The value of I or the "mass" moment of inertia is obtained after dividing by 386 or 32.2, as the case may be. On the blueprints of propellers, turbine rotors, and the like the WR^2 value, usually in lb-ft^2, is often given, rather than the value of I in lb-ft-sec^2. This is because the weights in pounds of the constituent parts are usually known, and it does not make sense to divide every weight by 32.2 in the tabulation. Fewer mistakes are made by keeping the weights themselves throughout all the multiplications and additions and making the conversion from WR^2 to I at the very end of the work by a single division by 32.2 or 386, as the case may be.

Problems 235 *to* 244.

CHAPTER XIII

DYNAMICS OF PLANE MOTION

42. Rotation about a Fixed Axis. We shall now study the rotational motion of a thin, plane, rigid body about a fixed axis perpendicular to its plane. The problem was started on page 214 (Fig. 188); it will now be taken up again in a more general form. Whereas in Fig. 188 the body had an angular acceleration but no angular speed, we now give it both acceleration $\dot{\omega}$ and speed ω. We ask for the external forces required to bring about this motion and for the reaction forces at the axis caused by those external forces and the motion. Figure 209 shows the body, and the coordinate system Oxy has been located with the fixed axis O as origin. One particle dm of the body is shown with its accelerations $\dot{\omega}r$ and $\omega^2 r$, and the corresponding inertia forces $\dot{\omega}r\,dm$ and $\omega^2 r\,dm$, directed against the acceleration. Similar inertia forces act on all the elements of the body. The reader is once more

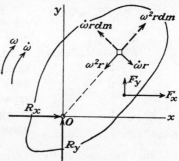

Fig. 209. A flat body pivoted at O, subjected to external forces F_x, F_y, showing the "inertia forces" on a particle.

reminded that these "inertia forces" are fictitious and have no actual existence, as explained on pages 211 and 215. The actual forces acting on the body consist of the external forces (of which one, F_x, F_y, is shown as a sample in the figure) and the reaction force R_x, R_y exerted by the axle on the body. Now, by d'Alembert's principle (page 213), the body is in equilibrium under the influence of the actual forces F and R and the (imaginary) inertia forces. The next step in the analysis is to replace the many inertia forces on the constituent particles by their resultant, which is a problem in statics. This is conveniently done by Cartesian components, and Fig. 210 shows the inertia forces of one particle resolved into those components. We note that the two shaded "force" triangles and the one location triangle x,y,r are similar, so that we can write for the Cartesian components f_x, f_y of the inertia

231

forces on a single particle

$$f_x = \omega^2 x \, dm - \dot{\omega} y \, dm$$
$$f_y = \omega^2 y \, dm + \dot{\omega} x \, dm$$

By the theorems of pages 9 and 34, the resultants of these inertia-force components are

$$F_{x \text{ inertia}} = \int(\omega^2 x \, dm - \dot{\omega} y \, dm)$$
$$= \omega^2 \int x \, dm - \dot{\omega} \int y \, dm = \omega^2 m x_G - \dot{\omega} m y_G$$

and similarly

$$F_{y \text{ inertia}} = \omega^2 m y_G + \dot{\omega} m x_G$$

The magnitudes of these inertia forces then are the same as if the entire mass of the body were concentrated at the center of gravity G, and as if the point G were a particle. Combining the x and y Cartesian components at G we see that they combine into one single force $m\omega^2 r_G$ directed centrifugally outward, and one single force $m\dot{\omega} r_G$ directed tangentially counterclockwise. Thus we have the *magnitudes* of the resultant inertia forces, but not yet their *locations* or points of action, which determine the value of their moment about the axis O. We could write the moment equation about O in Cartesian components (and the reader is advised to do this for exercise), but it is clear that in this case the radial and tangential components are more suitable than the Cartesian ones. The centrifugal force $\omega^2 r \, dm$ has no moment about O, and the tangential force $\dot{\omega} r \, dm$ has a moment $\dot{\omega} r^2 \, dm$, or, integrated over the whole body, the moment of all inertia forces about O is $\dot{\omega} I_O$. This moment is $\dot{\omega} m k_O^2$, where k_O is the radius of gyration (page 224). The moment of the inertia forces concentrated at the center of gravity would have been $\dot{\omega} m r_G$, which is different, because in general r_G differs from k_O. (This last fact can be understood most easily for the special case where the center of gravity G coincides with the axis O; then $r_G = 0$, but $k_O = k_G = \sqrt{I_G/m}$ is not zero.)

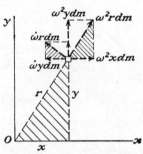

FIG. 210. Resolution of the inertia forces at a point into Cartesian components.

Therefore we see that for calculating the *magnitude* of the resultant inertia force we can replace the body by a particle at its center of gravity, but the moment of all the small inertia forces is determined by the formula $\dot{\omega} I_O$. Summarizing these results in one sentence:

The angular acceleration $\dot{\omega}$ of a thin, flat, rigid body constrained to move in rotation about a fixed axis O perpendicular to its plane is found from the formula:

$$\text{Moment of external forces about O} = I_o\dot{\omega} \qquad (15)$$

while the reaction forces from the axis on the body are determined by solving the static equilibrium equations of the body under the influence of the external forces and of two fictitious "inertia forces" $m\dot{\omega}r_G$ and $m\omega^2r_G$, as if the entire mass of the body were concentrated in its center of gravity.

From this general rule we can at once derive the special case that if the axis of rotation happens to coincide with the center of gravity, the axis reaction force is equal, parallel to, and opposite in direction to (but not in line with) the resultant of the external forces. If in addition to the coincidence of O with G, the external forces consist of a pure couple in the plane of rotation, then the axis reaction is zero, and we may remove the physical axis without changing the motion. Hence, if the center of gravity of a thin, flat body is at rest and we apply a pure couple to that body in its plane, the center of gravity will remain at rest, and the body will experience an angular acceleration about that center of gravity.

Now it remains to be seen how we can apply the above general theorem to an actual problem. In such a problem usually only the external forces are known, while $\dot{\omega}$, ω, and the axis reactions are to be determined. We can start by writing the moment equation about the axis, in which equation the unknown axis reaction and the unknown centrifugal force $m\omega^2r_G$ do not appear. Only the moment $I_o\dot{\omega}$ and the external forces enter, so that we can solve for the acceleration $\dot{\omega}$. With the given initial conditions of the problem this can be integrated to ω. At this stage only the axis-reaction components are unknown, and they can be found by writing the static equations of vertical and horizontal equilibrium or equivalent equations. This will now be done on a number of examples.

43. Examples of Fixed-axis Rotation. The following cases will be discussed:

    a. The pulley wheel
    b. The compound pendulum
    c. The falling trap door
    d. The adjustable-pitch aircraft propeller

a. The Pulley Wheel. A pulley wheel attached with its center to a
fixed ceiling and having tensile forces T_1 and T_2 in the two branches of
its string is shown in Fig. 211. The center of gravity of the wheel is
supposed to be in its axis, the wheel is a uniform disk of radius R and
weight W, and the bearing is assumed frictionless. What is its motion
under the influence of T_1 and T_2 and what is the axis bearing reaction?
The moment equation about O is

$$(T_2 - T_1)R = I_o\dot\omega = \frac{WR^2}{2g}\dot\omega$$

This determines the angular acceleration, which is proportional to
the difference in the two string forces. Since the resultant of all the
inertia forces is zero, the axis reaction is equal, parallel, and opposite

to the resultant external force $T_1 + T_2$. It is noted
that the external forces T_1, T_2 and the axis reaction
$(T_1 + T_2)$ together do not form a system in equilibrium, but they result in a clockwise couple. This is
the couple that accelerates the pulley angularly.
Equilibrium is obtained only by adding to this the
fictitious couple $-I_o\dot\omega$ of the inertia forces, as indicated by dotted lines in Fig. 211. Obviously if the
inertia of the wheel is zero $(I_o = 0)$ the string tensions on both sides are equal $(T_1 = T_2)$, which justi-

FIG. 211. Angu-
lar acceleration of
a pulley wheel.

fies the statement made on page 191 when we were discussing the pulley
wheel for the first time. To obtain an understanding of the numbers
involved, we now ask for the difference $T_2 - T_1$ in pounds, required
to accelerate a 6-in.-diameter uniform pulley weighing 2 lb with
$\dot\omega = 10$ rpm/sec. The answer is

$$T_2 - T_1 = \frac{WR}{2g}\dot\omega = \frac{2 \times 3}{2 \times 386}\left(\frac{10}{60}2\pi\right) = 0.0081 \text{ lb}$$

b. The Compound Pendulum. A compound pendulum is a rigid
body hinged about a horizontal axis not passing through its center of
gravity, and acted upon by its own weight as an external force. It
differs from the simple pendulum (page 184) in that the various constituent particles of the compound pendulum have different motions
and accelerations, whereas the simple pendulum consists of a single
particle only. Figure 212 shows the pendulum acted upon by its
weight W through the point G at distance a from the axis O. It is
shown in an arbitrary angular position φ, measured positive clockwise

starting from the vertical position of equilibrium. First we solve for
the motion by the moment equation about O:

$$I_O\ddot{\varphi} = -Wa \sin \varphi$$

The $-$ sign appears, because $\ddot{\varphi}$, by definition, is positive clockwise
and the weight couple tends to rotate the pendulum counterclockwise.
The equation can be written as

$$\ddot{\varphi} = -\frac{Wa}{I_O} \sin \varphi = -\frac{g}{I_O/ma} \sin \varphi$$

and, on comparing it with the equation of page 185, we conclude that
the motion φ of a compound pendulum is identical with the motion of a
simple pendulum of "equivalent" length l_{equiv}, if

$$l_{\text{equiv}} = \frac{I_O}{ma} \tag{16}$$

All conclusions drawn for the motion of the simple pendulum therefore
are applicable to the compound pendulum; in particular the frequency
[Eq. (9d), page 185] is

$$f = \frac{1}{2\pi} \sqrt{\frac{Wa}{I_O}} \tag{9e}$$

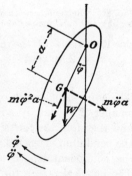

Suppose the pendulum of Fig. 212 had along
the line OG and its extension a large number of
small holes fitting the shaft O, so that the distance $OG = a$ can be changed by hooking
the pendulum on the shaft through various
holes O; the question comes up of how the
frequency f or the equivalent length varies
with a. To understand this relation we
notice that in Eq. (16) the quantity I_O depends on a, because we can write [Eq. (11a),
page 220]

$$I_o = I_G + ma^2 = m(k_G^2 + a^2)$$

Fig. 212. The compound
pendulum.

and I_G or k_G is independent of a. We can now rewrite Eq. (16).

$$l_{\text{equiv}} = \frac{k_G^2}{a} + a$$

or

$$\frac{l_{\text{equiv}}}{k_G} = \frac{a}{k_G} + \frac{1}{a/k_G}$$

The last form of the equation is called "dimensionless," the variables l_{equiv} and a appear as ratios in terms of a constant length k_G. The equation is plotted in Fig. 213, the first term on the right-hand side being a straight line through the origin, and the second term being a rectangular hyperbola, both shown in dashed lines. At point A, $l_{equiv}/k_G = a/k_G = 1$, the two curves both have 45-deg tangents, so that the fully drawn sum curve has a horizontal tangent at B. (This of course can be verified by differentiating.) We see that the smallest

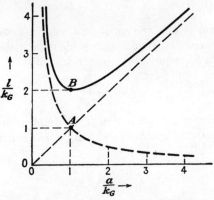

Fig. 213. Relation between the equivalent length l and the distance between the center of gravity and the peg a of a compound pendulum.

possible value of the equivalent length occurs at point B, where $l_{equiv}/k_G = 2$, so that

$$l_{equiv\ minimum} = 2k_G$$

The equivalent length of a given pendulous body can never be made smaller than this by adjusting the axis location; this is the shortest and therefore the *fastest* possible pendulum. The graph shows that if the pendulum is pegged very close to its center of gravity (a/k_G small), the equivalent length is very large, or the pendulum is very slow. The physical reason for this is that the accelerating gravity moment is very small while the inertia is finite. To get an idea of how a pendulum in the condition of point B (Fig. 213) looks, consider a uniform bar of length l. For such a bar, by Eq. (13) (page 224), we have $I_G = ml^2/12$ and $k_G = l/\sqrt{12} = 0.29l$. At point B we have $a = k_G = 0.29l$. The pendulum is shown in Fig. 214, and if the knife-edge is moved either up or down from the position shown, the pendulum will become slower. However, at this position a slight change in the location of the knife-edge will make practically no change at all in the frequency, because the tangent at B in Fig. 213 is horizontal. This fact has been utilized in the construction of extremely accurate pendulum clocks for astronomers, and after its inventor, the pendulum is known as a *Schuler* pendulum. The period of such a clock is not affected by wear of the knife-edge in the course of time.

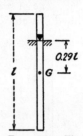

Now we are ready to return to Fig. 212 to calculate the bearing reaction force. By the general statement of page 233, this force is equal to the vector sum of the weight and of two inertia forces, which can be calculated as if the mass were concentrated at the center of gravity. This, as can be seen on page 186, is exactly the same as for the simple pendulum. Summarizing, we can say that a compound pendulum (Fig. 212) acts like a simple pendulum of length l_{equiv} [Eq. (16)] so far as its *motion* is concerned, but acts like a simple pendulum of length a in the matter of the axis reaction force.

Fig. 214. The *Schuler* pendulum is the fastest possible suspension of a given bar l in a given gravitational field g.

c. The Falling Trap Door. Figure 215 shows a uniform beam of weight W and length l, supported at both ends. The support at B is suddenly removed. What is the acceleration of point B and what is the reaction force at A during the first instant thereafter? The problem is that of a compound pendulum with $a = l/2$ and $\varphi = 90°$. However, we will solve it here from the beginning again. The moment equation about A at the first instant is

$$\frac{Wl}{2} = I_A \ddot{\varphi} = \frac{1}{3} \frac{Wl^2}{g} \ddot{\varphi} = \frac{1}{3} \frac{Wl}{g} (\ddot{\varphi}l) = \frac{1}{3} \frac{Wl}{g} \ddot{x}_B$$

Therefore

$$\ddot{x}_B = \frac{3}{2} g$$

The far end B accelerates downward 50 per cent faster than a freely falling particle, and since x varies linearly with the distance from A, the center of gravity has a downward acceleration $\frac{3}{4}g$. The inertia force $mr\dot{\omega}$ thus is $\frac{3}{4}W$ upward, and the inertia force $mr\omega^2$ is zero because the door has not yet acquired velocity. Thus the reaction from the support at A is $W/4$ upward on the beam. This result is independent of the speed ω. If the door had fallen from a higher

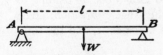

Fig. 215. The problem of the falling trap door.

position and were moving through the horizontal position with angular speed ω, there would be an additional reaction, at A to the left on the beam, of $(W/g)\omega^2(l/2)$.

   d. *The Adjustable-pitch Aircraft Propeller.* Figure 216 is a simplified sketch of a four-bladed airplane propeller rotating at constant

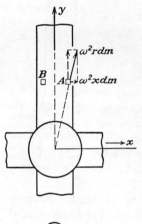

angular speed ω about the horizontal z axis. In the plane of the propeller, the x and y axes are along the center lines of two blades at a certain instant. Consider an element dm at point A in one of the blades. That point moves in a circular path in a plane described by the equation $z = $ constant, and applying d'Alembert's principle, the forces in the system can be calculated by statics if we apply the hypothetical "centrifugal" force $\omega^2 r\, dm$ to the element. This "force" can be resolved into the components $\omega^2 x\, dm$ and $\omega^2 y\, dm$ parallel to the axes (see Fig. 210, page 232). Assume a symmetrical blade as shown. Then the y component $\omega^2 y\, dm$ of particle A and the corresponding component of particle B, symmetrically located, are parallel and add up to a resultant along the y axis. Thus, integrated over the entire blade, this becomes

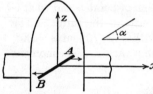

Fig. 216. Explaining the centrifugal twisting torque of an airplane propeller blade.

$$\omega^2 \int y\, dm = \omega^2 m y_G$$

the centrifugal force of the entire blade pulling up at the root of the blade, with which we are familiar. Next consider the smaller sidewise component $\omega^2 x\, dm$ at A and at B. These two forces cancel each other as far as magnitude goes, but they do form a couple tending to rotate the blade about the y axis into the "flat" direction. The moment of the force at A about the y axis is $\omega^2 x\, dm\, z$, and the total moment of the entire blade is

$$\text{Moment about } y \text{ axis} = \int \omega^2 xz\, dm = \omega^2 I_{xz}$$

Thus we see a twisting moment on the blade proportional to its product of inertia or centrifugal moment. In Fig. 206 (page 228) this moment was calculated for a flat rectangular blade. In an actual construction the value of I_{zz} has to be found by numerical or graphical methods. To get an idea of the magnitude of this "centrifugal twisting torque" on the blade, consider an aluminum airplane-propeller blade, about 1 in. thick, 10 in. wide, and 6 ft long, rotating at 1,500 rpm, and oriented at $\alpha = 30°$. Then

$$W = 720 \text{ cu in.} \times 0.10 \text{ lb/cu in.} = 72 \text{ lb}$$

$$I_{zz} = \frac{ml^2}{24} \sin 2\alpha = \frac{72 \times 100}{386 \times 24} \sin 60° = 0.67 \text{ lb-in.-sec}^2$$

$$\omega = \frac{1,500}{60} \times 2\pi = 156 \text{ radians/sec}$$

$$\text{Torque} = \omega^2 I_{zz} = 16,300 \text{ in.-lb} = 1,350 \text{ ft-lb}$$

Before leaving this example we do well to consider why it seems to be more involved than the previous three examples. In the theory of the previous article, culminating in the rule of page 233, we dealt with a "thin, flat" body, and the axis reaction was a single force in the plane of the thin body. In the applications on the pulley wheel, the pendulum, and the trap door the bodies may not have been thin, but at least they could be considered as built up of thin layers, each of which was an exact copy of all the others. The trap door, for example, might be much longer perpendicular to the paper than the dimension l shown, but every section parallel to the paper shows the same picture. This is no longer the case with the aircraft propeller. The sections for different values of z (Fig. 216) are different, and this is the reason for the greater complication of our last example.

Problems 245 to 254.

44. General Motion in a Plane. We will now consider the most general motion of a thin flat body in its own plane under the influence of forces in that plane. We have seen some simple cases of such motion before, and we will now once more examine those. The first case is the parallel acceleration under the influence of a force acting through the center of gravity, which was discussed on page 212 in connection with Figs. 185 and 186. The second case is that of a free body under the influence of a pure couple only. We saw on page 233 that then the center of gravity will not be accelerated and the body experiences an angular acceleration about the center of gravity.

Now we are ready to consider the general case of an unconstrained flat body under the influence of arbitrary forces in its plane, not limited to a force through G or to a pure couple. We approach the problem by resolving the general case into a sum of the two special cases, as illustrated in Fig. 217. In Fig. 217a only a single force F is shown to act on the body; this force can be considered to be the resultant of all forces acting on it, so that it represents the most general case. The

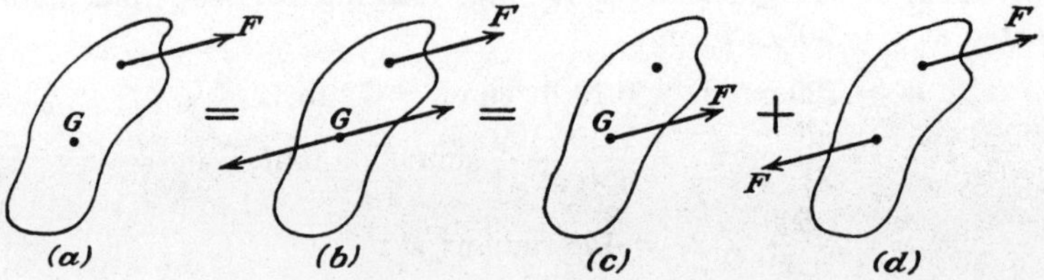

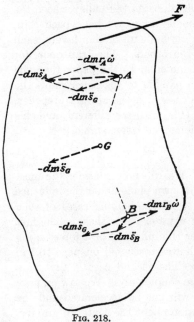

FIG. 217. The most general motion (a) of a body in a plane can be considered as the superposition of a parallel motion (c) and a rotation about the center of gravity (d).

step from Fig. 217a to 217b is the familiar one of "adding nothing" (Fig. 7, page 11), and then we see that the system can be considered as the sum of a parallel translatory acceleration (217c) and of a pure rotation about a fixed center of gravity (217d).

This is illustrated once more in Fig. 218 for the case of a body starting from rest, *i.e.* without centrifugal acceleration. The inertia force for an element dm at the center of gravity is directed opposite to the moving force F. At point A the inertia force is made up of two components: one being the translatory acceleration, which is the same all over the body and hence equal to $-dm\ddot{s}_G$, and another one due to the rotation about the point G, directed tangential to the circle about G opposite to the acceleration, which is $r\dot{\omega}$. These two components are vectorially added and the resultant must be the total inertia force of the particle, and hence must be $-dm\ddot{s}_A$ at location A. From the figure we see that there must be a particle in the body of which the inertia force is zero; it is located somewhat farther down than point B and is the acceleration pole for the motion. Writing this in the form

FIG. 218.

of a formula referred to an xy Cartesian coordinate system, we have

$$\left.\begin{array}{l} \Sigma F_x = m\ddot{x}_G \\ \Sigma F_y = m\ddot{y}_G \\ \Sigma M_G = I_G\ddot{\varphi} \end{array}\right\} \tag{17}$$

The Σ signs are written because there usually are several forces acting on the body. The symbol M_G means the moment of all external forces about the center of gravity, and if we lay the origin of coordinates in that point G, the moment is $\Sigma M_G = \Sigma(F_x y - F_y x)$.

Equations (17) are just about the most important ones in this book; they will be illustrated and interpreted by examples in the next section. A few general conclusions we can read out of the equations immediately:

In case the resultant of the external forces passes through the center of gravity, $\ddot{\varphi} = 0$, or integrated, $\dot{\varphi} = \omega = $ constant. The angular speed of the body remains unchanged, and the center of gravity of the body moves like a single particle of a mass m equal to that of the entire body. As an example of this, consider the parabolic trajectory of a particle (Fig. 145, page 167), and suppose that instead of a particle we had a flat circular disk in the plane of the trajectory, rotating about its center of gravity. The disk is affected only by its own weight, which passes through the center of gravity, so that the center of gravity will describe its familiar parabolic path and the disk will keep on rotating at uniform speed. The path of a point away from the center of gravity will be a complicated one, a wavy line about the parabola; nevertheless the disk or rather the point G of the disk, behaves like a particle. This is the most general case of plane motion, and returning to the statement of page 213, we understand that the word "or" in the middle of the sentence is correct; it should not be replaced by "and" as we would have been inclined to write at that time.

Another example of the particle-like motion of rotating bodies exists in the planets. They are attracted by the sun by gravitational forces, which pass through the center of gravity of each planet. Therefore the path of the center of gravity of a planet can be found from the laws of particle motion (page 175), although a steady rotation of the body of the planet takes place simultaneously about the center of gravity. Still another example is the diving athlete, whose center of gravity will describe a parabolic path independent of the motions of his body.

45. Examples on General Plane Motion. Four representative examples will now be discussed for illustration of the theory:

    a. The rolling wheel
    b. The center of percussion
    c. The bifilar pendulum
    d. The sliding ladder

    a. The Rolling Wheel. A wheel of radius r, weight W, and moment of inertia I_G about the center of gravity, which is in the center of the

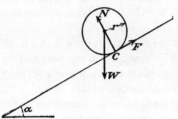

circle, rolls without slipping down an incline α (Fig. 219). What is the downward acceleration?

    The forces acting on the wheel are the weight W and the force from the incline, which can be resolved into a normal pressure N and a friction force F. The latter force may have any value between $-fN$ and

Fig. 219. A cylinder rolling down an incline without slipping.

$+fN$. We take as coordinates the distance x along the incline, and y perpendicular to it; we further let φ be the angle of rotation of the wheel. Then $\ddot{y} = 0$, because the wheel does not leave the plane. The Newton equations [Eqs. (16)] are

In the y direction: $N - W \cos \alpha = m\ddot{y} = 0$
In the x direction: $W \sin \alpha - F = m\ddot{x}$
In the φ direction: $Fr = I_G \ddot{\varphi}$

    The first of these equations tells us that the normal force is

$$N = W \cos \alpha,$$

independent of the motion. In the last two equations there are three unknowns, $\ddot{x}$, $\ddot{\varphi}$, and F. Thus we lack one equation, and that equation is supplied by the geometry of "pure rolling without sliding," which requires that $x = r\varphi$, or $\ddot{x} = r\ddot{\varphi}$. Solving the equations for the unknowns gives

$$\ddot{x} = \frac{g \sin \alpha}{1 + (I_G/mr^2)}$$

$$F = \frac{I_G}{r^2} \ddot{x}, \qquad \ddot{\varphi} = \frac{\ddot{x}}{r}$$

This solves the problem, but there are several interesting angles to the solution, which we will now start discussing.

    In the first place, we remember from page 179 that a sliding particle (without rolling) has an acceleration $g \sin \alpha$ down the incline. We

thus see that the rolling slows down the acceleration by a factor
$1 + (I_G/mr^2)$, which is always larger than unity, but which differs
for wheels of different mass distributions. If the wheel consists of
a heavy rim with light spokes we have $I_G = mr^2$ and the motion is
slowed down by a factor 2, as compared to sliding. If the wheel is a
solid cylinder with $I_G = \frac{1}{2}mr^2$ [Eq. (14)], it is slowed down by a

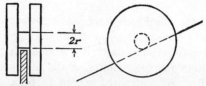

Fig. 220.   A wheel of a large moment of inertia in comparison to its rolling radius r will
go down the incline at a very slow rate.

actor $1\frac{1}{2}$.   We can construct an artificial wheel (Fig. 220) with a
radius of gyration much larger than r, in which the ratio I_G/mr^2 can
be made 10 or larger; in that manner we can make $\ddot{x}$ easily as low as
1 per cent of g.   In this way Galileo studied the laws of motion of
freely falling bodies, as was discussed on page 180.

A special case occurs when we do away altogether with the inclined
plane and hang the wheel from a string (Fig. 221).
This is a toy, called "yo-yo," which was very popular for a time.   The yo-yo problem is exactly the
same as that of an inclined plane with $\alpha = 90°$, and
the shape of the yo-yo is such that it descends with
an acceleration $g/10$ or slower.

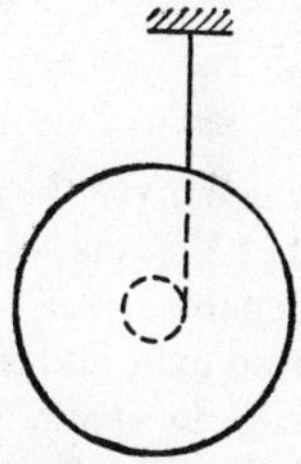

Next we have to investigate under which conditions the wheel refuses to roll and starts to slip.
That occurs when

Fig.   221.   The
yo-yo is a wheel
rolling down a
vertical inclined
plane.

$$\frac{F}{N} = \frac{I_G}{r^2} \frac{g \sin \alpha}{[1 + (I_G/mr^2)]W \cos \alpha} = \frac{\tan \alpha}{1 + (mr^2/I_G)} \geqslant f$$

If the angle α is sufficiently steep and if I_G/mr^2 is large, the required
friction force F may have to be larger than the maximum possible
fN and slipping will occur.   In that case we have to repeat the analysis
with a new assumption.   Then $F = fN$, so that F is no longer an
unknown and the three equations of Newton

$$N - W \cos \alpha = 0$$
$$W \sin \alpha - fN = m\ddot{x}$$
$$fNr = I_G\ddot{\varphi}$$

have only the three unknowns N, $\ddot{x}$, and $\ddot{\varphi}$ and can be solved. The solution is

$$\ddot{x} = g \,(\sin \alpha - f \cos \alpha)$$
$$\ddot{\varphi} = \frac{rfW \cos \alpha}{I_G}$$
$$N = W \cos \alpha$$

which means that the wheel goes down with $\ddot{x}$, but also starts rotating, with $\ddot{\varphi}$. Under no circumstances can $r\ddot{\varphi}$ become larger than $\ddot{x}$, because then the wheel would overspin itself; it would slip in the uphill direction. The condition for the validity of the above solution thus is $r\ddot{\varphi} \leqslant \ddot{x}$, which after some algebra, is seen to be identical with the previous result:

$$f \leqslant \frac{\tan \alpha}{1 + (mr^2/I_G)}$$

In the solution of the problem of Fig. 219 quite often a short cut is proposed, in which it is said that the contact point C is the instantaneous center of rotation and therefore equivalent to a fixed axis. Thus we can apply Eq. (15) (page 233) directly with respect to point C:

$$Wr \sin \alpha = I_C \ddot{\varphi} = (I_G + mr^2)\ddot{\varphi}$$

which gives the correct result for $\ddot{\varphi}$ and hence for $\ddot{x}$ with less work. It is true that the answer so obtained is correct, but nevertheless the method is questionable. An instantaneous center of rotation or a velocity pole is not a fixed axis, and Eq. (15) is not applicable to it. A fixed axis, being completely fixed in space, obviously has no velocity, nor acceleration, nor any higher derivative of the displacement. A velocity pole has no velocity at this instant; in general it does have velocity a little later, hence it does have acceleration. Figure 151 of page 173 shows that the instantaneous center of rotation of a rolling wheel at constant speed has an upward acceleration, while the wheel center, moving on a straight line has no acceleration. If the wheel were pegged to an axis at C (Fig. 219), it would dig into the track after a small rotation; in that case C would have no acceleration and the wheel center would have an acceleration towards C if ω were constant. The difference between the two cases thus consists of a centripetal acceleration $\omega^2 r$ in the direction NC and we will see on page 300 that we have obtained a correct answer for our short-cut analysis only because the inertia force of this "relative acceleration" happens to have no moment about point G. An example where this is not the case and where the proposed short cut leads to an incorrect answer is shown in Fig. 222a. The center of gravity of the wheel is off center by distance a, and the correct application of Eqs. (17) is

$$F = m\ddot{x}_G$$
$$N - W = m\ddot{y}_G$$
$$Na - Fr = I_G\ddot{\varphi}$$

or, eliminating the unknowns N and F,

$$(W + m\ddot{y}_G)a - m\ddot{x}_G r = I_G\ddot{\varphi}$$

The coordinates of G can be expressed in terms of φ by geometry:

$$x_G = r\varphi - a(1 - \cos \varphi)$$
$$y_G = r - a \sin \varphi$$

Differentiation of these expressions gives

$$\dot{x}_G = r\dot{\varphi} - a \sin \varphi \, \dot{\varphi}$$
$$\dot{y}_G = -a \cos \varphi \, \dot{\varphi}$$
$$\ddot{x}_G = r\ddot{\varphi} - a \cos \varphi \, \dot{\varphi}^2 - a \sin \varphi \, \ddot{\varphi}$$
$$\ddot{y}_G = a \sin \varphi \, \dot{\varphi}^2 - a \cos \varphi \, \ddot{\varphi}$$

Substituting these into the Newton equation leads to an equation for $\ddot{\varphi}$, which is too difficult to solve in the general case. We simplify the question by ask-

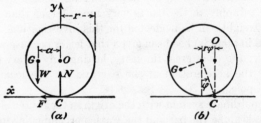

Fig. 222. A rolling wheel of which the center of gravity is off center.

ing for the acceleration at the start only, when $\varphi = 0$, and hence $\sin \varphi = 0$, $\cos \varphi = 1$, but $\dot{\varphi}$ and $\ddot{\varphi}$ are not zero. Then

$$\ddot{x}_G = r\ddot{\varphi} - a\dot{\varphi}^2, \qquad \ddot{y}_G = -a\ddot{\varphi}$$

and

$$Wa = ma^2\ddot{\varphi} + mr^2\ddot{\varphi} - mra\dot{\varphi}^2 + I_G\ddot{\varphi}$$
$$= [I_G + m(a^2 + r^2)]\ddot{\varphi} - mra\dot{\varphi}^2$$
$$= I_C\ddot{\varphi} - mra\dot{\varphi}^2$$

The short-cut application of Eq. (15) about the instantaneous center of rotation C gives immediately, with much less work,

$$Wa = I_C\ddot{\varphi}$$

which, on comparison with the above result, is seen to be incorrect, since the second term is missing. This second term represents the inertia force $mr\dot\varphi^2$ multiplied by the moment arm a, as will be explained more fully on page 300.

We see from the above example in small type that **the application of Eq. (15) about an instantaneous center of rotation is not correct; that equation holds only for rotation about a permanently fixed axis.**

b. The Center of Percussion. Consider the straight solid bar of mass m (Fig. 223), which is supposed to be outside the field of gravity.

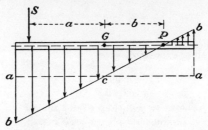

Fig. 223.   The center of percussion S and the center of rotation P.

This may be accomplished by assuming the bar to lie on a horizontal plane of ice without friction, but in practice we mean that the force S we propose to apply to the bar is very much larger than its weight, so that the weight can be neglected for this investigation. The force S (measured in pounds) has been given that letter because we visualize it as a shock; a force of great intensity, lasting only a very short time. We ask for the acceleration of the bar and for the location of the acceleration pole.

Take a coordinate system with the origin in the center of gravity G, with the x axis along the bar and the y axis perpendicular to it. Then Newton's equations [Eqs. (17)] are

$$S = m\ddot y_G$$
$$Sa = I_G\ddot\varphi$$

The acceleration $\ddot y$ of a point at location x along the bar is thus made up of two parts: $\ddot y = S/m$, which is the same at all points, and $x\ddot\varphi = Sax/I_G$, varying from point to point. At a point $x = b$ the two terms cancel each other:

$$\frac{S}{m} = \frac{Sab}{I_G}$$

so that

$$ab = k_G^2 \tag{18}$$

Here a is the distance between the center of gravity G and the center of percussion S; b is the distance between G and the center of rotation P, while k_G is the radius of gyration of G.

We thus define the center of percussion S as the point that can be struck without causing an acceleration at another designated point P, the center of rotation. It is clear from Eq. (18) that there is no such thing as *the* center of percussion of a body. Its location depends on the location of P; to each arbitrarily chosen center of percussion there belongs a center of rotation. In Eq. (18) the letters a and b are interchangeable; hence the points P and S bear a reciprocal relation to each other, which means that if S is the center of percussion for no acceleration at P, then P is the center of percussion for S as the center of rotation. It is instructive to follow the motion of P when S moves in from far away toward G.

The concept of center of percussion has many practical applications: the most obvious of them is a hammer, which is so shaped that when the center of shock S is located in the hammer head, the center of percussion is at the handle. Although the inventor of the modern hammer undoubtedly was blissfully ignorant of Eq. (18), and good hammers are of great antiquity, some of the tools of primitive tribes that can be seen in museums violate the relation (18), and are said to be "badly balanced."

c. *The Bifilar Pendulum.* The bifilar pendulum shown in Fig. 224 consists of a heavy mass, suspended from two parallel strings AC and BD and capable of swinging in its plane under the action of gravity. The geometry of the device is such that if AC and BD swing sideways, turning about A and B as fixed points, the

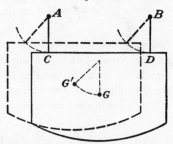

Fig. 224. The bifilar pendulum remains parallel to itself during the motion and hence acts like a particle.

whole line CD remains horizontal: in fact every point of the block, including the center of gravity G, describes a circular arc of radius l. Thus, by page 241, the block behaves like a particle, and the pendulum is a simple one (page 185) with a frequency

$$f = \frac{1}{2\pi} \sqrt{\frac{g}{l}} \qquad (9d)$$

The length, or equivalent length, of the pendulum can be made as short as we please and is not limited at all by the dimensions of the mass m, as it was in the compound pendulum (page 236). This property has been put to use in the loose counterweights in the crankshafts of modern reciprocating aircraft engines, where it was necessary (for the purpose of limiting the torsional vibrations in the engine) to have a pendulum weighing at least 5 lb of a length not exceeding ¼ in. This is obviously impossible with a compound pendulum and could be accomplished only with a bifilar one.

 d. *The Sliding Ladder.* The sliding-ladder problem is illustrated in Fig. 225. The friction is assumed to be negligible and the center of gravity is assumed to be in the center. There is obviously no fixed axis of rotation; therefore we write Newton's equations [Eqs. (17)] with respect to the center of gravity:

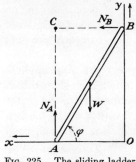

FIG. 225. The sliding ladder and its velocity pole C.

$$N_B = m\ddot{x}_G$$
$$N_A - W = m\ddot{y}_G$$
$$N_B \frac{l}{2} \sin\varphi - N_A \frac{l}{2} \cos\varphi = I_G\ddot{\varphi}$$

In writing the last equation the moment on the left-hand side has been written with such signs as tend to increase the angle φ. The observation that φ decreases when the ladder slides down is irrelevant; eventually we will find a negative answer for $\ddot{\varphi}$, but in writing the moment equation we know that $I\ddot{\varphi}$ equals the moment tending to *increase* φ. In solving for the motion, we first eliminate from these equations the reactions N_A and N_B in which we are not interested at this time.

$$m\frac{l}{2}\sin\varphi\,\ddot{x}_G - (W + m\ddot{y}_G)\frac{l}{2}\cos\varphi = I_G\ddot{\varphi}$$

This equation contains three unknowns, $\ddot{x}_G$, $\ddot{y}_G$, and $\ddot{\varphi}$, and hence we need two additional equations. They are furnished by the geometrical relation between x_G, y_G, and φ.

$$x_G = \frac{l}{2}\cos\varphi, \qquad y_G = \frac{l}{2}\sin\varphi$$

For substitution into the Newton equations we have to differentiate these.

$$\dot{x}_G = -\frac{l}{2}\sin\varphi\,\dot{\varphi} \qquad \dot{y}_G = \frac{l}{2}\cos\varphi\,\dot{\varphi}$$

In the next differentiation we have to remember that not only the angle φ is a function of time, but the angular speed $\dot{\varphi}$ also varies with the time. Thus

$$\ddot{x}_G = -\frac{l}{2}\cos\varphi\,\dot{\varphi}^2 - \frac{l}{2}\sin\varphi\,\ddot{\varphi}$$

$$\ddot{y}_G = -\frac{l}{2}\sin\varphi\,\dot{\varphi}^2 + \frac{l}{2}\cos\varphi\,\ddot{\varphi}$$

Substitution of these expressions into the above equation leads to an expression several lines long, which, however, simplifies very nicely under the algebraic operations. The result is

$$-Wl\cos\varphi = \left[I_G + m\left(\frac{l}{2}\right)^2 \right]\ddot{\varphi}$$

The square bracket can be interpreted as the moment of inertia of the ladder about point O or point C in Fig. 225 and the left-hand side is the moment of *all* forces about point C (since N_A and N_B have no moment about that point).

Therefore our last result could have been obtained much quicker by applying Eq. (15) about C as a fixed center of rotation. The answer happens to be correct only by the accidental fact (page 300) that the acceleration vector of point C passes through G, something that is not at all easy to recognize. In general, the application of Eq. (15) about the velocity pole leads to an erroneous result; for example, if the center of gravity should lie off the center of the ladder, the result so obtained would be incorrect.

Problems 255 *to* 266.

CHAPTER XIV

WORK AND ENERGY

46. Kinetic Energy of a Particle. In Chap. VIII the concept of work was defined and discussed. From pages 133 to 137 of that chapter we take the following propositions:

1. Work is defined as the product of a force and that component of the displacement of its point of action, which is in line with the force.
2. This definition is equivalent to that of the product of the entire displacement and the in-line component of the force, both definitions being expressed by the formula $W = Fs \cos \alpha$.
3. From the definition it follows that work is considered positive if the force and the in-line displacement are in the same direction; negative when these directions are opposite to each other.
4. Work is measured in foot-pounds or inch-pounds and is a "scalar" quantity, not a "vector" quantity.
5. The algebraic sum of the amounts of work done by two or more concurrent forces on the displacement of the point of intersection equals the work done by the resultant of those forces on the same displacement.
6. The work done by a pure couple M on a displacement is zero; on a rotation through φ radians the work done is $M\varphi$.
7. The work done by a set of forces acting on a rigid body for an arbitrary displacement of that body is equal to the work done by another set of forces, which is statically equivalent to the first set.

From these propositions we deduced in Chap. VIII (page 140) the theorem that the work done by all external forces and reactions on an arbitrary small displacement of a system in equilibrium is zero (provided that the internal forces do no work). Here we are dealing with dynamics, where the bodies are not in equilibrium, *i.e.*, they are not at rest, and we will now consider how this theorem is to be generalized to apply to moving bodies. We start with a single particle, on which

250

a single force is acting, and this single force is understood to be the resultant of all external forces and reactions acting on the particle (Fig. 226). At first we assume that the force has the same direction as the velocity of the particle, which therefore moves in a straight line. By Newton's law the magnitude of the force is $m\ddot{x}$, and while the particle moves through a small distance dx, the work done is

$$dW = F\,dx = m\ddot{x}\,dx = m\,\frac{d\dot{x}}{dt}\,dx = m\,d\dot{x}\,\frac{dx}{dt}$$
$$= m\dot{x}\,d\dot{x} = md(\tfrac{1}{2}\dot{x}^2) = d(\tfrac{1}{2}mv^2) = dT$$

or

$$\mathbf{F\,dx = d(\tfrac{1}{2}mv^2) = dT} \qquad (19)$$

The quantity $\tfrac{1}{2}mv^2$ is called the *kinetic energy* of the particle and is usually denoted by the letter T; it is a scalar quantity, which is either zero or positive, but can never be negative. Equation (19) can be integrated to take care of displacements of finite magnitude:

$$\int_1^2 F\,dx = \int_1^2 d(\tfrac{1}{2}mv^2) = \tfrac{1}{2}m(v_2^2 - v_1^2) = T_2 - T_1$$

Expressed in words: **The work done by the resultant of all forces acting on a particle equals the increment in the kinetic energy of the particle.**

This theorem holds in general, although so far we have proved it only for the case of rectilinear motion, where the force has the same direction as the velocity. For the general case (Fig. 227), where the force has a different direction, we choose a coordinate system with the x axis parallel to the velocity at a certain instant, and resolve the force F into an in-line component F_x and an across component F_y. By the fifth proposition of page 250 the work done by F is the sum of that done by F_x and F_y. The work done by F_x is the same as for the case of Fig. 226, and we repeat the calculation.

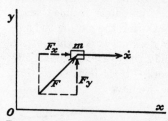

FIG. 227. The work done by a force on a particle equals the increment in kinetic energy of the particle.

$$dW_1 = F_x\,dx = m\ddot{x}\,dx = \cdots = d(\tfrac{1}{2}m\dot{x}^2)$$

Similarly

$$dW_2 = F_y\,dy = m\ddot{y}\,dy = \cdots = d(\tfrac{1}{2}m\dot{y}^2)$$

Adding

$$dW = dW_1 + dW_2 = d[\tfrac{1}{2}m(\dot{x}^2 + \dot{y}^2)]$$

Velocities can be added vectorially (page 162), and since the $\dot{x}$ vector is perpendicular to the $\dot{y}$ vector, we have

$$\dot{x}^2 + \dot{y}^2 = v_x^2 + v_y^2 = v^2$$

and

$$dW = d(\tfrac{1}{2}mv^2) = dT$$

This proves the theorem for the general plane case. For three dimensions, where the force has three components F_x, F_y, F_z for a velocity $\dot{x}$, the proof is the same.

We shall now illustrate the theorem with a few simple examples. First we consider the freely falling body. The force acting on it is $W = mg$ and while it falls through a height h, the work done by the force is mgh. By the theorem this equals the increment in kinetic energy, and if the body starts from rest ($T = 0$), we have

$$mgh = \tfrac{1}{2}mv^2 \quad \text{or} \quad v = \sqrt{2gh}$$

If the body starts with a velocity v_1 at height h_1 and falls to a lower height h_2, we have

$$mg(h_1 - h_2) = \tfrac{1}{2}m(v_2^2 - v_1^2)$$

from which the terminal velocity v_2 can be calculated.

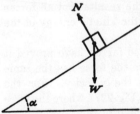

FIG. 228. The work theorem applied to the problem of the inclined plane.

The next example is the particle sliding down a frictionless inclined plane (Fig. 228). During the motion the normal force N does no work, while of the weight W only the in-line component $W \sin \alpha$ does work. If the body starts from rest and slides a distance s along the plane, we have

$$(W \sin \alpha)s = \tfrac{1}{2}mv^2$$

and

$$v = \sqrt{2(g \sin \alpha)s}$$

which is the sliding velocity at the end of distance s.

If there is friction on the plane, the component of force doing work is $W \sin \alpha - fN = W \sin \alpha - fW \cos \alpha$, and the terminal velocity becomes

$$v = \sqrt{2g(\sin \alpha - f \cos \alpha)s}$$

If we compare these two examples with the solutions obtained for them previously by direct application of Newton's law (pages 176 and 179), we notice that those previous solutions gave us the accelerations only, and that in order to find the velocities we had to integrate. By the application of the work-energy theorem we thus save one integration, because that integration has already been performed once and for all cases, when the theorem was derived.

A third example is that of the stone whirled overhead at the end of a string at high speed in a horizontal plane, neglecting gravity (page 211). In this case the only force acting on the particle is the string tension, which is directed perpendicular to the circular path and hence does no work. Therefore the kinetic energy of the particle remains constant, so that it whirls around with constant speed.

The only way in which the string tension mv^2/r can be made to do work is by pulling in on the string, *i.e.*, by making r shorter. Suppose we permit r to increase by the small amount dr, sufficiently slowly for the string tension to remain practically constant during the process. Then the work done by the string force on the particle is $-m(v^2/r)\, dr$, negative because the particle was permitted to displace itself in a direction opposite to that of the string force. By the theorem we have

$$-m \frac{v^2}{r} dr = d \left(\frac{1}{2} mv^2 \right) = mv\, dv$$

or

$$-\frac{dr}{r} = \frac{dv}{v}$$

Integrated:

$$\log v = -\log r + \text{constant}$$
$$\log vr = \text{constant}$$
$$vr = \text{constant}$$

Therefore if the string is pulled in, the work so performed is transformed into additional kinetic energy of the whirling particle, which goes faster and faster the shorter the radius becomes. This case will be discussed from a different viewpoint on page 280.

The next example is the simple pendulum (Fig. 158, page 184), which is released from a position φ_0, starting from rest. We ask for the velocity of the particle at the bottom of its path $\varphi = 0$. The forces acting on the particle are the string tension and the weight. The string tension, being always perpendicular to the path, does no work on the particle. The weight $W = mg$ has a component tangent to the path and hence it does perform work, but to calculate it we do

better to consider the whole force mg and multiply it with the in-line
component of displacement. That displacement is partly horizontal
and partly vertical, but W performs work only on the vertical component. The work done is $mg\ \Delta h$, where Δh is the vertical descent of
the particle. Thus, if starting from rest $(v = T = 0)$, we have

$$mg\ \Delta h = \tfrac{1}{2}mv^2$$

The descent $\Delta h = l(1 - \cos \varphi_0)$, between the extreme position $\varphi = \varphi_0$
and the mid-position $\varphi = 0$, so that

$$v = \sqrt{2gl(1 - \cos \varphi_0)} = \sqrt{2g\ \Delta h}$$

In this example the velocity was found very simply from the work-energy
theorem. To derive the same result from Newton's laws directly is much
more complicated, as follows:

We start from the last line of page 184, giving the acceleration

$$\ddot{\varphi} = -\frac{g}{l}\sin \varphi$$

To integrate this, we write

$$\ddot{\varphi} = \frac{d\dot{\varphi}}{dt} = \frac{d\dot{\varphi}}{dt} = \frac{d\dot{\varphi}}{d\varphi}\frac{d\varphi}{dt} = \dot{\varphi}\frac{d\dot{\varphi}}{d\varphi}$$

Substituting

$$\dot{\varphi}\ d\dot{\varphi} = -\frac{g}{l}\sin \varphi\ d\varphi$$

$$\int_0^{\dot{\varphi}} \dot{\varphi}\ d\dot{\varphi} = -\frac{g}{l}\int_{\varphi_0}^0 \sin \varphi\ d\varphi$$

$$\frac{\dot{\varphi}^2}{2} = +\frac{g}{l}[\cos 0 - \cos \varphi_0]$$

$$v = \dot{\varphi}l = l\sqrt{2\frac{g}{l}(1 - \cos \varphi_0)} = \sqrt{2gl(1 - \cos \varphi_0)}$$

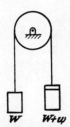

W $W+w$

FIG. 229. At-
wood's ma-
chine.

The last example to be discussed in this article is
Atwood's machine (Fig. 229), which is a system consisting of two particles of weight W and $W + w$, connected by a weightless string slung over a weightless pulley. We
ask the same question as in all previous examples: *what is the
velocity of the system after a given displacement?* To apply the
energy theorem to this case we remark that on each particle two
forces are acting: the weight and the string tension. The string tensions on both weights are equal, because the pulley is supposedly massless (page 234), and if the system is displaced, one weight goes up as
much as the other goes down. Therefore the amounts of work done by

the two string tensions are equal and opposite and their sum is zero. Applying the theorem to one mass, we find

$$-Ws + \text{work of string tension} = \Delta\left(\frac{1}{2}\frac{W}{g}v^2\right)$$

and for the other mass, we have similarly

$$+(W + w)s + \text{work of string tension} = \Delta\left(\frac{1}{2}\frac{W + w}{g}v^2\right)$$

When adding the two equations, the string tension drops out, as explained before:

$$ws = \Delta\left(\frac{1}{2}\frac{2W + w}{g}v^2\right)$$

or, when starting from rest,

$$v = \sqrt{2\left(g\frac{w}{2W + w}\right)s}$$

The parentheses in the square root have been placed there to indicate the relation between this result and the one obtained on page 191 with Newton's equations.

Problems 267 and 268.

47. Potential Energy; Efficiency; Power. As we have seen in the preceeding examples, the forces that may act on a particle arise from various causes. They can be listed as follows:

a. Gravity forces or weights
b. Spring forces or other elastic forces
c. Normal forces from a wall or other guide
d. Friction forces
e. All other forces, such as pull in a rope, steam pressure on a piston, etc.

It is useful to distinguish between some of these, and in this article we propose to prove that **work done in overcoming the first two kinds of forces is recoverable and is stored in the form of potential energy; work done by the third kind is always zero, while work done in overcoming the fourth kind is not recoverable and is said to be "dissipated" in the form of heat.**

First we consider a body acted upon by its weight and by an upward force P equal and opposite to the weight, arising, for instance, from a string (Fig. 230). Suppose the weight to be moved at slow speed from one position to another one, h higher, by way of any devious path and

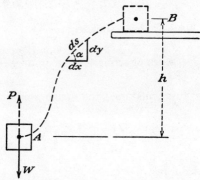

Fig. 230. The work done by the string force in bringing the weight from position A to position B along any path is Wh.

to be deposited on a table at the end. The work done by the string force, which overcomes the weight, is equal to

$$W = \int_0^s P(ds \sin \alpha) = \int_0^h P \, dy = \int_0^h W \, dy = Wh = V \quad (20)$$

The work so done is called the **potential energy** and is usually denoted by the letter V. The name suggests that this work is not lost to us, but can be recovered, for instance, by removing the table, and letting the weight descend slowly while pulling up another equal weight on the other end of a pulley, or by letting it drop freely through distance h

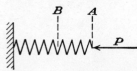

Fig. 231. The work done by a force compressing a spring is stored in the form of elastic energy.

and using the kinetic energy at the end of the fall for a drop-forging operation.

Next, let us look at the coil spring of Fig. 231, which is being compressed slowly by a force P, overcoming the spring force. As explained before on page 181, the force P is proportional to the deflection, and if x is measured from the unstressed condition of the spring, the force $P = kx$, when k is the spring stiffness measured in pounds per inch. The work done by P in compressing the spring x inches then is

$$W = \int_0^x P \, dx = \int_0^x kx \, dx = k \frac{x^2}{2} \Big|_0^x = \frac{kx^2}{2} = V \quad (21)$$

This is the potential energy stored in a spring, sometimes also called **elastic energy** or **resilience**, and denoted by the same symbol V as the gravitational potential energy.   The work done by P is recoverable because the spring is ready to push back against the force P and thus to return the same amount of work to us.

With a friction force (Fig. 232), the situation is entirely different. Let the body or particle lying on a rough horizontal plane be displaced

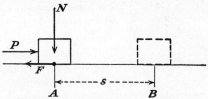

Fig. 232.   Work done against a friction force is dissipated in heat.

through distance s by a force P, overcoming the friction force F and hence equal to it.   The work done by P is $Ps = Fs$, and at the end of the process, the particle stays where it is and has no tendency whatever to return to its initial position or to give us any of the work back. The work is irrecoverable and is said to be "dissipated" in heat. Gravity forces and spring forces are said to be **potential forces,** while friction forces are called **dissipative forces.**   Work done in overcoming a potential force is like money in the bank: it is payable on demand; work done against friction is like money dissipated: the bartender will not pay it back.   The term "dissipated" is sometimes criticized on the grounds that the work is not destroyed; it is only changed into another form, heat energy.   A small part of that heat energy can be restored to mechanical energy, potential or kinetic, by means of a complicated heat engine.   But

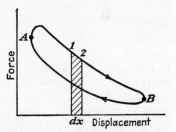

Fig. 233.   The area of an engine-indicator diagram represents the work done during one cycle.

money spent in a bar is not destroyed either; it has only changed hands, and on rare occasions a small part of it is restored to the former owner on the morning after.

The difference between a potential force and a dissipative force can be seen clearly in a force-displacement diagram as in Fig. 233, where the force acting on a point and the displacement of that point in the direction of the force are plotted against each other.   The work done by the force between points 1 and 2 on a small displacement dx is $F\,dx$, represented by the shaded region. If we start with the force at point A and permit the point to be displaced to B, the work done by the force is the entire area under the upper branch of the curve. If we then go back from B to A, and the force during this return displacement is different from the one going from A to B, as shown, the negative work done from B to A is the area under the lower branch of the curve. Thus the net work done during the round trip A to B to A is represented by the area of the closed loop. Diagrams like Fig. 233 are very useful as "indicator diagrams" in engines, where the piston force is plotted vertically and the piston displacement horizontally. The piston force, of course, is greater going one way than the other, because the object of the engine is to perform work.

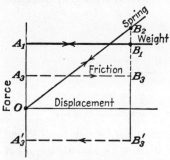

Fig. 234. Force-displacement diagrams for a weight force, an elastic spring force, and a friction force.

Now let us examine the potential forces of Figs. 230 and 231, and the dissipative force of Fig. 230 in the light of such an "indicator diagram." These diagrams are shown in Fig. 234. The weight force W is independent of the location or displacement, and is represented by a horizontal line. Going from A to B back to A in Fig. 232 corresponds to going back and forth horizontally along the line A_1B_1 in Fig. 234. The work done going up is equal to the negative work going down: all the work is restored. The spring force kx of Fig. 231 is represented by a straight line through the origin in Fig. 234 and again the spring force going forward is the same as that going backward, and the round trip ABA in Fig. 231 corresponds to OB_2O in Fig. 234; again with the negative work during the backstroke equal to the positive work during the forward stroke.

The friction force in Fig. 232, however, reverses its sign with the motion, and in Fig. 234 is represented by OA_3 when going to the right A_3B_3, and by $OA_3' = -OA_3$ during the backstroke to the left. Thus the total work done by force P (Fig. 232) during a round trip is the area $A_3B_3B_3'A_3'$ in Fig. 234, and since we are back to the starting position, all this work is disspated. Thus

If a potential force acts on a particle, which is given a round-trip displacement returning to its original position, the loop area enclosed

in the force-displacement diagram of Fig. 233 is zero. If the force is purely dissipative, there will be a loop, and its area above the zero force line will be equal to that below the zero force line.

Now we are ready to write the rule of page 251 in a few different forms. We remember that the work done by the force *overcoming* the weight or spring force equals the gain in potential energy, and that thus the work by the weight itself is the negative gain in potential energy. (A weight going *up* gains V, but does negative work.) Therefore we can rewrite the rule of page 251 as follows:

a. The work done on a particle by the external forces, exclusive of the potential forces, equals the increment in the sum of the potential and kinetic energies. Or

b. The work done on a particle by the external forces exclusive of the potential and dissipative forces equals the increment of the sum of the kinetic and potential energies plus the work dissipated. Or again

c. If the external forces doing work on a particle are all potential forces, the total energy of the particle, being the sum of the potential and kinetic energies, remains constant.

Examples of the last statement are the freely falling stone, the particle sliding down a frictionless plane, or Atwood's machine. In all those cases the total energy remains constant; the potential energy decreases and the kinetic energy increases. In the simple pendulum the energy is all potential in the extreme positions, all kinetic in the bottom position, and mixed in between. As long as no friction occurs, the total energy remains constant. Figure 235 shows a simple pendulum OA, which is impeded in its motion by a fixed pin P. The pendulum is shown in four positions, and when it is let go at A from rest, it will ultimately swing out to the position B of equal height with A, because the string

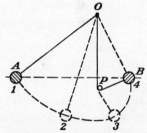

Fig. 235. A pendulum, released from rest in position A, will rise to the position B of equal height, even if it strikes the pin P.

tension does no work and the weight is a potential force, so that the total energy is constant.

In the example of the rough inclined plane the forces acting are the normal one, doing no work, the weight, which is a potential force, and the friction, which is a dissipative one. Applying statement *b* above, there are no external forces that do work other than the potential and dissipative ones. Thus the total energy is diminished by the work dissipated. Suppose $\alpha = 30°$, so that $\sin \alpha = dh/ds = \frac{1}{2}$, and suppose the friction force to be 25 per cent of the weight, then the work dissipated is half the loss of potential energy; the other half of the potential energy is transformed to kinetic energy; hence v^2 is half as large and v is 0.707 times as large as it would be in the absence of friction.

Now we are ready to give a satisfactory definition of the term *efficiency*, which was used before on pages 88, 91, and 98. In a mechanical machine, such as a hoist, a certain amount of work is done by us (the input), with the object of getting most of it back in some other form (the output). In the process some friction occurs and part of the work we do is dissipated in friction. The **efficiency η is defined as:**

$$\eta = \frac{\text{work output}}{\text{work input}} = \frac{\text{work input} - \text{work dissipated}}{\text{work input}} \qquad (22)$$

Power is defined as the rate of doing work; hence it is expressed in foot-pounds per second or similar units. The usual unit of power in mechanical engineering is the horsepower (hp), defined by James Watt to be considerably more than the power of a strong healthy horse (because James wanted the purchasers of his new fangled engines to be well satisfied).

$$1 \text{ hp} = 33,000 \text{ ft-lb/minute} = 550 \text{ ft-lb/sec} \qquad (23a)$$

The unit of power commonly used in electrical engineering is the kilowatt (kw).

$$1 \text{ kw} = 1.34 \text{ hp} \qquad (23b)$$

Since the horsepower or kilowatt signifies work or energy per unit of time, the horsepower-hour or kilowatt-hour (kw-hr) represent just work again:

$$1 \text{ kw-hr} = 60 \text{ kw-minutes} = 60 \times 1.34 \text{ hp-minutes}$$
$$= 60 \times 1.34 \times 33,000 \text{ ft-lb} = 2,650,000 \text{ ft-lb}$$
$$= 1,325 \text{ ft-tons}$$

It is impressive that such an enormous amount of work is available at any time in our private homes at a cost of 2 or 3 cents.

Problems 269 to 277.

48. Energy of Plane Bodies. Since energy, potential as well as kinetic, is a scalar quantity, the energy of a large body or system of bodies is equal to the sum of the energies of its constituent particles.

To find the potential energy of gravitation of a large body we have to sum the potential energies of all particles dw, located at various heights:

$$V = \int z \, dw = Wz_G = mgz_G$$

by Eq. (2) (page 34).

The gravitational potential energy of a two- or three-dimensional body is the same as that of a particle located at the center of gravity of the body, in which the entire weight of the body is concentrated, or

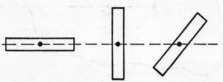

FIG. 236. The gravitational potential energy of a body is determined by the vertical location of its center of gravity.

shorter, **the potential energy of a body equals that of its center of gravity.** As an example, Fig. 236 shows a body in three positions of equal potential energy.

The kinetic energy of a particle dm is $\frac{1}{2}v^2 \, dm$, and if a large body is in a state of parallel motion, having equal velocities in all of its points, it acts like a single particle, and

The kinetic energy T of a body in parallel, non-rotating motion is $\frac{1}{2}mv^2$, equal to the kinetic energy of its center of gravity, or of any other point of the body having the total mass concentrated in it.

A body rotating at speed ω about a fixed axis has a speed $r\omega$ at distance r from the axis. The kinetic energy of one particle thus is $\frac{1}{2}(r\omega)^2 \, dm$, and the total kinetic energy of the body is

$$T = \int \frac{1}{2} r^2 \omega^2 \, dm = \frac{\omega^2}{2} \int r^2 \, dm = \frac{1}{2} I_o \omega^2 \qquad (24)$$

The kinetic energy of a body rotating about a fixed axis O, not necessarily its center of gravity, **is $\frac{1}{2}I_o\omega^2$.** This rule is not restricted to flat bodies, but applies to three-dimensional ones as well, as follows from its derivation.

Now we are ready to find the kinetic energy of a body in general two-dimensional motion. The motion is described as the superposition of a parallel translation of speed v_G of the center of gravity, and a rotation ω around the center of gravity. Choose a coordinate system with the origin in the center of gravity and the x axis along the velocity v_G of the center of gravity (Fig. 237).   The velocity of any particle dm then is the vector sum of v_G and $r\omega$; the angle between those two com-

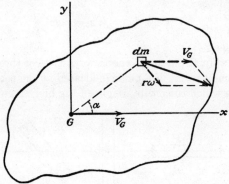

Fig. 237.   To calculate the kinetic energy of a particle in a translating and rotating rigid body.

ponent vectors being $90° - \alpha$ (Fig. 237).   Then, by the cosine rule of triangles, we find for the resultant velocity of particle dm

$$v_{\text{total}}^2 = v_G^2 + (r\omega)^2 + 2v_G r\omega \cos (90 - \alpha)$$
$$= v_G^2 + r^2\omega^2 + 2v_G\omega r \sin \alpha$$
$$= v_G^2 + r^2\omega^2 + 2v_G\omega y$$

The kinetic energy of the entire body then is

$$T = \tfrac{1}{2}\int v_{\text{total}}^2 \, dm = \tfrac{1}{2}v_G^2\int dm + \tfrac{1}{2}\omega^2\int r^2 \, dm + v_G\omega\int y \, dm$$

By Eq. (2) (page 34) the last integral is zero, because we have taken the center of gravity for our origin, so that

$$\mathbf{T} = \tfrac{1}{2}\mathbf{m}v_G^2 + \tfrac{1}{2}\mathbf{I}_G\omega^2 \qquad (25)$$

The kinetic energy of a body is the sum of the kinetic energies of and about the center of gravity.   This rule has just been proved for two-dimensional motion, but it is true for motion in space as well.

Now we are ready to extend the rule of page 251 from a particle to a larger rigid body, or to a system of rigid bodies, connected together by hinges or ropes, by an argument very closely related to that of page 139.   The body, or system of bodies, is thought of as built up of many particles.   On each constituent particle there act such external forces of the system as may happen to be on that particle *plus* the push-pull forces from the neighboring particles.   For any one such particle the rule of page 251 holds, and considering proposition 5 of page 250, we can say that the work done by all external and internal forces on that one particle equals the increment in its kinetic energy. Now we sum this statement over all the particles of the system and conclude that the work done by all external and internal forces of the entire system equals the increment in kinetic energy of the system.

The internal forces obey the law of action equals reaction, so that for any internal force acting on a particle a, there is an equal and opposite force acting on a neighboring particle b, and if those two forces do not move with respect to one another, the work done by the one cancels that of the other. For a single rigid body that is always the case, and it is also the case for systems of rigid bodies where there is no friction in the hinges and where there is no stretch in the ropes or springs holding the rigid bodies together (page 141).

Thus we come to the general rule:

The work done by all external forces on a rigid body or on a system of rigid bodies without friction in their internal connections equals the increment in the kinetic energy of the system.

A special case of this rule occurs when the system is in equilibrium: no motion takes place by itself and a small imaginary displacement imposed from the outside will create no kinetic energy: the work is then zero, which fact is the foundation for the method of work in statics (Chap. VIII).

An important special case of the above theorem, corresponding to the statement c of page 259 is

If all external forces doing work on a single rigid body (or on a system of rigid bodies without internal friction and without elasticity in their joints) are potential forces, then the total energy of the system, being the sum of the potential and kinetic energies, remains constant.

This theorem applies particularly to systems moving under the influence of gravity without friction in their guides, such as multiple compound pendulums, cylinders rolling down inclines, and systems of frictionless pulleys.

If there is friction between the various rigid parts of a system, for instance, when a cylinder partly rolls and partly slips over an inclined or curved track, then the rule is

The work done by all external forces on a system of rigid bodies equals the increment in kinetic energy in the system plus the work dissipated in friction in the joints or connections.

We will now illustrate these theorems on a number of examples. *Problems* 278 *and* 279.

49. Applications. The foregoing theories will be applied to six problems:

    a. A rolling cylinder
    b. The compound pendulum
    c. A system of pulleys
    d. The sliding ladder
    e. A dynamometer
    f. An engine flywheel

a. A Rolling Cylinder. The problem of the cylinder of mass m, radius r, and moment of inertia I_G, rolling down an incline α was discussed on page 242 (Fig. 219) by means of Newton's equations. With the energy method the solution is obtained with less effort. We note that the normal force N does no work. Neither does the friction force F perform work, because we have assumed no slipping. The only force doing work is the weight, and its work is transformed into kinetic energy. Let the cylinder roll down a distance s along the plane, starting from rest. The work done by the weight (or the loss in potential energy V) is $Ws \sin \alpha$. Let the velocity of the center of the cylinder at the end of the path s be v. Then the angular speed $\omega = v/r$, if there is no slipping. Thus the kinetic energy, by Eq. (25), is

$$T = \frac{1}{2} mv^2 + \frac{1}{2} I_G \frac{v^2}{r^2} = Ws \sin \alpha$$

But $W = mg$, and for I_G we can write mk_G^2. Thus

$$v^2 + \frac{k_G^2}{r^2} v^2 = 2gs \sin \alpha$$

or

$$v^2 = 2 \frac{g \sin \alpha}{1 + k^2/r^2} s$$

This formula relates the velocity v to the distance s traveled down the incline. The formula is seen to be of the form $v^2 = 2gh$, indicating that the motion is a uniformly accelerated one with the acceleration $g \sin \alpha/(1 + k^2/r^2)$. This result was found previously on page 242.

b. The Compound Pendulum. Let the compound pendulum of Fig. 212 be released from position φ_0 without initial angular speed. What is the angular speed at the bottom position $\varphi = 0$? We note that the question is worded like all the others in this chapter: *what is the velocity for a given displacement?* A question of this type is usually answered in the simplest manner by the energy method.

Of the two forces acting on the pendulum, one, the axis support, does no work, while the other one, the weight, is a potential force. Hence the sum of the potential and kinetic energies is constant. The loss in potential energy is

$$\Delta V = W \, \Delta h = mga(1 - \cos \varphi_0)$$

We set this equal to the gain in kinetic energy by Eq. (24).

$$mga(1 - \cos \varphi_0) = \tfrac{1}{2} I_0 \omega^2 = \tfrac{1}{2} m(k_G^2 + a^2) \omega^2$$

and

$$\omega^2 = 2 \frac{ga}{k_G^2 + a^2} (1 - \cos \varphi_0)$$

For reasonably small angles φ_0, we may write

$$\cos \varphi_0 = 1 - \frac{\varphi_0^2}{2} + \cdots$$

and the result is

$$\omega \approx \varphi_0 \sqrt{\frac{ga}{k_G^2 + a^2}}$$

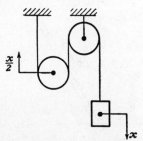

c. A System of Pulleys. In Fig. 238 we have a system of three equal weights W, two of which are in the form of uniform pulley disks of radius r. Starting from rest, what is the velocity of the system after a given descent x? What is the acceleration?

FIG. 238. To calculate the accelerations of a system of two pulleys and a weight by the energy method.

The only force doing work is gravity. If the weight goes down distance x, then the floating pulley goes up distance $x/2$. The fixed pulley rotates through angle x/r and the floating pulley through $x/2r$. The loss in potential energy is

$$W x \text{ (for the weight)} - W \frac{x}{2} \text{ (for the pulley)} = \frac{Wx}{2}$$

The gain in kinetic energy of the weight is $\tfrac{1}{2} m v^2$; that of the fixed pulley is $\tfrac{1}{2} I_G v^2 / r^2$, and of the floating pulley, $\tfrac{1}{2} m(v/2)^2 + \tfrac{1}{2} I_G(v/2r)^2$. Therefore

$$\frac{Wx}{2} = \frac{5}{8} m v^2 + \frac{5}{8} I_G \frac{v^2}{r^2}$$

By Eq. (14) we have $I_G = mr^2/2$; further $W = mg$, so that

$$v^2 = \frac{8}{15}\, gx = 2\left(\frac{4}{15}\, g\right) x$$

The latter answer is in the form $v^2 = 2gh$ for the freely falling body, and we conclude that the acceleration of the descending weight (of which the displacement is x) is $\frac{4}{15}g$. The acceleration of the floating pulley then is $\frac{2}{15}g$ upward.

These answers have been found without calculating the tensile forces in the various sections of rope. If Newton's equations had been used, as in Problem 256, the three rope tensions would have appeared as unknowns and the solution would have been much more cumbersome. If we want to know the rope tensions, we can still find them easily after the energy analysis. Knowing that the weight acceleration is $\frac{4}{15}g$, the force moving the weight must be $\frac{4}{15}W$, and hence the rope above the weight must hold back with a force $\frac{11}{15}W$.

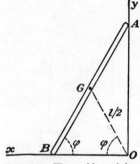

FIG. 239. The problem of the sliding ladder.

d. The Sliding Ladder. Figure 239 shows a sliding ladder without friction anywhere. Starting from the upright position $\varphi = 90°$ at zero speed, what is the angular speed $\dot\varphi$ of the ladder and what is the velocity of the bottom point B as a function of φ?

The normal forces at A and B do no work and there is no friction. Hence the only force doing work is the weight. The loss in potential energy is the loss in V of the center of gravity (page 261 and Fig. 236) or

$$\Delta V = W\left(\frac{l}{2} - \frac{l}{2}\sin\varphi\right) = \frac{mgl}{2}(1 - \sin\varphi)$$

The kinetic energy is that *of* and *about* G. The velocity of G is found easily after we recognize from the figure that OG is a constant length $l/2$ and that therefore G describes a circle about O as center with a speed $\dot\varphi l/2$. Hence

$$\Delta T = \frac{1}{2}\, m\left(\frac{\dot\varphi l}{2}\right)^2 + \frac{1}{2}\, I_G \dot\varphi^2$$

$$= \frac{ml^2}{8}\, \dot\varphi^2 + \frac{1}{2}\left(\frac{1}{12}\, ml^2\right)\dot\varphi^2 = \frac{1}{6}\, ml^2 \dot\varphi^2$$

Equating this to the loss in V leads to

$$\dot{\varphi} = \pm \sqrt{\frac{3g}{l}\,(1 - \sin\,\varphi)}$$

This expression contains a $\pm$ ambiguity, which is due to the fact that the kinetic energy depends on $\dot{\varphi}^2$ only and is equally large for a ladder sliding uphill or downhill. Here we know that it must be sliding downhill so that $\dot{\varphi}$ must be negative. The $+$ sign before the square root, signifying uphill rotation, must be discarded. This then answers the first question. For the second question we see that $OB = x_B = l\cos\,\varphi$. Hence

$$\dot{x}_B = -l\,\sin\,\varphi\,\dot{\varphi} = \sqrt{3gl\,\sin^2\varphi(1 - \sin\,\varphi)}$$

an expression which is zero for $\varphi = 90°$ and also for $\varphi = 0$, as it should be. What is the velocity of point A at position $\varphi = 0$?

  e. A Dynamometer. A dynamometer is a device for measuring either forces or torques. Figure 240 shows one form of it, consisting of two half rings lined on the inside with brake lining, loosely clamped around a rotating shaft or wheel. When no weight W is placed in the pan at the end of the arm, the friction torque of the shaft turns the arm so that it rests against the upper stop S. When too large a weight is placed

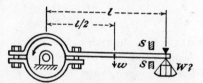

Fig. 240. From a measurement of the weight W and of the speed of the engine the horsepower can be calculated.

at W, the arm will rest on the lower stop, and for a certain definite value of W, it will play between the two stops. Given the horsepower HP of the engine, which is dissipated in the dynamometer, the speed N of the shaft expressed in rpm, the arm length l and the arm weight w, find the necessary weight W.

  The moment equilibrium of the half rings and arm demands that $Wl + (wl/2)$ equals the engine torque. If that torque is M_t, then the work $M_t\varphi$ per revolution is $M_t 2\pi$, and the work per minute is

$$2\pi N M_t = 33{,}000\,HP$$

by Eq. (23a). Thus

$$M_t = \frac{33{,}000\,HP}{2\pi N} = Wl + \frac{wl}{2}$$

For a 60-hp engine at 1,200 rpm, the torque $M_t = 262$ ft-lb. With an arm length $l = 3$ ft and an arm weight of 20 lb, the moment of the arm itself is 30 ft-lb, so that a 232-ft-lb torque is to be carried by the end weight W. Therefore $W = 233$ ft-lb/3 ft $= 78$ lb.

f. An Engine Flywheel. A flywheel on an engine serves the purpose of keeping the speed of the engine more or less constant when the torque applied to it is subject to variations. A gasoline engine has

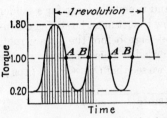

large torque variations: shortly after a cylinder sparks, the torque is large; midway between explosions, some cylinder is compressing gas and the torque is small. Figure 241 shows the approximate torque curve for a four-cylinder four-cycle spark-plug engine. The torque variations are 80 per cent of the mean torque, and occur twice per

FIG. 241. The torque curve of a four-cylinder, four-cycle spark-plug engine.

revolution, with explosion frequency. The torque is approximately represented by

$$M_t = M_{t\,\text{mean}} \left(1 + 0.80 \sin \frac{\pi}{15} Nt \right)$$

where N is the rpm of the engine, and t is measured in seconds. The reader should verify this by observing that for $t = 1$ sec the angle of the sine is $\pi N/15 = 4\pi N/60$ or 4π times the revolutions. Let this engine be coupled to an electric generator of which the counter torque is constant $= M_{t\,\text{mean}}$. Then the torque variations will cause speed variations.

For a 60-hp engine running at 1,200 rpm, we now propose to calculate the required WR^2 of the flywheel effect if the speed is to be kept between 1,200 rpm $\pm$ 1 per cent, *i.e.*, between 1,188 and 1,212 rpm.

To solve this problem, we observe that at point A in Fig. 241 the torque has just been larger than normal so that the speed will be maximum. Between A and B the electric counter torque is larger than the engine torque and the unit will slow down, till at B the lowest speed of 1,188 rpm occurs. The mean speed then will occur midway between A and B.

Now let us calculate the work done by the excess torque between time B and the next time A. That work is

$$\int M_{t\,\text{excess}} \, d\varphi$$

The angle between B and A is ¼ revolution or $\pi/2$ radians. The "average height" of a sine wave is $2/\pi$ times its peak height, which the reader should check by performing the integration. Thus the above integral is

$$\frac{2}{\pi} M_{t \text{ excess (peak)}} \frac{\pi}{2} = M_{t \text{ excess (peak)}} = 0.8 M_{t \text{ mean}}$$

This work is used to increase the kinetic energy of the engine, generator, and flywheel, or

$$\Delta(\tfrac{1}{2} I_{\text{combined}} \omega^2) = I_{\text{combined}} \omega \, \Delta \omega = 0.8 M_{t \text{ mean}}$$

Therefore

$$I_{\text{combined}} = \frac{0.8 M_{t \text{ mean}}}{\omega \, \Delta \omega}$$

In this formula the numerical values are

$$M_{t \text{ mean}} = \frac{33,000 HP}{2\pi N} = \frac{33,000 \times 60}{2\pi \times 1,200} = 263 \text{ ft-lb}$$

$$\omega = \frac{1200}{60} 2\pi = 126 \text{ radians/sec}$$

$$\Delta \omega = 2\% \text{ of } \omega = 2.52 \text{ radians/sec}$$

Hence:

$$I_{\text{combined}} = \frac{0.8 \times 263}{126 \times 2.52} = 0.662 \text{ lb-ft-sec}^2$$

$$= 7.95 \text{ lb-in.-sec}^2$$

This is the required moment of inertia of the combined engine-generator-flywheel. To get a picture of what this represents, let us calculate the diameter of a solid steel disk of 1-in. thickness having this combined moment of inertia:

$$I = \frac{1}{2} \frac{W}{g} r^2 = \frac{1}{2} \left(\frac{0.28}{386} \pi r^2 \right) r^2 = 0.00114 r^4 \text{ lb-in.-sec}^2$$

Hence

$$r^4 = \frac{7.95}{0.00114} = 6,950 \text{ in.}^4$$

and

$$r = 9.1 \text{ in.} \quad \text{or} \quad d = 18.2 \text{ in.}$$

Problems 280 *to* 288.

CHAPTER XV

IMPULSE AND MOMENTUM

50. Linear Momentum. Let us consider once more the simplest case of particle motion, that of a force F acting on a particle in the same direction as its velocity v. Taking the x axis of coordinates to coincide with this direction, we can write Newton's equation in a form slightly different from the one we have used so far:

$$F = m\ddot{x} = m\frac{d\dot{x}}{dt} = \frac{d}{dt}(m\dot{x}) = \frac{d}{dt}(mv)$$

The quantity mv is called the *momentum* of the particle, or sometimes also the "linear" momentum, to distinguish it from "angular" momentum, which will be introduced on page 276. In words Newton's law states that **the force equals the rate of change of linear momentum.** Integrating with respect to time leads to

$$\int_{t_1}^{t_2} F\,dt = \int_1^2 d(mv) = mv_2 - mv_1 = \Delta(mv)$$

The integral expression at the left is called the *impulse* (or linear impulse) and it is measured in pound-seconds.

The linear impulse equals the increment in linear momentum.

Since the impulse is measured in pound-seconds, so must be the momentum, which is verified as follows:

$$mv = \frac{W}{g}v = \left[\frac{\text{lb}}{\text{in.}/\text{sec}^2} \times \frac{\text{in.}}{\text{sec}}\right] = [\text{lb-sec}]$$

The equation so far has been stated only for the case that the direction of the force coincides with that of the velocity of the particle. We now investigate the more general case where these two directions are *not* the same. This is illustrated in Fig. 242 in two dimensions only, but our reasoning will be in three dimensions. Newton's law states that the acceleration of a particle has the same direction as the force acting on it, and this fact can also be expressed by resolving both the force and the acceleration into three Cartesian coordinates [Eq. (7), page 176]. These equations, written in the new form, are

270

$$F_x = \frac{d}{dt}\,(mv_x)$$
$$F_y = \frac{d}{dt}\,(mv_y) \qquad (26)$$
$$F_z = \frac{d}{dt}\,(mv_z)$$

or, rolled together into a single vector equation,

$$\mathbf{F} = \frac{d}{dt}\,(m\mathbf{v}) \qquad (26)$$

and

$$\int_1^2 \mathbf{F}\,dt = m\mathbf{v}_2 - m\mathbf{v}_1$$

Force $\mathbf{F}$ and velocity $\mathbf{v}$ are vector quantities. Impulse is the vector force multiplied by the scalar time; so is momentum the vector $\mathbf{v}$

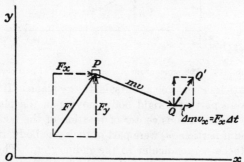

Fig. 242. The vector QQ', being the increment of the momentum vector PQ, has the same direction as the force F acting on the particle P.

multiplied by the scalar m. Hence both impulse and momentum are vector quantities that can be resolved into components or compounded into a resultant just like force or velocity.

In Fig. 242 the vector PQ is the momentum $m\mathbf{v}$, having the direction of the instantaneous velocity $\mathbf{v}$ at time $t = 0$. The vector QQ', parallel to the force $\mathbf{F}$, is the increment in momentum $m\,\Delta\mathbf{v} = \mathbf{F}\,\Delta t$. The vector PQ' represents the momentum at time Δt and is the vector sum of $m\mathbf{v}$ and $m\,\Delta\mathbf{v}$.

It is important to note that while impulse and momentum are vector quantities, energy is not. The kinetic energy of a moving particle does **not** have the direction of the velocity; it is a scalar, a number equal to $\frac{1}{2}mv^2$, not having any direction at all.

So far we have seen nothing new. The importance of the concept of momentum lies in its applications to larger bodies and systems of bodies. We therefore now proceed to calculate the linear momentum of a conglomerate of particles moving in space. **The linear momentum of a conglomerate is defined as the vector sum of the linear momenta of the constituent particles.**

Figure 243 shows three particles m_1, m_2, m_3 of different sizes and velocities lying in the xy plane, but we imagine that there are many

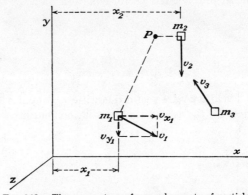

FIG. 243.   The momentum of a conglomerate of particles.

more such particles in, as well as outside, the xy plane. If the particles m_1 and m_2 were part of a rigid body moving in the plane, point P would be the instantaneous center of rotation of the body (Fig. 171, page 197), and if particle m_3 were part of the same body, its velocity v_3 would have to be perpendicular to the radius Pm_3. The velocity v_3 is purposely drawn quite differently to indicate that these particles move like an unconnected swarm in the most arbitrary manner in three-dimensional space.

Let us first calculate the x component of the instantaneous location of the center of gravity G of this swarm. It is found, by Eq. (2) (page 34), from

$$m_1 x_1 + m_2 x_2 + \cdots + m_n x_n = \Sigma m_r x_n = x_G \Sigma m_n = m x_G$$

where m denotes the total mass of the swarm. Of course similar equations hold for the y and z directions. Let us now differentiate these equations with respect to time. All the x_n, y_n, z_n vary with time but the individual masses m_n of the particles are constant. Thus

$$\Sigma m_n \dot{x}_n = m \dot{x}_G$$
$$\Sigma m_n \dot{y}_n = m \dot{y}_G$$
$$\Sigma m_r \dot{z}_n = m \dot{z}_G$$

The left-hand sides are the sums of the x, y, z components of momentum of all the individual particles. The right-hand sides can be interpreted as the three components of momentum of an imaginary particle having a mass equal to the total mass of the swarm and having the same velocity as the center of gravity of the swarm. Thus we have proved that

The linear momentum of any conglomerate of particles equals the linear momentum of its total mass concentrated at the center of gravity. Of course the "conglomerate" includes the rigid body as a special case.

Now we are ready to generalize the statement of page 270 from a single particle to a swarm (and incidentally to a rigid body or to a system of interconnected rigid bodies). For each constituent particle the resultant force acting on it equals the rate of change of its linear momentum. As always, there are two kinds of forces acting on a particle: (*a*) external ones and (*b*) the pushes or pulls from neighboring particles, called "internal forces." The internal forces obey Newton's law of action equals reaction. Summing Newton's statement over all the particles of the swarm, we see that the vector sum of all external forces acting on all particles plus the vector sum of all internal forces on all particles equals the rate of change of momentum of the system of all particles. But, on account of the law of action equals reaction each internal force is balanced by an equal and opposite one on the neighboring particle so that their total sum is zero. This leads us to the result that

The resultant of all external forces acting on a conglomerate of particles equals the rate of change of momentum of the center of gravity.

$$\mathbf{F} = \frac{d}{dt}\,(m\mathbf{v}_G) \qquad (26a)$$

This is an important and far-reaching theorem.

Its most spectacular applications are in connection with exploding bodies. Let a shell fly in a parabolic path, and let it explode in mid-air. The explosion is brought about by internal forces, whose total impulse is zero (by action equals reaction); the external force is the weight, before as well as after the explosion; therefore the center of gravity of the conglomerate of splinters and combustion gases continues in its parabolic path. The truth of this can be visualized easily for an explosion breaking the shell into two equal parts, flying away in opposite directions, one to the right, one to the left of the parabola, but the theorem is true no matter into how many parts the projectile breaks up.

A similar situation exists with an athlete diving off a high spring-board. The center of gravity of the diver's body describes exactly a parabola, irrespective of the twists and turns and somersaults he may make in the meantime. All these contortions are brought about by internal forces; while the only possibility of deviating the center of gravity from its parabola lies in an external force, other than the weight, of which the only conceivable one, that from the surrounding air, is altogether too small to be effective.

Another application of the theorem is in the recoil of guns. Let the loaded gun be at rest with horizontal bore and free to roll on wheels in the direction of its bore. The firing explosion creates a large force pushing forward on the projectile and pushing back equally hard on the gun. By Eq. (26a) the momentum of the combination remains zero, because there is no external force. Hence, when the projectile leaves the gun, the combined center of gravity is still at rest and the velocities of the projectile and the gun must be inversely proportional to their masses. The gun recoils, and this recoil motion is taken up by springs and friction, and its kinetic energy is dissipated in friction. Another way of expressing the situation is that the momentum $m_p V$ of the projectile is equal and opposite to the momentum of the gun $m_g v$, so that $v/V = m_p/m_g$, the total momentum being zero all the time. What happens to the kinetic energy? That was obviously zero before the firing and no longer zero after the firing, because we remember from page 251 that kinetic energy (unlike momentum) is a scalar and is always positive. The energies of the projectile and of the gun must be added to give the system energy. Is the energy of the projectile equal to that of the gun, or if not equal, which is the larger energy? We have

$$T_p = \tfrac{1}{2}m_p V^2 = \tfrac{1}{2}V(m_p V)$$
$$T_g = \tfrac{1}{2}m_g v^2 = \tfrac{1}{2}v(m_g v)$$

The two quantities in parentheses are the momenta, which are equal, so that the kinetic energies are in ratio of the velocities, or in inverse ratio of the masses, which is lucky because the enemy will have greater difficulty in stopping the projectile than we have in taking the recoil of the gun.

A device that was often used for determining the velocity of a bullet (before the days of high-speed cameras or oscillographs) is the *ballistic pendulum*, shown in Fig. 244. It consists of a bag of sand of weight

W suspended from a string of length l. The bullet of weight w is
fired into the sand at the unknown speed V,
and the pendulum swings out to an angle α,
which is observed. From α it is possible
to determine the bullet speed V.

Just before striking, the momentum of
the system is wV/g, and during the period
of embedding of the bullet into the sand-
bag no external forces act, because it all
takes place in so short a time that the
pendulum remains vertical. Then the system momentum remains constant during

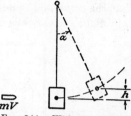

FIG. 244. With the ballistic
pendulum the velocity V of a
bullet can be measured.

the impact, and the sandbag-bullet combination acquires a speed v so
that

$$\frac{W + w}{g} v = \frac{w}{g} V$$

or

$$v = \frac{w}{W + w} V$$

Immediately after the impact the pendulum swings out and comes to
rest at angle α. The calculation of α from the initial speed v is a
typical question for an energy analysis rather than for a momentum
analysis, because it expresses a relation between velocity and displacement, rather than between velocity and time.

$$\frac{1}{2} \frac{W + w}{g} v^2 = (W + w)h \qquad \text{or} \qquad v^2 = 2gh$$

and by geometry

$$h = l(1 - \cos \alpha)$$

so that

$$V = \frac{W + w}{w} v = \frac{W + w}{w} \sqrt{2gl(1 - \cos \alpha)}$$

is the desired relation between the bullet velocity V and the angle of
swing α.

In the example of Fig. 245 many of the foregoing effects are combined. A man in a closed boxcar (without friction in the wheel axles)
shoots a bullet into a target at the other end. The reader should
reason out for himself what happens during the three stages of the
process: first while the bullet is traveling inside the gun barrel, then
while it is in the air, and finally while it is embedding itself in the

target. It is instructive to do this with the momentum theorem [Eq. (26a)] as well as with Newton's laws directly. The conclusion is that the combined center of gravity never moves, that while the bullet is in flight the car rolls back slowly, that at the end the car is stopped

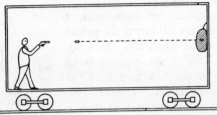

FIG. 245. Linear momentum relations in the problem of the man firing a bullet inside a closed car.

again and is slightly behind its first position. During the flight of the bullet its forward momentum equals the rearward momentum of the car, but the bullet's kinetic energy is very much greater than that of the car.

Problems 289 *to* 293.

51. Angular Momentum. The moment of a force about an axis, oblique in space with respect to the line of action of the force, is found by first resolving the force into two components, one parallel to the moment axis and the other one in a plane perpendicular to that axis, then multiplying the latter component with the normal distance between its line of action and the moment axis (page 103). The component of force parallel to the moment axis has no moment about that axis. In the same manner we can form a moment about an axis of any other vector quantity instead of force. We could take the moment of a displacement, velocity, or acceleration, but such moments have never been found of much use. However, the moment of the momentum vector about an axis is of great utility and is denoted by the letter $\mathfrak{M}$.

The unfamiliar Gothic $\mathfrak{M}$ is used for moment of momentum, M is used for moment, and m for mass. Unfortunately, too many words in mechanics start with the letter m.

For a rigid body rotating about a fixed axis O at angular speed ω, a particle dm at radius r has a speed $r\omega$ and a momentum $r\omega\,dm$ perpendicular to the axis of rotation. The **moment of** this **momentum,** also called the **angular momentum** of the particle is $r\,r\omega\,dm = r^2\omega\,dm$, and the angular momentum of the entire body is the integral of that expression or is ωI_0. Newton's law for the rotation of a rigid body about a fixed axis was [Eq. (15), page 233]

$$M = I_0 \ddot{\varphi} = I_0 \frac{d\omega}{dt} = \frac{d}{dt} (I_0 \omega) = \frac{d\mathfrak{M}}{dt} \qquad (27a)$$

or in words

The moment of all external forces acting on a rigid body pivoted about an axis fixed in space equals the rate of change of angular momentum $I\omega$ about that axis.

The next case we consider is that of the general two-dimensional motion of a body. The velocity of such a body can be considered (Fig. 169, page 196) as the superposition of a parallel motion and a rotation. The angular momentum of the parallel motion about the center of gravity is zero, by the definition of center of gravity. (By page 33 it is the point through which passes the resultant of a set of parallel "forces" proportional to the masses dm, or the point about which those parallel "forces" have no moments. Instead of "forces" $g\,dm$ we have here quantities $v\,dm$ of the same nature.) Therefore the angular momentum about G is due to the rotational motion only and is $I_G \omega$. With the help of this we can rewrite Newton's equations [Eqs. (17)] of page 241 as follows:

$$\left. \begin{aligned} F_x &= \frac{d}{dt} (m\dot{x}_G) \\[2mm] F_y &= \frac{d}{dt} (m\dot{y}_G) \\[2mm] M_G &= \frac{d}{dt} (I_G \omega) \end{aligned} \right\} \qquad (27b)$$

or in words

For the plane motion of a rigid body the resultant of all external forces equals the rate of change of the linear momentum of the center of gravity G, and the moment of the external forces about G equals the rate of change of angular momentum about G.

These are very useful theorems, but they tell us nothing new; they are the old equations (17) of Newton's law, stated in terms of new words. However, the concept of angular momentum owes its great significance in mechanics not so much to applications to single rigid bodies, but rather to composite systems, such as precessing gyroscopes or rotating and oscillating governor flyballs, for which the new theory is very useful. With the goal of a gyroscope theory in view, we now return once more to a single particle, having not only a tan-

gential speed about the moment axis, but having radial and longitudinal speed components as well, and subjected to a force likewise
having all three components. For such a particle we shall prove that

**The moment about an arbitrary line fixed in space of the force
acting on a particle equals the rate of change of the angular momentum
of that particle about that axis, or $M_F = d\mathfrak{M}/dt$.**

This proposition is very important and is by no means as obvious
as it might seem at first reading. For its proof we start without a
figure but imagine the moment axis in space and the particle not on
that axis with an arbitrary velocity vector and an arbitrary force
vector in different directions. We resolve the velocity as well as the

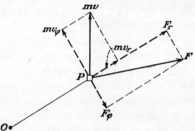

Fig. 246. To prove that the moment of a force acting on a particle about a line in
space equals the time rate of change of angular momentum of the particle about that
line.

force into components parallel to the moment axis and in a plane
perpendicular to that axis. The parallel (longitudinal) components
of force or velocity have no moment about the axis. By Newton's
law a force along the axis will change the particle velocity only in that
direction, but the perpendicular velocity component is not affected
by it. Therefore, in our derivation of the moment relations we can
disregard the longitudinal components of force and velocity (or
momentum) and deal with the perpendicular components only. Now
we are ready to draw a figure. Figure 246 is the plane perpendicular
to the moment axis, which appears as a point O only. The particle
P has a momentum $m\mathbf{v}$ and the force $\mathbf{F}$ is acting on it. We resolve the
force as well as the momentum into radial and tangential components,
as shown in the figure. By Varignon's theorem (page 103), the
moment of a vector about an axis is the sum of the moments of its
components. But the radial components, F_r or mv_r, have zero moment
about O, so that the moments of F and mv are equal to those of F_φ and
mv_φ, respectively. With this the problem is reduced to rotation about
a fixed axis, for which it is expressed by Newton's law $F_\varphi = d(mv_\varphi)/dt$.

in which both sides of the equation are multiplied by the common moment arm *OP* of Fig. 246.

Another proof of the theorem is by resolution into Cartesian coordinates F_x, F_y, F_z, mv_x, mv_y, and mv_z as shown in Fig. 247. The coordinate system has been so chosen that the *x* axis coincides with the moment axis.

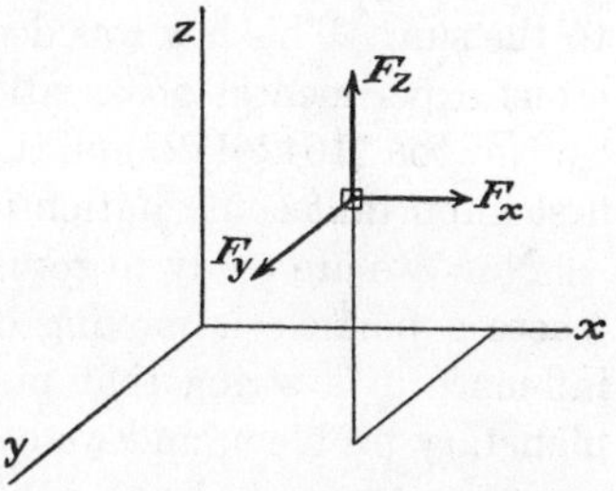

The moment of the total force about the *x* axis is, by geometry,

$$M_F = F_y z - F_z y$$

which by Newton's law is

$$m\ddot{y}z - m\ddot{z}y = m(\ddot{y}z - \ddot{z}y)$$

The moment of the momentum vector about the *x* axis is, like the moment of the force,

$$\mathfrak{M} = mv_y z - mv_z y = m(\dot{y}z - \dot{z}y)$$

Fig. 247.

Then, by differentiation:

$$\frac{d\,\mathfrak{M}}{dt} = m[(\ddot{y}z - \dot{y}\dot{z}) - (\ddot{z}y - \dot{z}\dot{y})]$$
$$= m(\ddot{y}z - \ddot{z}y)$$

which is the same as the expression for the moment of the force M_f, so that the theorem is proved. In this proof Varignon's theorem has been freely used although it was not mentioned. The reader is advised to point out where in the proof this was done.

The simplest and most beautiful application of this angular-momentum theorem of a particle is to a problem in astronomy, to Kepler's law of equal areas in equal times. In Fig. 248 let *S* be the

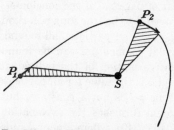

FIG. 248. A planet in its path around the sun sweeps out equal areas in equal times.

sun and P_1 and P_2 be two positions of a planet moving around the sun (we do not have to know that the path is an ellipse). Newton assumed that the sun and planet, both considered to be particles, attract each other with a force along their connecting line. The attractive force on the planet has no moment about *S*, and hence $d\mathfrak{M}/dt$ is zero, or $\mathfrak{M}$, the angular momentum, is constant. On page 14 we saw that a moment can be represented by twice the area of the shaded triangle, erected on the *mv* vector as a base. If we multiply the $m\mathbf{v}$ vector by a small time Δt so that the base is short, then all the triangles for equal times Δt must have equal areas. For longer times t the base of the figure becomes curved, and the figure can be considered as the sum or integral of many small triangles. Hence equal areas are swept out in equal times, and the velocity is larger when the planet gets closer to the sun. This law was deduced by Kepler (1571–1630) from numerous experimental observations on the planets, and its explanation by Newton (1642–1727) on the basis of his own laws of motion was the first important confirmation for the correctness of those laws.

Now we are ready to return to the problem discussed on page 253 where a particle is moving on a horizontal smooth table under the influence of a string that pulls it. This is exactly the same as the planetary problem, and we can write immediately

$$\mathfrak{M} = r\,r\dot{\varphi} = r^2\omega = \text{constant}$$

a result which was found much more elaborately on page 253.

Our next step is to extend the theorem of page 278 from a single particle to a conglomerate of particles. We follow the same reasoning as in all such cases. Each particle of the swarm is subjected not only to an external force, but to internal forces from its neighbors as well. By action equals reaction the momenta caused by these internal forces cancel each other in pairs, so that the total momentum and hence the moment of momentum caused by the internal forces is zero. Therefore

The vector sum of the moments of all external forces acting on a conglomerate of particles about an arbitrary axis fixed in space equals the rate of change of the angular momentum of that conglomerate about the axis.

This statement is identical with that of page 277 if the conglomerate is a rigid body, but we have now proved it to be true for a non-rigid body as well. In the general case of a swarm of particles, all moving independently of each other, the angular momentum can be found only by summation. In case the relative motion of the individual particles is radial only, the angular speed ω is the same for all particles and the formula $\mathfrak{M} = I\omega$ applies, but now I is not a constant, but varies with the time. Thus the theorem furnishes the explanation for the familiar experiment in physics (Fig. 249), where a man stands on a rotating platform without external forces. The angular momentum about the vertical axis must remain constant, and if the man's internal motions consists of radial displacements of his arms only, this can be expressed by

$$\frac{d\mathfrak{M}}{dt} = \frac{d(I\omega)}{dt} = 0$$

in which I is variable. The product of I and ω is constant, so that ω increases when the man pulls his arms together. This is a trick commonly practiced by figure skaters or ballet dancers to increase their vertical angular speed at will in the movement known as the "pirouette."

If the man on the rotatable platform starts without angular speed and does not push against anything outside the system (including the surrounding air), he cannot acquire permanent angular speed by whatever contortions he might make. He can make the platform rotate by swinging a stick or his arms con-
tinuously above his head in a horizontal circle, but as soon as he stops that motion, the platform stops rotating. If he happens to have a bicycle wheel with a couple of handles on its axle, he can acquire an angular speed starting from rest by holding the wheel in a horizontal plane and then start-
ing it spinning by hand. The total angular momentum about a vertical axis must remain zero, so that if the wheel spins clock-
wise, he himself spins counterclockwise.

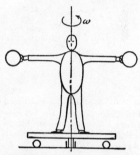

FIG. 249. The system of the man and the frictionless turntable cannot change its angular momentum about a vertical axis by motions of the man's arms.

If, while this is going on he turns the wheel upside down, while spin-
ning, his own angular speed will be reversed. But when he stops the wheel, he will stop rotating himself.

Problems 294 to 297.

52. Applications. Work is force times distance; impulse is force multiplied by time. When the **velocity** of a system is asked after a certain **distance**, we do well to **use the work-energy theorem.** Similarly, when the **velocity** after a certain **time** is wanted, we **use the impulse-momentum theorem.**

The following problems will be discussed:

a. The block sliding down an incline
b. The cylinder rolling down an incline
c. Mixed sliding and rolling
d. The water jet against a flat wall
e. The Pelton water turbine
f. Reaching the moon with a rocket

a. The Sliding Block.   A block slides down an incline α.   What is the speed after a given time *t* from the start?   If there is no friction, the force in the direction of the incline is $W \sin \alpha$, constant in time. Equating the impulse to the increment in momentum, we write

$$(W \sin \alpha)t = mv$$

or

$$v = (g \sin \alpha)t$$

which is the desired relation.   In case there is friction, the retarding

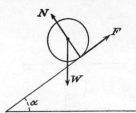

friction force is $fN = fW \cos \alpha$, so that the net force down the incline is $W(\sin \alpha - f \cos \alpha)$.   The analysis is the same and the answer is

$$v = gt(\sin \alpha - f \cos \alpha)$$

b. The Rolling Cylinder.   What is the speed of a cylinder rolling down an incline

Fig. 250.   To find the velocity after a certain *time* for a cylinder rolling down a rough incline.

α at a time *t* after starting from rest? From Fig. 250 we see that the net force down the incline is $W \sin \alpha - F$, in which

F is not known, except that it must be smaller than $fW \cos \alpha$.   We apply Eqs. (27*b*) along the plane and about the center of gravity:

$$(mg \sin \alpha - F)t = mv$$

$$Frt = I_G \omega = \frac{I_G v}{r} \qquad \text{(for pure rolling)}$$

In this pair of equations there are two unknowns, F and t.   Eliminating F and solving for *t*, we find

$$v = \frac{g \sin \alpha}{1 + I_G/mr^2} t$$

verifying the result obtained on page 242.   In this general form the solution holds for solid or hollow cylinders, spheres, or yo-yo wheels like that of Fig. 221 (page 243), all depending on the value we substitute for I_G.

c. Mixed Sliding and Rolling.   A billiard ball is given a central kick by the cue.   We shall see on page 288 that this results in a parallel velocity v_0 of the ball without rotation.   During the first part of the path (Fig. 251), slipping takes place and the friction force starts a rotation, which after a while becomes so large that the slip stops and the ball is purely rolling.   After what time t_1 does this occur?

The only forces acting on the ball are the weight W, the normal force (also equal to W), and the friction force, which equals fN during the time interval $0 < t < t_1$. The equations [Eqs. (27b)] are

$$fNt_1 = m(v_0 - v_1)$$
$$fNrt_1 = I_G\omega_1$$

At the time t_1, pure rolling has set in, so that

$$v_1 = \omega_1 r$$

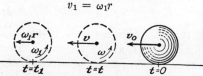

$$t=t_1 \qquad\qquad t=t \qquad\qquad t=0$$

FIG. 251. A billiard ball starting at time $t = 0$ with parallel sliding motion gradually comes to a purely rolling motion at $t = t_1$.

In these three equations t_1, ω_1, and v_1 are unknown, so that we can solve for them:

$$v_1 = \frac{v_0}{1 + I_G/mr^2}$$

$$\omega_1 = \frac{v_1}{r}$$

$$t_1 = \frac{v_0}{fg(1 + mr^2/I_G)}$$

which solves the problem.

d. The Water Jet. A jet of water of cross section A and speed v strikes a flat plate (Fig. 252), which has a cone-shaped guide attached to it, so that the water stream spreads out evenly in all directions. What is the force exerted on the plane by this jet?

The volume of water striking the plate each second is Av, and its mass is $Av\rho$, where ρ is the mass per unit volume, or the density of the water. The water in the jet before striking the plate has momentum all in one direction, whereas after striking,

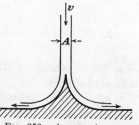

FIG. 252. A water jet striking a plate exerts a force $Av^2\rho$.

the particles have velocities equally distributed in all directions, so that the momentum is zero. The loss of momentum per second of the water is

$$Av\rho \, v = Av^2\rho,$$

and by Eq. (26a), this is associated with a retarding force of that

magnitude exerted by the plate on the water, and hence by the water
on the plate.   Suppose $A = 1$ in.^{2} and $v = 100$ ft/sec.   Water weighs
62.4 lb/cu ft, so that

$$\rho = \frac{62.4 \text{ lb/cu ft}}{32.2 \text{ ft/sec}^2} \approx 2 \text{ lb sec}^2 \text{ ft}^{-4}$$

The force is

$$F = \rho v^2 A = 2 \times (100)^2 \times \tfrac{1}{144} = 139 \text{ lb}$$

FIG. 253. A
water jet strik-
ing a Pelton
wheel bucket ex-
ercises a force
$2Av^2\rho$.

If the jet strikes a Pelton wheel bucket, shaped as in
Fig. 253, so that the velocity is reversed without any
loss in kinetic energy, the momentum is reversed and
the change in momentum is twice that of Fig. 252 or $F = 2A\rho v^2$.

   e. The Pelton Water Turbine.   A Pelton wheel consists of a large
number of buckets, each shaped like Fig. 253, arranged on the periphery of a wheel (Fig. 254).   Let the speed of the water jet be V.

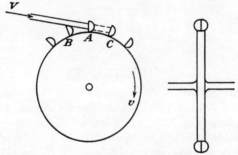

FIG. 254.   In a Pelton wheel two buckets are simultaneously struck by the water jet
during part of the time.

First let us consider a single bucket, moving at velocity v.   During
each second a length $V - v$ of water strikes it, with a relative velocity
$V - v$.   If the bucket is well constructed, this water reverses its speed
and leaves the bucket, with a reverse speed $V - v$ relative to the
bucket.   The force exerted on the bucket then is

$$F = 2A\rho(V - v)^2$$

Since the bucket moves with speed v, the power or work done per
second, W_1, by the water on the bucket is

$$W_1 = Fv = 2A\rho v(V - v)^2 = 2A\rho V^3 \frac{v}{V}\left(1 - \frac{v}{V}\right)^2$$

The expression has been written in the last form, because we wish to investigate how this power varies with the ratio v/V. For $v/V = 0$, there is no power, because the bucket does not move; for $v/V = 1$, no water catches the bucket and the power is again zero. The relation is plotted in Fig. 255, and the reader should verify by differentiation that the maximum power or rate of work is being delivered if the bucket moves at one-third jet speed. Then the water strikes

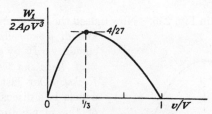

Fig. 255. The power delivered by a single Pelton bucket as a function of the speed ratio.

the bucket at two-thirds jet speed, and leaves the bucket at one-third jet speed in a direction opposite from the original one. It thus retains one-ninth of its kinetic energy and gives up eight-ninths doing work on the bucket.

One would expect that if the bucket speed were half the jet speed and hence the final speed of the water zero, then each drop of water would give up all of its kinetic energy and hence more work would be done by it. This is indeed so, but, for $v = V/2$, less water strikes the bucket per second than for $v = V/3$, and the maximum rate of work per bucket is at $v = V/3$, as shown in Fig. 255.

Now we return to the multibucket wheel of Fig. 254, and remark that for each individual bucket the result of Fig. 255 applies. What is now the power of the wheel? It is more than the power of a single bucket, because sometimes two buckets receive the water stream simultaneously. For example, Fig. 254 shows the instant that the stream is just beginning to be interrupted by bucket B. During the short time that follows, bucket B receives the full stream, while bucket A is using up the portion BA of the stream simultaneously. The relative speed of catching up is $V - v$, and if s is the bucket pitch, or length of the column AB, the time of double exposure is $s/(V - v)$. The time of one pitch advance is s/v, so that double exposure occurs during a fraction of the total time expressed by

$$\frac{s/(V - v)}{s/v} = \frac{v}{V - v}$$

The power of the wheel thus is the power of a single bucket multiplied by

$$1 + \frac{v}{V-v} = \frac{V}{V-v} = \frac{1}{1-v/V}$$

or

$$W_1 = 2A\rho V^3 \frac{v}{V}\left(1 - \frac{v}{V}\right) \quad \text{(for the complete wheel)}$$

a relation shown in Fig. 256.

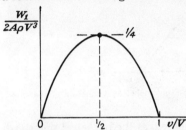

FIG. 256. The power delivered by a complete multibucket Pelton wheel as a function of the speed ratio.

Now indeed the maximum power occurs at half jet speed, as the reader should verify by differentiation.

f. Reaching the Moon with a Rocket. Our last topic in this article is a discussion by means of the momentum theorem of the possibility of sending a rocket to the moon. A rocket operates by ejecting from behind at high speed some parts of its own substance. In order to impart high speed to that expelled material the rocket machinery has to give it a large rearward acceleration, and hence a large push. By the fact that action equals reaction, the rocket pushes itself forward against the inertia of the expelled material. The rocket has only a definite amount of matter aboard for expelling purposes, and it does best by ejecting it at the highest possible speed. This ejecting speed is limited by energy considerations, and the best that can be done by ordinary chemical combustion or by explosives is about 5,000 ft/sec or 4½ times the speed of sound. Any substantially higher ejecting speeds await the atomic age. Given this jet speed, the rocket push can be regulated by the rate at which matter is expelled, and the question comes up as to the best rate for the purpose of getting to the moon. The answer to that question is simple: the best we can do is to throw out the total available charge in the shortest possible time at the start. This can be recognized by observing that if we should expel matter at such a moderate rate that the upward push equals the rocket weight only, then the rocket would just hover as long as it was burning, and would get nowhere. By ejecting the total charge at the beginning, the maximum possible speed is attained, which sends the rocket farthest away. The only limitation is the maximum acceleration the rocket structure can withstand.

Now we are ready to calculate the maximum speed V of the rocket, after the available charge has been expelled. Let V_{jet} be the expelling speed, m_{full} the mass of the rocket including fuel at the start, m_{empty} the mass at the end of combustion, and let μ be the mass ejected per second. Then the momentum thrown off during time dt is $\mu V_{jet}\, dt$, which is also the gain in forward momentum of the rocket itself. The mass of the rocket at time t is $m_{full} - \mu t$, so that

$$\mu V_{jet}\, dt = (m_{full} - \mu t)\, dV$$

$$V_{jet} \int_0^t \frac{\mu\, dt}{m_{full} - \mu t} = \int_0^V dV$$

$$-V_{jet} \log_e (m_{full} - \mu t)\Big|_0^t = V \Big|_0^V$$

$$V_{jet} \log_e \frac{m_{full}}{m_{full} - \mu t} = V$$

This is the relation between the variable rocket speed V and the variable time t during the ejection period. But this period only lasts until $m_{full} - \mu t$ becomes m_{empty}, so that

$$\frac{V_{(max)\ rocket}}{V_{jet}} = \log_e \frac{m_{full}}{m_{empty}}$$

This simple relation must be satisfied irrespective of the internal construction of the rocket.

Now we must calculate what initial speed $V_{(max)\ rocket}$ is required to reach the moon. This is an exercise in integration that is left to the reader as Problems 299 and 300 (page 430), with the result that the required speed is 37,000 ft/sec or about 34 times the speed of sound. With the best available V_{jet} of 5,000 ft/sec, we find that

$$\log_e \frac{m_{full}}{m_{empty}} = \frac{37,000}{5,000} = 7.4$$

$$\frac{m_{full}}{m_{empty}} = e^{7.4} = 1,640$$

Thus, for every pound of rocket structure, machinery, fuel tanks, instruments, and controls, the rocket must carry 1,639 lb of fuel or similar ejectable material, which shows that the moon is still far away.

Problems 298 *to* 303.

53. Impact. When two elastic bodies impinge on one another, like two billiard balls, a hammer and a piece of steel, or a golf club

and its ball, it has been found that the contact is of extremely short
duration and the contact force is of very large magnitude. It is
common experience that with a light blow of a peening hammer we
can make a permanent dent in a piece of steel, whereas in order to
make that same dent with static loading we have to put thousands of
pounds on the hammer head. Figure 257 shows a photograph of a
golf ball being hit by a club, from which the enormous magnitude of

Fig. 257. A golf ball being struck by the club. This photograph with an exposure
of 10^{-6} sec was taken by Prof. Harold E. Edgerton with one of the marvelous high-
speed cameras he developed.

the contact force is apparent. The duration of contact is of the order
of one-thousandth of a second. With these phenomena Newton's
laws are true as always, the accelerations are very large, but the time
interval during which the acceleration applies is so small that the
struck body hardly displaces itself. The momentum equation still is

$$\int_{t_1}^{t_2} F \, dt = m(v_2 - v_1)$$

but in the impulse integral we do not care to distinguish the details of
the force-time relationship, considering the "impulse" as a whole only.
In Fig. 258 two impulses are shown, which are entirely equivalent for
our purpose as long as the areas under the curves are the same. These

two impulses will "instantaneously" change the momentum of the struck object by the same amount.

Now we propose to study the general case of central impact between two bodies, and after obtaining the general formulas, we will interpret the results to various practical cases. By "central" impact we mean that the line of action of the contact force is perpendicular to the contact surface and passes through the centers of gravity of both bodies. The masses m_1 and m_2 and the velocities V_1 and V_2 before impact are known; it is required to calculate v_1 and v_2 after the impact (Fig. 259). All velocities are considered positive in the same direction; the quantities m_1, m_2, V_1, V_2 may have arbitrary values,

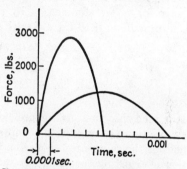

FIG. 258. The important property of an impulse is the area under the curve, while the shape of the curve itself has little significance.

but V_1 must be larger than V_2, otherwise the two bodies will not meet. The impact force is an internal force in the system of two masses; hence the total momentum remains the same under all circumstances, whether the balls are made of lead or of ivory. Thus

$$m_1V_1 + m_2V_2 = m_1v_1 + m_2v_2 \qquad (28)$$

This is only one equation in the two unknowns v_1 and v_2. The second equation is an energy equation and expresses the difference between lead and ivory. If a perfectly elastic hard steel ball from a ball bearing is dropped from a certain height on a hard steel plate, the impact

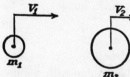

FIG. 259. Two balls over-taking each other, just before impact.

takes place without loss of energy; *i.e.*, in a force-displacement diagram like Fig. 233 or 234 the forward and return curves lie on top of each other and do not form a loop. Then the steel ball rises to the same height as that it fell from, and the velocity of the ball relative to the plate reverses its direction during the impact. On the other hand a lead ball just stays down and does not bounce at all. With Newton we define **the coefficient of restitution** e as the ratio between the relative speeds after and before impact: for a purely elastic impact $e = 1$;

for a purely inelastic one $e = 0$; all actual cases are in between. This is expressed by

$$e(V_1 - V_2) = v_2 - v_1 \qquad (28a)$$

which furnishes the second equation for the solution of v_1 and v_2. To solve for v_2, we eliminate v_1 by multiplying the last equation by m_1 and by adding it to Eq. (28), with the result

$$\left.\begin{aligned} v_2 &= \frac{m_2 V_2 - m_1 e V_2 + m_1 V_1(1 + e)}{m_1 + m_2} \\ v_1 &= \frac{m_1 V_1 - m_2 e V_1 + m_2 V_2(1 + e)}{m_1 + m_2} \end{aligned}\right\} \qquad (29)$$

It is seen that these two expressions can be found one from the other simply by reversing the subscripts 1 and 2; which must be so because neither the momentum equation [Eq. (28)] nor the restitution equation changes by reversing the subscripts. Two special cases of Eq. (29) are interesting. The first one is that for purely elastic impact when $e = 1$.

$$v_1 = \frac{(m_1 - m_2)V_1 + 2m_2 V_2}{m_1 + m_2}$$

$$v_2 = \frac{(m_2 - m_1)V_2 + 2m_1 V_1}{m_1 + m_2}$$

which for the special case of equal masses $m_1 = m_2$ reduces to

$$v_1 = V_2$$
$$v_2 = V_1$$

which means that the velocities just reverse. If the front ball stands still and the rear ball of equal weight strikes it, the rear ball stops dead and the front ball runs on with the same speed as the rear ball at first. This can be observed with billiard balls or marbles.

The second interesting special case is inelastic impact: $e = 0$. Then Eq. (29) reduces to

$$v_1 = v_2 = \frac{m_1 V_1 + m_2 V_2}{m_1 + m_2}$$

which means that after the impact the two bodies run on with a common speed. A piece of putty falling on the floor illustrates this case.

The loss or dissipation of energy in the impact can be calculated from

$$\Delta T = (\tfrac{1}{2}m_1 V_1^2 + \tfrac{1}{2}m_2 V_2^2) - (\tfrac{1}{2}m_1 v_1^2 + \tfrac{1}{2}m_2 v_2^2)$$

When Eqs. (29) are substituted into this expression, the energy loss can be written in terms of the known quantities m_1, m_2, V_1, V_2. This, however, involves a page of algebra before the expression is reduced to its simplest form, which is

$$\Delta T = \frac{1 - e^2}{2} \frac{m_1 m_2}{m_1 + m_2} (V_1 - V_2)^2 \tag{30}$$

The reader who has time to spare may verify this, but we will now derive the result in a simpler manner. We reason that the loss of energy during the impact does not depend on V_1 and V_2 individually, but rather on their difference $V_1 - V_2$, because only this difference determines the intensity of the shock. Adding equal amounts to V_1 and V_2, which means moving the pair of balls at an additional uniform speed, should not change the energy loss. Then we choose V_1 and V_2 so as to make the situation as simple as possible, namely, by setting the combined center of gravity at rest. The two particles move toward each other, and after the impact the point G still remains at rest, so that the two particles individually reverse their speeds, only diminished by the factor e. The center of gravity is at rest when

$$m_1 V_1 + m_2 V_2 = 0$$

or

$$V_1 = \frac{m_2}{m_1 + m_2} (V_1 - V_2) \qquad \text{and} \qquad -V_2 = \frac{m_1}{m_1 + m_2} (V_1 - V_2)$$

The incoming energy of m_1 is $\frac{1}{2} m_1 V_1^2$; the outgoing energy is $\frac{1}{2} m_1 (e V_1)^2$. A similar formula holds for the other mass m_2. Therefore

$$\begin{aligned}
\Delta T &= \frac{1 - e^2}{2} (m_1 V_1^2 + m_2 V_2^2) \\
&= \frac{1 - e^2}{2} \left[m_1 \left(\frac{m_2}{m_1 + m_2} \right)^2 + m_2 \left(\frac{m_1}{m_1 + m_2} \right)^2 \right] (V_1 - V_2)^2 \\
&= \frac{1 - e^2}{2} \frac{m_1 m_2}{m_1 + m_2} (V_1 - V_2)^2
\end{aligned}$$

which is Eq. (30).

We can draw some practical conclusions from these formulas, or rather we can show that our ancestors who developed hammers, anvils, and pile drivers by a process of common sense and trial and error succeeded in adopting dimensions which, by the formulas, prove to be the most suitable ones. First let us consider the forging operation where m_1 is the hammer and m_2 is the combination of anvil and forging. The object is to forge, *i.e.*, to transform as great a part as possible of the kinetic energy of the hammer into dissipation by changing the shape of the forging. In this case, when the forging is sufficiently hot, the restitution coefficient e is very small, practically zero, so that by Eq. (30), the dissipated energy is

$$\frac{1}{2}\frac{m_1 m_2}{m_1 + m_2}V_1^2$$

This can be written as

$$\frac{m_2}{m_1 + m_2}\frac{1}{2}m_1 V_1^2$$

or in words, $m_2/(m_1 + m_2)$ times the original energy of the hammer. In order to dissipate as large a fraction as possible of this energy, we must make the ratio $m_2/(m_1 + m_2)$ as close to unity as possible, which means that the anvil m_2 must be made as large as possible with respect to the hammer m_1.

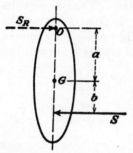

The opposite is the case in a pile driver. There the pile m_2 takes the place of the anvil, and the object is to transform the kinetic energy of the hammer or ram into kinetic energy of the pile, which then shoots a few inches further into the ground, changing its own kinetic energy into work overcoming side friction from the ground. Here, it is of advantage to make the hammer m_2 large in comparison to the pile m_1. Also it is of advantage to make e as large

FIG. 260. A compound pendulum subjected to a shock S experiences a reaction shock S_R from its support.

as possible. The reader should verify from Eq. 29 that for purely elastic impact, $e = 1$, the pile acquires four times as much kinetic energy as it would for inelastic impact $e = 0$.

We now turn to an entirely different problem, illustrated in Fig. 260. A compound pendulum, suspended at O at rest, is given a sudden impulse or shock S at distance b below point G, and directed perpendicular to OG. What does the pivot axle at O feel? The shock $S = \int F\, dt$ consists of a very large force during a very short time, as explained in Fig. 258; hence if O feels anything at all, it must also be a shock. Assume that reaction shock to be S_R, directed parallel to S. (We can reason that there cannot be a reaction shock component along OG, because that would impart a velocity along OG to the body, which is impossible). We write three equations

$$S - S_R = mv_G$$
$$Sb + S_Ra = I_G\omega$$
$$v_G = a\omega$$

The first two of these are the integrated Newton equations [Eqs.(17) of page 241]; the third equation is a geometric one expressing that point O is fixed. These three equations contain the three unknowns v_G, ω, and S_R. The quantities v_G and ω are the velocities of the system immediately after the shock. Eliminating them and solving for the reaction shock S_R, we find

$$S_R = S\frac{I_G - mab}{I_G + ma^2} = S\frac{k_G^2 - ab}{k_O^2}.$$

We see that the reaction shock is zero if $k_G^2 = ab$, that is, when the shock S is given in the center of percussion (page 247) with respect to the point of suspension O.

Problems 304 *to* 306.

CHAPTER XVI

RELATIVE MOTION

54. Introduction. In all previous statements in this book about displacements, velocities, or accelerations, these quantities were expressed in terms of a coordinate system "at rest." By that we tacitly meant that the coordinate system is at rest with respect to what Newton called "absolute space," which is the space of the "fixed" stars. Newton's law of the proportionality of force and acceleration is found to agree very well with experiment when the acceleration is referred to a coordinate system "at rest in absolute space." The earth rotates with respect to that absolute space, so that a coordinate system fixed to our earthly surroundings is not strictly at rest, and Newton's laws do not apply quite as well, but for almost all our engineering applications we can say that an earthly coordinate system is sufficiently close to being "at rest." Only for a few devices, of which the gyroscopic ship's compass is the most notable one, does the rotation of the earth become of engineering interest.

In many practical cases a motion can be described more simply in terms of a moving coordinate system than in terms of an absolute one or an earthly one. Take for example the motion of a point on the periphery of a rolling wheel (Fig. 151, page 173). The path of that point is a cycloid, and the determination of the velocity and acceleration is complicated. If, however, we set up a coordinate system with the origin in the wheel center, moving with it, and with the x axis horizontal and the y axis vertical, then the path of a peripheral point becomes a circle, and its velocity and acceleration appear very much simpler. Or consider a Watt flyball engine governor (Fig. 166, page 193), of which the balls oscillate up and down while rotating. The actual path in space of a ball is very complicated, and the accelerations are difficult to determine. We are very much tempted to place ourselves as observers on the rotating governor spindle and describe the ball motion with respect to the rotating coordinate system. The motion is then a simple up and down oscillation, and the acceleration is easily found. However this acceleration relative to the rotating system is different from that relative to the surroundings at rest,

294

and only the latter acceleration equals F/m. Newton's law is true for coordinate systems at rest; in general it does not hold for moving coordinate systems.

The science of mechanics did not start with engineering, but with astronomy, and naturally the ancient astronomers described their observations in terms of a coordinate system of which the earth was the origin. The paths they found for the planets were awful, hypo- and epicycloids, and it was a great accomplishment when Copernicus

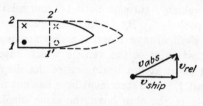

FIG. 261. The absolute velocity is the vector sum of the relative velocity and the vehicle velocity.

(1473–1543) and Kepler (1571–1630) recognized that these paths could be described more simply as ellipses in terms of a coordinate system with the sun as origin. This discovery was one of the starting points for Newton's great work.

In engineering there are many cases where motions with respect to a rotating coordinate system are simpler than those with respect to absolute or terrestrial space. In the counterweights of aircraft engines there are loose pendulous masses, whose motion in absolute space is very complicated indeed, but which only oscillate with respect to a moving coordinate system attached to the crankshaft. The motion of fluid or gas particles in the blades and passages of turbines or rotating pumps are other examples of this kind.

Thus we recognize the desirability of finding out what we have to do in order to make Newton's laws applicable to moving coordinate systems, and that is the object of this chapter. We shall make no change in Newton's law itself, but we shall find rules by which the actual or absolute acceleration can be deduced from the simpler relative acceleration with respect to the moving coordinate system.

Some of the rules of relative motion are extremely simple, almost obvious, and they have been applied here and there in the previous pages already. Consider, for example, Fig. 261, where a ship moves with respect to the shore, which, being at rest, is an absolute coordinate system. The captain walks across the deck from starboard to port, from point 1 to point 2. While he does that, point 1 of the ship goes to position 1' and point 2 of the ship goes to 2', and, of course, the captain ends up in position 2'. We call 1-2 the relative displacement and 1-1' the vehicle displacement. The word "vehicle" will be used throughout for the moving coordinate system; in this case the vehicle is a ship; in further examples the vehicle will be a turbine rotor, an elevator cab, a rotating table, the earth, or a crankshaft. In the example of Fig. 261 the captain is the "moving point," which moves relative to the "vehicle" through path 1-2 and relative to "absolute space" through path 1-2'. By "vehicle" displacement or velocity or acceleration we shall always mean the displacement or velocity or acceleration of that point of the vehicle which happens to coincide with the moving point at the beginning of the displacement. This in our case is point 1. The statement here has not much significance because all points of the ship have the same displacement, but in future cases of rotating vehicles it is important to keep this definition in mind.

From Fig. 261 we draw the conclusion that **the absolute displacement is the vector sum of the relative displacement and the vehicle displacement.**

We shall see in the next article that this vector addition of a relative and a vehicle quantity resulting in an absolute quantity holds not only for displacements, but also for velocities. It even holds for accelerations, provided the vehicle does not rotate. But when the vehicle rotates, we shall see that the absolute acceleration of a point is *not* equal to the vector sum of the relative and vehicle accelerations.

55. Non-rotating Vehicles. We first investigate the case where the vehicle moves parallel to itself, but not necessarily in a straight path. The path may be curved, but all points of the vehicle move in congruent and parallel paths like the bifilar pendulum of Fig. 224 (page 247). We choose one point of the vehicle for the origin O' of the moving coordinate system and lay the x' and y' axes fixed in the vehicle (Fig. 262). Then the $O'x'y'$ coordinate system moves parallel to itself in a curved path with the vehicle. Let the distance 1-1' be the distance traveled by the vehicle point 1 in time Δt in its curved path. If Δt be made small the piece of path 1-1' becomes almost straight, and the distance 1-1' can be written as $v_v \Delta t$ where v_v is the average velocity of point 1 of the vehicle during the time Δt.

Similarly the distance 1-2 can be written $v_r \Delta t$ and 1-2' becomes $v_a \Delta t$. We have seen that these displacements satisfy the vector equation

$$\mathbf{v}_r \, \Delta t + \mathbf{v}_v \, \Delta t = \mathbf{v}_a \, \Delta t$$

Now we divide by Δt, and let Δt become zero in the limit, so that the average velocities become true instantaneous velocities. This leads to the result

$$\mathbf{v}_r + \mathbf{v}_v = \mathbf{v}_a$$

or in words:

For a non-rotating vehicle the **absolute velocity is the vector sum of the relative and vehicle velocities.**

This sentence is only partly printed in bold-face type, because we shall see later that the statement holds true for rotating vehicles as well, although we have not proved it at this time.

We now proceed to consider accelerations, which are rates of change of velocities. In Fig. 262 the vehicle is shown in two positions, time

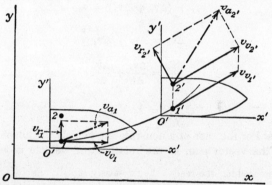

Fig. 262. The vehicle and its various velocities shown in two consecutive positions, at time $t = 0$ and at time $t = \Delta t$.

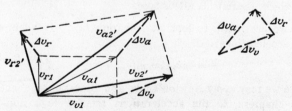

Fig. 263. The velocities of Fig. 262 reassembled into one figure.

Δt apart. The vehicle is moving through a curved path, and the velocity of its point 1 at the two positions 1 and 1' is different in direction as well as in magnitude. The same is true of the relative speed; at time $t = 0$ the captain walks to portside, but at time $t = \Delta t$ he runs towards the port aft corner of his ship. The two consecutive positions of the "moving point," the captain, are 1 and 2' and the corresponding velocities are plotted in the figure. The vehicle velocity at time Δt is the velocity of point $2'$ of the vehicle, which is the same as the velocity of point $1'$ because the vehicle does not rotate. Thus $\mathbf{v}_{v2'} = \mathbf{v}_{v1'}$. In Fig. 263 the vectors of Fig. 262 have been drawn once more, the velocities at time $t = 0$ in light lines, the velocities at time $t = \Delta t$ in heavy lines, and the differences, which are $\dot{\mathbf{v}}\,\Delta t$, in dashed lines. We see that the directions of the accelerations $\dot{\mathbf{v}}$ in general are totally different from the directions of the velocities. From the geometry of Figs. 262 and 263 we deduce that $\dot{\mathbf{v}}_a\,\Delta t$ is the vector sum of $\dot{\mathbf{v}}_r\,\Delta t$ and $\dot{\mathbf{v}}_v\,\Delta t$, by the following process:

$$\mathbf{v}_{r1} + \Delta\mathbf{v}_r = \mathbf{v}_{r2'}$$
$$\underline{\mathbf{v}_{v1} + \Delta\mathbf{v}_v = \mathbf{v}_{v1'} = \mathbf{v}_{v2'}}$$
$$(\mathbf{v}_{r1} + \mathbf{v}_{v1}) + (\Delta\mathbf{v}_r + \Delta\mathbf{v}_v) = (\mathbf{v}_{r2'} + \mathbf{v}_{v2'}) \quad \text{(add)}$$
$$\mathbf{v}_{a1} + (\Delta\mathbf{v}_r + \Delta\mathbf{v}_v) = \mathbf{v}_{a2'}$$

Therefore

$$\Delta\mathbf{v}_a = \Delta\mathbf{v}_r + \Delta\mathbf{v}_v$$

$$\frac{\Delta\mathbf{v}_a}{\Delta t} = \frac{\Delta\mathbf{v}_r}{\Delta t} + \frac{\Delta\mathbf{v}_v}{\Delta t}$$

and

$$\dot{\mathbf{v}}_a = \dot{\mathbf{v}}_r + \dot{\mathbf{v}}_v \tag{31}$$

or in words: **For the case of a non-rotating vehicle the absolute acceleration is the vector sum of the relative and vehicle accelerations.**

Before proceeding, the reader should satisfy himself that the validity of this proof depends on the fact that in Fig. 262 the vehicle velocities of points $1'$ and $2'$ are the same. If these velocities are different, which is the case for a rotating vehicle, the formula (31) is false.

Now we are ready to look at Newton's law. It holds only for absolute accelerations:

$$\mathbf{F} = m\dot{\mathbf{v}}_{\text{abs}} = m(\dot{\mathbf{v}}_{\text{rel}} + \dot{\mathbf{v}}_{\text{veh}}) \tag{31a}$$

Therefore we may apply Newton's law, and all its consequences of the previous chapters, to the accelerations relative to a non-rotating moving coordinate system, provided we add vectorially to these accelerations the parallel field of accelerations of the moving coordinates.

Before discussing examples of this theorem, we will express it in a somewhat different manner yet. The equation for a moving particle can be written

$$\mathbf{F} - m\dot{\mathbf{v}}_{\text{veh}} = m\dot{\mathbf{v}}_{\text{rel}} \tag{31b}$$

or

Newton's law may be applied to the relative accelerations of a moving, non-rotating coordinate system, if only we add to each mass element dm **a fictitious or supplementary force of magnitude** $-\dot{\mathbf{v}}_{\text{veh}}\, dm$.

As an example consider the space inside an elevator cab rising with an upward acceleration $g/2$. In the cab we, as observers, are looking at a 110-lb lady who stands on a scale, holding a pendulum in one hand and dropping her purse out of the other hand. What is (a) the scale reading, (b) the period of the pendulum, (c) the acceleration of the purse, and (d) the telephone number of the lady?

Applying statement (31a) we observe that the lady has zero acceleration, but we must add $\frac{1}{2}g$ upward to that before applying Newton's law. The scale thus reads 110 lb for the weight and an additional 55 lb to push the lady up. By statement (31b) we add a force $\frac{1}{2}mg$ *down*ward to the 110-lb weight of the lady, who thus tips the scale at 165 lb. By statement (31a) the pendulum swings like a pendulum that is accelerated upward at $g/2$, although to me, the observer, no such acceleration is visible. By statement (31b) the pendulum swings under the influence of the gravity force mg plus a fictitious force $\frac{1}{2}mg$ downward. Thus it acts the same way as an ordinary pendulum in a field of $1\frac{1}{2}g = 48.3$ ft/sec². The purse goes down in absolute space with acceleration g. By statement (31a) we have to add to our observed acceleration an acceleration $g/2$ upward; hence we observe $\frac{3}{2}g$ downward. By statement (31b) the purse is acted upon by its own weight mg and by an additional downward force $\frac{1}{2}mg$; its mass is m, hence it goes down with acceleration $\frac{3}{2}g$. Thus questions (a) to (c) have been elucidated. The answer to question (d) is left to the initiative and ingenuity of the reader.

An important conclusion that can be drawn from the theorems (31a) and (31b) is that

Newton's laws apply without any correction to coordinate systems moving at uniform velocity, because the vehicle acceleration for such a coordinate system is zero.

This places us in a position to clear up the question, discussed on page 233, concerning the applicability of the formula $M = I\ddot{\varphi}$ to two-dimensional motion. It was proved that this formula holds for the center of gravity and also for a fixed axis of rotation. We suspected that it might be applicable to the instantaneous center of rotation, *i.e.*, the velocity pole, and possibly also to the acceleration pole. Neither of these two latter points is a fixed center; the velocity pole has acceleration and the acceleration pole has velocity. Let us now view the system from some suitably chosen moving coordinates. First we take a coordinate system moving at uniform speed with the speed of the acceleration pole P_a. Newton's laws apply directly to this vehicle, and with respect to this vehicle the acceleration pole not only has zero acceleration, but zero velocity as well. It therefore is a fixed center, and **the formula** $M = I\ddot{\varphi}$ **is applicable to the acceleration pole of a system moving in a plane.**

Next we consider a vehicle with zero velocity but with an acceleration $\ddot{x}_P$ equal to that of the velocity pole. With respect to this coordinate system the velocity pole is a fixed axis, because it has neither velocity nor acceleration. Newton's law is applicable in this coordinate system *only* after we have added to the system a set of imaginary forces $-\ddot{x}_P\, dm$. If these forces have a moment about the velocity pole they will affect the angular acceleration, and we will find a different answer for $\ddot{\varphi}$. If however these supplementary forces have no moment about the velocity pole, we find the correct answer for $\ddot{\varphi}$. The supplementary forces are a parallel field $\ddot{x}_P\, dm$, and their resultant passes through the center of gravity. This force has no moment about the velocity pole if the direction of $\ddot{x}_P$ passes through G. **Thus we find that the formula** $M = I\ddot{\varphi}$ **is applicable to the velocity pole of a two-dimensional motion only when the acceleration vector of that velocity pole passes through the center of gravity.**

Another application of the theorems (31a) and (31b) is to the situation inside Jules Verne's projectile traveling to the moon. Looking at this projectile from a terrestrial or absolute coordinate system, we say that the outside shell as well as the passengers and objects inside are all subject to the same acceleration due to the attraction of the earth. Hence everything is floating inside the shell and nothing appears to have weight. Looking at it from a coordinate system moving with the shell, we first establish that the shell has an acceleration towards the earth equal to the local g (which is less than 32.2 ft/sec² at some distance). Then we apply to all objects inside the shell the forces mg toward the earth, being the weight, and the supplementary force mg away from the earth. Hence the objects, to an observer inside the shell, behave as if no forces at all were acting on them.

Even when a physical body is rotating, the theory of this article can be applied. The limitation is that the vehicle or coordinate system should not rotate. Consider for example an airplane moving at high speed through a curve, so that the center of gravity has a centripetal acceleration of $5g$, while the airplane is turning in space.

We can describe the motion with reference to a vehicle or coordinate system with its origin in the center of gravity of the plane and with its *xyz* axes pointing north, west, and up. With respect to this system Newton's laws hold, provided that supplementary forces $5g \, dm$ are applied centrifugally. The center of the plane appears at rest, and the plane turns with respect to our coordinate system. It is only when we insist on choosing a coordinate system with the axis directions fixed to the plane instead of to space that the more complicated theory of the next article must be applied.

Problems 307 *to* 310.

56. Rotating Vehicles; Coriolis' Law. When the vehicle translates and rotates, as in Fig. 264, the total or absolute displacement 1-2' can still be considered to be the vector sum of a vehicle displacement 1-1' and a relative displacement 1'-2'. Again considering those displacements to take place during the short time Δt and letting Δt go to zero, **the absolute velocity is** seen to be **the vector sum of the relative velocity and the vehicle velocity,** even for the rotating vehicle. In Fig. 264 we could have reversed the procedure, and instead of going from 1 to 2' via 1', we could have gone via 2. Still the

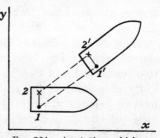

FIG. 264. A rotating vehicle.

above statement holds verbatim, the direction 1-2 is different from 1'-2' and the direction of 1-1' is different from 2-2', but when we go to the limit $\Delta t = 0$, all these distances become small, the directions of 1-2 and 1'-2' come closer and closer together and in the limit coincide.

Thus for velocities, rotating coordinate systems are no more complicated than non-rotating ones, but when we proceed to accelerations we broach a subject that is more difficult than anything we have seen so far in this book. An analytical treatment is apt to hide the physical relations behind mathematical operations. We therefore adopt a geometrical manner of proof, which, although much longer than the analytical one, brings out the physical significance more clearly. The proof will be given for a number of simple special cases, from which the general case will be built up gradually.

We first consider a table rotating at uniform speed ω about its center O (Fig. 265), and a point moving at constant speed v_r along a radial track attached to the table. The path of that point in space will be a spiral curve and the determination of the acceleration not simple. We now look upon the table as the vehicle, so that the origin of the moving coordinate system O is at rest and the axes rotate. The

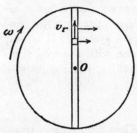

vehicle acceleration is $\omega^2 r$ toward the center; the relative acceleration is zero, because v_r is constant. Therefore, if Eq. (31) (page 298) would apply here, the absolute acceleration would also be $\omega^2 r$ directed centripetally. We can see at once that this is not correct by considering the tangential velocity of our point in absolute space. That velocity is ωr, and it is not constant, because the point moves to larger radii, into a region of greater tangential speed.

FIG. 265. The first special case of the proof of Coriolis' theorem: a rotating table with a radial track.

A point of which the tangential speed increases with time has a tangential acceleration, which Eq. (31) fails to disclose.

Now let us calculate the acceleration of the particle, which is shown again in Fig. 266, in two consecutive positions 1 and 2'. The absolute velocity at point 1 is the vector sum of the relative velocity v_r at point 1 and the vehicle velocity ωr of point 1. The same is true for point 2', but the vehicle velocity there is $\omega(r + \Delta r) = \omega(r + v_r \Delta t)$. The absolute acceleration is the difference between the two absolute velocities divided by Δt. We calculate this difference in components: in the directions parallel to $O1\text{-}2$ and perpendicular to it. The angle $\omega \Delta t$ is small, so that $\sin \omega \Delta t \approx \omega \Delta t$ and $\cos \omega \Delta t \approx 1$, in which terms of the second and higher powers of Δt have been neglected. Then, in the direction parallel to $O12$, we have

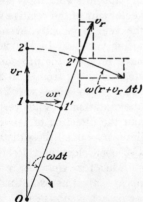

FIG. 266. Toward the proof of Coriolis' theorem. A point moving along a radial track on a rotating table is shown in two consecutive positions 1 and 2' with all its velocity components.

$$\Delta v = [v_r - \omega(r + v_r \Delta t) \, \omega \, \Delta t]$$
$$- v_r = -\omega^2 r \, \Delta t$$

and

$$\frac{\Delta v}{\Delta t} = -\omega^2 r$$

so that

$$\dot{v}_{\text{radial}} = \lim \frac{\Delta v}{\Delta t} = -\omega^2 r$$

In the direction perpendicular to $O12$ we have

$$\Delta v = [\omega(r + v_r \, \Delta t) + v_r \, \omega \, \Delta t] - \omega r$$
$$= \omega v_r \, \Delta t + v_r \omega \, \Delta t = 2\omega v_r \, \Delta t$$

and

$$\dot{v}_{\text{tang}} = \lim \frac{\Delta v}{\Delta t} = 2\omega v_r$$

The absolute acceleration is thus seen to consist of two components: an outward radial one of magnitude $-\omega^2 r$ (which is a centripetal one of $+\omega^2 r$), and a tangential one to the right of $2\omega v_r$. The first of these is the vehicle acceleration; the second one is something new; it is known as the "Coriolis acceleration," after its inventor Coriolis (1792–1843). Thus, we see that this special case satisfies the following rule:

The absolute acceleration is the vector sum of three components: the relative acceleration, the vehicle acceleration, and the Coriolis acceleration. The Coriolis acceleration has the magnitude $2\omega v_{r\perp}$, where $v_{r\perp}$ is the component of the relative velocity perpendicular to the axis of vehicle rotation. The Coriolis acceleration is directed perpendicular to the v_r vector and also perpendicular to the ω vector of the vehicle.

The rule could have been stated much more simply for this special case; however the reader should verify that, as stated, it is correct for Fig. 266, and we shall presently prove that, in the above form, it applies to the most general case as well.

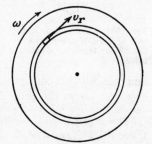

The next simple system to be considered is shown in Fig. 267. Again the vehicle or moving coordinate system is a table rotating at uniform speed ω. Instead of the radial track of Fig. 265, we now have a circular track. Let us imagine a toy locomotive running over this track with constant speed v_r, while the table rotates.

FIG. 267. The second special case in the proof of Coriolis' theorem: a rotating table with a concentric circular track.

Then the absolute velocity of the locomotive is still tangential and equal to $\omega r + v_r$. Its path in absolute space is still the same circle, so that its absolute acceleration is centripetal and of magnitude [Eq. 6a, page 165]

$$\dot{v}_a = \frac{1}{r}(\omega r + v_r)^2 = \omega^2 r + 2\omega v_r + \frac{v_r^2}{r}$$

an algebraic sum of three terms, all directed centripetally.

The first of these three terms is seen to be the vehicle acceleration, or the acceleration of that point of the track just under the locomotive. The last term is the relative acceleration of the locomotive with respect to an observer rotating with the vehicle. The middle (Coriolis) term is extra; it has the magitude $2\omega v_r$ and is directed perpendicular to the relative speed as well as to the ω vector, which, as before, is perpendicular to the table. Thus the result in this case again obeys the general rule.

The third special case to be considered is illustrated in Fig. 268; the rotating table is the same as before, but the track this time is

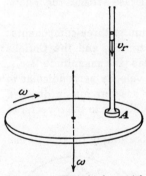

not radial or circular, but perpendicular to the table, and consists of a tube through which the particle is made to move at constant speed v_r. The absolute velocity of the particle in space consists of a vertical component v_r and a tangential one ωr. Only the latter velocity changes with time and is the same as the velocity of point A of the vehicle. Thus the absolute acceleration is equal to the vehicle acceleration only. There does not seem to be a Coriolis term and, by the general rule, there should not be any, because the relative speed is parallel to the axis of rotation

FIG. 268. The third special case of Coriolis' theorem: a perpendicular track on a rotating table.

and has no component perpendicular to it. Therefore, this special case also obeys the general rule.

The reader should now repeat the reasoning for the three special cases (Figs. 265, 267, and 268), dropping the assumption that ω and v_r are constants, and introducing the accelerations $\dot{\omega}$ and $\dot{v}_r$ in addition to the velocities ω and v_r. He should verify that the results in all cases conform to the general rule of page 303.

After the general rule thus has been proved for the three special cases of radial, tangential, and vertical relative velocity, we proceed to a particle of which the velocity relative to the rotating table has all three components simultaneously. The absolute acceleration of that point then in general will have nine components: the relative, vehicle, and Coriolis components for each of the radial, tangential, and vertical cases. The three relative acceleration components add up vectorially to the combined relative acceleration, and the same is the case with the vehicle acceleration. Only for the Coriolis acceleration must we satisfy ourselves that the resultant of the three components still conforms to the rule for magnitude $(2\omega v_{r\perp})$ and direction. Since the case of Fig. 268 has no Coriolis acceleration, and the other two components (for Figs. 265 and 267) lie in the plane of rotation, the resultant Coriolis acceleration is at least perpendicular to the ω vector.

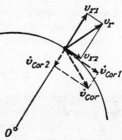

Fig. 269. Compounding the Coriolis accelerations arising from radial and tangential relative speeds.

Figure 269 shows in full lines the two components of relative velocity in the plane of rotation, in dotted lines the corresponding Coriolis accelerations. In each case the accelerations contain the common factor 2ω and are further proportional to the v_r component and perpendicular to it. Then the resultant Coriolis acceleration is also perpendicular to the resultant relative velocity and proportional to it, because the dotted rectangle is similar to the fully drawn one.

Thus the general rule of page 303 is proved for the most general case of a rotating vehicle of which the center point is at rest.

For a non-rotating vehicle of which the origin moves, the general rule of page 303 reduces to the special one of page 298, because the Coriolis acceleration is zero. For a vehicle which not only rotates, but of which the origin moves at the same time, we have a superposition of the two previous cases and the general rule of page 303 still holds, although we will not prove it here.

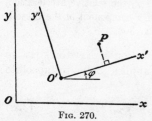

Fig. 270.

The analytical proof for the two-dimensional case is shorter than the geometrical proof just given. In Fig. 270 let Oxy be a coordinate system at rest and $O'x'y'$ be a moving coordinate system. A point P has the absolute coordinates x,y and the relative coordinates x',y', and the relation between these can be found from geometry (see also Fig.197, page 222):

$$x = x_{O'} + x' \cos \varphi - y' \sin \varphi$$
$$y = y_{O'} + x' \sin \varphi + y' \cos \varphi$$

Differentiation gives

$$\dot{x} = \dot{x}_{O'} + \dot{x}' \cos \varphi - x' \sin \varphi \, \dot{\varphi} - \dot{y}' \sin \varphi - y' \cos \varphi \, \dot{\varphi}$$
$$\dot{y} = \dot{y}_{O'} + \dot{x}' \sin \varphi + x' \cos \varphi \, \dot{\varphi} + \dot{y}' \cos \varphi - y' \sin \varphi \, \dot{\varphi}$$

In these expressions $\dot{x}_{O'}$ and $\dot{y}_{O'}$ are the absolute velocities of the moving origin O'; for $\dot{\varphi}$, which is the angular speed of the vehicle, we may write ω. Rearranging the terms somewhat, we write

$$\dot{x} = [\dot{x}_{O'} - \omega(x' \sin \varphi + y' \cos \varphi)] + (\dot{x}' \cos \varphi - \dot{y}' \sin \varphi)$$
$$\dot{y} = [\dot{y}_{O'} + \omega(x' \cos \varphi - y' \sin \varphi)] + (\dot{x}' \sin \varphi + \dot{y}' \cos \varphi)$$

Examining the brackets on the right of the $\dot{x}$ and $\dot{y}$ expressions, we see that they mean the absolute velocities of a point P, when P is fixed with respect to $O'x'y'$. These then are what we have called the "vehicle velocities," by the definition of page 296. The parentheses in the above expressions are the velocities of point P with respect to the coordinates $O'x'y'$. Thus the above equations state in words that the absolute velocity components are the sums of the vehicle and relative velocity components.

Now we differentiate once more. We will do it here only for $\dot{x}$, leaving the similar $\dot{y}$ analysis to the reader.

$$\ddot{x} = \ddot{x}_{O'} - \dot{\omega}x' \sin \varphi - \omega \dot{x}' \sin \varphi - \omega x' \cos \varphi \, \omega - \dot{\omega}y' \cos \varphi - \omega \dot{y}' \cos \varphi$$
$$+ \, \omega y' \sin \varphi \, \omega + \ddot{x}' \cos \varphi - \dot{x}' \sin \varphi \, \omega - \ddot{y}' \sin \varphi - \dot{y}' \cos \varphi \, \omega$$

Rearranging,

$$\ddot{x} = [\ddot{x}_{O'} + \omega^2(y' \sin \varphi - x' \cos \varphi) - \dot{\omega}(x' \sin \varphi + y' \cos \varphi)]$$
$$+ \, (\ddot{x}' \cos \varphi - \ddot{y}' \sin \varphi)$$
$$- \, 2\omega(\dot{x}' \sin \varphi + \dot{y}' \cos \varphi)$$

An examination of this expression shows that the bracket is the x component of the absolute acceleration of point P, when P is fixed to $O'x'y'$, and thus the bracket is the vehicle acceleration. The second line is the x component of the acceleration of P relative to $O'x'y'$, the relative acceleration. The third line upon inspection is seen to be the x component of the Coriolis acceleration, as defined in the general rule of page 303.

Now we are ready to consider the application of Newton's law to rotating coordinate systems. The law applies only to absolute accelerations, or

$$\mathbf{F} = m\dot{\mathbf{v}}_a = m(\dot{\mathbf{v}}_{\text{rel}} + \dot{\mathbf{v}}_{\text{veh}} + \dot{\mathbf{v}}_{\text{Cor}}) \qquad (32a)$$

in which the additions must be understood to be in a vectorial sense. This equation can also be written as

$$\mathbf{F} - m\dot{\mathbf{v}}_{\text{veh}} - m\dot{\mathbf{v}}_{\text{Cor}} = m\dot{\mathbf{v}}_{\text{rel}} \qquad (32b)$$

or in words:

Newton's law applies in a moving coordinate system if only we add to each mass element two fictitious supplementary forces: the vehicle force $-\dot{v}_{\text{veh}}\,dm$, and the Coriolis force $-\dot{v}_{\text{Cor}}\,dm$.

Thus, the term "Coriolis force" means a force equal to the mass times the Coriolis acceleration, directed opposite to that acceleration. It is a fictitious force, of the same nature as the inertia or centrifugal force.

57. Applications. The theory of Coriolis will now be illustrated by applications to

a. Easterly and westerly deviations of projectiles
b. Bending in the arms of a flyball engine governor
c. The man on the turntable
d. The fluid drive of automobiles

a. Easterly and Westerly Deviations of Projectiles. Imagine a vertical mine shaft a mile deep, located near the equator. If a plumb line is hanging in the shaft and a stone is dropped from rest next to it, the stone will not fall parallel to the plumb line, but will deviate in an easterly direction. The reason for this appears in Fig. 271, which shows the earth when looked down upon from the North Pole. The sun appears to us to run from east to west; hence the earth rotates from west to east. Looking on the phenomenon from an outside or absolute coordinate system, we see that the stone, before falling, moves easterly with the peripheral speed of the equator. The bottom of the mine pit moves a little slower, being closer to the center of the earth. When the stone is dropped, the only force acting on it is

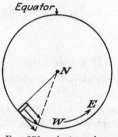

Fig. 271. A stone dropping down a deep mine shaft at the equator.

mg downward, so that the (absolute) acceleration in any direction but the downward one is zero. Hence the stone keeps on going easterly at its original speed and overtakes the bottom of the pit.

An observer on the earth, a rotating vehicle, would reason as follows: If the stone should slide down a purely vertical or radial track parallel to the plumb line, it would go to a region of smaller easterly tangential speed; hence it would experience a Coriolis force to the west from the guide. Since there really is no guide, this force is absent and the stone will deviate towards the east. The Coriolis acceleration is $2\omega v$, where v is the velocity of the stone $= gt$, and ω is the angular speed of the earth: 2π radians/24 hours. Let the y axis

point east, then
$$\ddot{y} = 2\omega g t$$
and integrated twice
$$y = \omega g \frac{t^3}{3} + C_1 t + C_2 = \omega g \frac{t^3}{3}$$

The integration constants C_1 and C_2 are zero because at time $t = 0$ we call $y = 0$ and the easterly speed $\dot{y}$ is also zero. The time t is found from
$$x = \frac{1}{2} g t^2$$
where x points downward. Eliminating the time between the two equations, we find
$$y = \frac{\omega g}{3} \left(\frac{2x}{g} \right)^{3/2} \quad \text{or} \quad y^2 = \frac{8}{9} \frac{\omega^2}{g} x^3$$

If the depth $x = 1$ mile $= 5{,}280$ ft, and
$$\omega = 2\pi / (24 \times 3{,}600) \text{ radians/sec,}$$
we find for the easterly deviation
$$y = 4.6 \text{ ft} \qquad (\text{for } x = 1 \text{ mile})$$

If a projectile is shot straight up in the air at the equator, the Coriolis acceleration is opposite to that of the falling stone; the deviation will be westerly, and at the top of the trajectory there will be a westerly velocity. On reaching the earth again there will be a westerly deviation. The calculation is exactly like that of the falling stone, only $-v = v_0 - gt$, instead of $v = gt$.

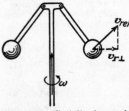

FIG. 272. Coriolis forces in the arms of a hunting engine governor.

If a projectile is shot horizontally from a gun at the equator toward the north or south, there is no Coriolis effect. (Why not?) If a projectile is shot at the equator 45 deg upward to the north, the deviation will be westerly, the same as if it were shot purely upward with 0.707 times its initial velocity. (Why?) A bullet fired horizontally toward the north at 45° northern latitude will reach the ground with an easterly deviation.

b. *Bending in the Arms of a Governor.* Suppose that the balls of an engine governor (Fig. 272) "hunt" or oscillate up and down while rotating at constant angular speed ω. What force does that cause on the arms? Consider the rotating spindle as the vehicle. Then the arms and balls oscillate in a plane containing the axis of rotation of the vehicle and hence the Coriolis forces are perpendicular to that plane. They do not, therefore, influence the motion in the plane of the arms. Referring to Eq. (32b) (page 306), we see that the supplementary force $-m\dot{\mathbf{v}}_{veh}$ is the centrifugal force of the balls. Hence the balls move in their plane under the influence of the centrifugal force, the weight, and the tension force of the arms. The Coriolis force has the magnitude $2\omega v_{r\perp}$ and is directed perpendicular to the

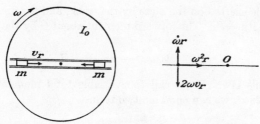

Fig. 273. Shows the accelerations of the arms of the man standing on the turntable of Fig. 249, page 281.

plane. While the balls are going *up*, *i.e.*, in the direction of increasing tangential speed, the arms must push them forward, and by action equals reaction, the balls push back on the arms. Thus the arms are bent in a direction against the rotation while the balls are swinging up, and the arms are bent with the rotation while the arms swing downward. This alternating bending moment has to be taken by the hinges at the top and it becomes quite large for high speeds of rotation ω.

c. The Man on the Turntable. In Fig. 249 (page 281) it was shown that a man on a rotating turntable can increase his angular speed by pulling in his arms. We now propose to examine this question from a somewhat different viewpoint. Consider as the system the turntable plus the man without his arms. This is a rigid body, and when the angular speed of this rigid body changes, a couple must be acting on it. That couple evidently must originate in the moving arms, and in order to understand how, we idealize the system to that of Fig. 273. Here the turntable plus the armless man has the constant moment of inertia I_o, and the arms are replaced by masses m moving inward with constant velocity v_r in a radial track fixed to I_o. For our vehicle we choose the turntable I_o, which has a speed ω and an acceleration $\dot{\omega}$, as yet unknown. One of the masses m is the moving point. The absolute acceleration of this point is the vector sum $\dot{\mathbf{v}}_{veh} + \dot{\mathbf{v}}_{Cor}$, because we have assumed the relative velocity constant. The vehicle acceleration has a radial component $\omega^2 r$ and a tangential one $\dot{\omega}r$, as shown in Fig. 273. The Coriolis acceleration is $2\omega v_r$ and is directed against the rotation, because m moves to a region of diminishing tangential speed. Hence the forward tangential force that must be acting on the mass m is

$$F = m\dot{\omega}r - 2m\omega v_r$$

This force is exerted on the mass by the track, and the mass reacts on the track with $-F$. The forward moment about O by the two masses on I_o thus is

$$M_O = 4m\omega v_r r - 2m\dot{\omega}r^2 = I_o\dot{\omega}$$

by Newton's law. If we call the variable total moment of inertia $I = I_o + 2mr^2$, we can combine two terms.

$$4m\omega v_r r = (I_o + 2mr^2)\dot{\omega} = I\dot{\omega} = I\frac{d\omega}{dt}$$

Also

$$v_r = -\frac{dr}{dt} \qquad \text{and} \qquad dI = 4mr\,dr$$

Substituting, we obtain

$$-\omega\,dI = I\,d\omega$$

and

$$\frac{d\omega}{\omega} + \frac{dI}{I} = 0 \qquad \text{or} \qquad \log\omega + \log I = \text{constant}$$

$$\log\omega I = \text{constant} \qquad \text{and} \qquad \omega I = \text{constant}$$

This result, in which we note that I is variable, is that of page 280, the theorem of conservation of angular momentum. From this example we see that the torque accelerating the turntable is the torque of the Coriolis forces of the inward-moving arms.

d. The Fluid Drive of Automobiles. The most important element of an automobile fluid drive is the *hydraulic coupling* sketched in Fig. 274, consisting of two equal halves, each of which is keyed to a shaft. Each shaft runs in its own bearings and has no mechanical connection with the other. The angular speeds of the two shafts are independent of each other and may have any ratio, positive or negative. On each shaft is keyed a coupling half A,B that can best be described as having the shape of half a doughnut, subdivided into sections by thin radial plates like an orange. The torus-like space of the doughnut is filled with light oil or another liquid, held in place by a cover C, bolted to one coupling half and free to rotate without friction around the other shaft at D. The cover C has no function other than holding the oil inside the coupling. During starting and stopping of the car, the driver and follower shafts (attached respectively to the engine transmission and to the driving wheels of the car) have widely different speeds, but during ordinary running the driving

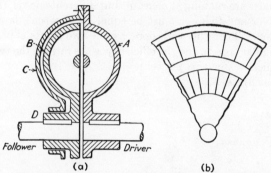

Fig. 274. The hydraulic coupling or "fluid flywheel" of an automobile drive.

shaft has a speed only a few per cent faster than the follower shaft. The torque is transmitted from the driver or engine shaft to the follower shaft by the action of the Coriolis force of the liquid in the doughnut, which will now be explained.

The radial vanes carry the liquid around with shaft speed and thus the liquid is subjected to centrifugal force. This centrifugal force is slightly larger in the driving shaft than in the following one because its angular speed is slightly higher. Thus the liquid trapped in any one of the orange segments will go into a circulatory motion, going outward in the driving shaft and inward in the following one. If a fluid particle moves through a full circle, the centrifugal force will do positive work on it and the oil will accelerate until the friction resistance makes equilibrium with the centrifugal driving force. This occurs in practice with a "slip" of about 2 per cent, *i.e.*, with the speed of the follower shaft being about 98 per cent of the speed of the driver.

Now, the liquid in the driver moves radially outward, into a region of greater tangential speed and has to be accelerated by the vanes of the driver. Thus the moving oil exerts a reactive counter torque on the driver, holding it back. Similarly, in the follower the oil moves inward, to a region of smaller tangential speed, and its inertia pushes the follower forward. Thus the torque is transmitted by action of the Coriolis forces. If the torque demanded by the follower shaft is suddenly increased, as when the car comes to a hill, the follower shaft will slow down, thus increasing the difference in centrifugal forces, thus increasing the circulatory relative speed and with it the Coriolis forces. Therefore the torque transmitted by the oil from the driver to the follower is roughly proportional to the slip.

The device is a torque transmitter of very high efficiency. When the two shafts are running at constant but different speeds, the torques on the driver and follower must be equal by Newton's law of action and reaction. Then the work done by the two shafts is proportional to their angular speeds, and the efficiency is the ratio of the two angular speeds, which is about 98 per cent.

Problems 311 *to* 323.

CHAPTER XVII

GYROSCOPES

58. Theorems on Rotation in Space. In the previous chapters a complete theory of the dynamics of bodies moving in two dimensions has been given, and only occasionally mention was made of three-dimensional motion. Most of the theorems derived heretofore are valid only for plane motion and that fact has been stated explicitly in those cases. In a few instances, however, a theorem happened to be generally true for three-dimensional motion as well, and we succeeded in proving it.

The most notable property so found is the one expressed by Eq. (26a) (page 273), which in slightly different words states that **the motion of the center of gravity of any system (rigid or non-rigid) in three-dimensional space is the same as the motion of a particle, in which the total mass of the system is concentrated, under the influence of all external forces of the system, displaced parallel to themselves to act in the center of gravity.**

This theorem enables us to calculate the motion of the center of gravity of a rigid body under all circumstances where the external forces are known. In order to find the complete motion of a rigid body we still have to determine the rotation of the body about its center of gravity. The theory of the rotation of rigid bodies in the general case is extremely complicated and consequently is outside the scope of this book. In fact, many problems in this category have never been solved yet. However, the object of this chapter is to give a theory of the gyroscope, sufficient to explain and predict the performance of most technically important applications, and this can be done with comparatively simple means. For an understanding of the technical applications, a study of the remainder of this article and of the next one is not essential, so that the reader may proceed directly to page 322. However, in order to indicate what is difficult and what is simple in the general theory we will now state a number of theorems on three-dimensional rotation of rigid bodies. After the statements a discussion follows, in which some of these theorems will be proved and others merely made plausible and illustrated by examples. The twelve statements are as follows:

I. Finite angles of rotation of a rigid body about different axes in space, all intersecting in one point, cannot (repeat not) be compounded vectorially (Fig. 275).

II. Very small angles of rotation of a rigid body about different axes in space, all intersecting in one point, can be compounded vectorially into a resultant angle of rotation about the resultant axis. Hence, if we divide these small angles by a time Δt and go to the limit $\Delta t = 0$, we find

III. Angular velocities of a rigid body about various axes in space, all intersecting in a point, can be compounded vectorially into a resultant angular speed about THE axis of rotation. This theorem will be proved on page 316.

IV. When the moment of a force in space about an axis is laid off as a vector along that axis, and the moments of one such force are so laid off on three mutually perpendicular axes through a point O, then the vector sum of these three vectors represents the moment of that same force about the resultant vector line. The moment of the force about this resultant line is greater than about any other line in space, the resultant line being perpendicular to the plane through the force and the point O. The resultant vector is called **the moment vector M** of the force about the point O.

V. When the angular momentum of a rigid body or of a non-rigid system about an axis is laid off as a vector along that axis, then the vector sum of three mutually perpendicular angular-momentum vectors through a point represents the angular momentum about the resultant vector line. The resultant then is THE angular momentum vector $\mathfrak{M}$ of the system, being greater than the angular momentum about any other line through the point.

It is noted that theorems IV and V are limited to three mutually perpendicular axes, while II and III are true for any number of axes with any angles between them.

VI. A rigid body has within it three mutually perpendicular directions (through its center of gravity or through any other point) for which all products of inertia are zero. These directions are called the "principal axes of inertia" for that point of the body. One of the principal moments of inertia is the maximum moment of inertia for any axis through that point; another principal moment of inertia is minimum; the third is necessarily "intermediate."

VII. The angular momentum of a rigid body about an axis O fixed in space is expressed by $I_o \omega_o$ ONLY if O coincides with one of the three principal axes of inertia of the body or also if O happens to be the axis of rotation. If the axis O is not a principal one and if the angular speed ω_2 about any axis perpendicular to O is not zero, then the angular momentum $\mathfrak{M}$ is **not** equal to $I_0\omega_0$ (see Fig. 282, page 321).

VIII. If a rigid body rotates with speed ω_1 about a principal axis of inertia, and with $\omega_2 = \omega_3 = 0$ about the other two principal axes, **then the angular-momentum vector $\mathfrak{M}$ has the same direction as the angular-speed vector** ω_1 (which is along the axis of rotation).

IX. If a rigid body with two or three different principal moments of inertia rotates about an axis not coinciding with one of the principal axes, then the angular-velocity vector ω and the angular-momentum vector $\mathfrak{M}$ have different directions.

X. The formula $M = d\mathfrak{M}/dt$, **is true about any axis fixed in space** (this is the theorem of page 280); **it takes the form** $M = d(I\omega)/dt$ **ONLY if the axis is a principal axis of inertia of a rigid body or if the rigid body is constrained by bearings to rotate permanently about an axis fixed to the body.**

XI. Applying the scalar formula $M = d\mathfrak{M}/dt$ about three mutually perpendicular axes through a point, and performing vector additions of the three components of the moment **M** as well as of the three components of rate of change of angular momentum $\mathfrak{M}$, which is permissible by theorems IV and V, we find the result

$$\mathbf{M} = \frac{d}{dt}\,(\mathfrak{M}) \qquad (33)$$

This vector equation states that for a rigid body the vector representing the moment of the external forces (either about the center of gravity or about a point of the body that is fixed in space) equals the time rate of change of the angular-momentum vector. The definition of "moment about a point" is given in theorem IV. Equation (33) is the fundamental formula by which gyroscopic phenomena will be explained.

XII. A rigid body free in space without any constraints can rotate permanently only about a principal axis of inertia.

These are the twelve statements. We now proceed to discuss them one by one.

59. Discussion of the Theorems. The first theorem states that finite angular rotations cannot be added vectorially. This is illustrated in Fig. 275, where a book is shown in four positions, each found from the previous one by a 180-deg rotation about some axis; first vertical, then fore-aft, and finally left-right. We see that after these three rotations the book ends up in the original position, so that the resultant rotation is obviously zero. A vector addition of the three rotation vectors shows a rotation of $180° \times \sqrt{3}$ about a diagonal of the cube, which is an incorrect answer. This case is brought up only to show that not every directed quantity can be compounded vectorially and that for each case where it *is* permissible we have to

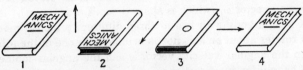

FIG. 275. Three consecutive 180-deg rotations about perpendicular axes bring a rigid body back to its starting position.

prove it. In particular, Fig. 275 is an apology for the tedious proof of theorems II and III, which now follows.

We start in Fig. 276 with a single angular-speed vector, which is drawn in the plane of the paper. We consider a point P at height h above this plane, and of which the projection on the plane is at normal distance n from the vector, so that the distance r between point P and the vector is $r = \sqrt{h^2 + n^2}$. The velocity of the point P is then ωr, perpendicular to its radius r, and this velocity can be resolved into a component ωn perpendicular to the plane of the paper and a component ωh parallel to that plane, which is illustrated in Fig. 276

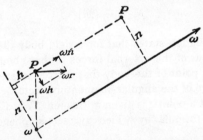

FIG. 276. Shows the velocity components caused by the angular velocity ω vector of a point P lying at distance h above the plane of the drawing.

Now we proceed to Fig. 277, showing two angular speed vectors ω_1 and ω_2 in the plane of the paper. The point P, at distances n_1,h and n_2,h, is under the influence of both rotations simultaneously. By what we just saw in Fig. 276, the velocity of P has four components, two parallel to the plane and two perpendicular to it. The two velocity components of P parallel to the plane are sketched to the left of Fig. 277, and are $\omega_1 h$ and $\omega_2 h$. They add up to a resultant $\omega_{12}h$ in the plane, perpendicular to the resultant angular-speed vector ω_{12}, because the P-velocity parallelogram to the left is similar to the angular-speed diagram to the right, all corresponding lines being perpendicular to each other. The two velocity components perpendicular to the plane are $\omega_1 n_1$ and $\omega_2 n_2$ and add up algebraically, being both in the same direction. Now $\omega_1 n_1$ can be looked upon as twice the area of triangle POA_1, and $\omega_2 n_2$ is twice the area of triangle POA_2. Next we look at these two triangles as having the common base PO and having for heights the perpendiculars dropped from A_1 and A_2 on PO. The sum of the two areas is a triangle with base PO and a

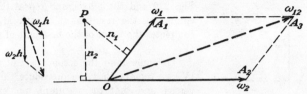

Fig. 277. The vector sum of the velocities of point P caused by ω_1 and by ω_2 separately equals the velocity of P caused by ω_{12} alone.

height equal to the sum of the heights, *i.e.*, the sum of the projections A_1 to PO and A_2 to PO, which is A_3 to PO. Thus the velocity of P normal to the plane of the paper is twice the area of triangle POA_3 which is ω_{12} multiplied by n_{12}, the normal from P on ω_{12}. Thus the two components of the speed of P parallel and perpendicular to the plane are $\omega_{12}h$ and $\omega_{12}n_{12}$, so that by Fig. 276 we can regard the velocity of P as being generated by the single rotation vector ω_{12} only.

The theorems II and III have thus been proved for two rotations. A third angular speed vector ω_3 can be laid through O in Fig. 277, this time not in the plane of the paper. We now lay a plane through ω_3 and the resultant ω_{12}, and repeat the proof. In this manner the theorem is seen to be true for three and hence for any number of angular-speed vectors.

We now proceed to the proof of theorem IV and the closely related theorem V.

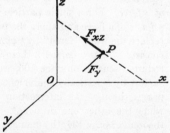

Fig. 278. Toward the proof of theorem IV.

Let in Fig. 278 the three mutually perpendicular directions through point O be x, y, z and let the force F pierce the xz plane in point P. We resolve the force into components F_y parallel to the y axis and F_{xz} in the xz plane. We aim to prove the theorem first for two axes, x and z, proceeding to the third axis later. We note that the component F_{xz} has no moment about either the x or the z axis, since it intersects both. Thus the moment of F about those axes is due to F_y only, and we now turn to Fig. 279, which is the xz plane of Fig. 278 and shows the force F_y perpendicular into the paper at point P. The moments about the x and z axes are $M_x = -F_y a$ and $M_z = F_y b$ plotted with the usual right-hand-screw convention of Fig. 95 (page 104). It is seen in Fig. 279 that the vector sum $(M_x + M_z = M_{zz} = F_y r)$ of these two moments indeed represents the moment of forces F_y about it, so that we have proved that the vector sum of the moments of force F about *two* perpendicular axes, x and z, represents the moment of force F about the resultant vector as axis. For the next step we lay a new plane coordinate system through the $F_y r$ direction of

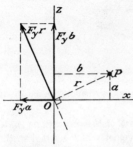

FIG. 279. Vectors representing the moments of a force at P about the x and z axes.

Fig. 279 and the y axis, and repeat the proof, combining the resultant M_{zz} with M_y. This completes the proof of the first part of theorem IV.

To understand the second part of theorem IV, we remark that the procedure can be reversed and that the resultant moment M_{xyz} just found can be resolved again into three moments about three perpendicular axes. In doing this we do not need to take the same xyz system we started from, but can take any other; in this way sweeping through all possible directions in space through point O. Now, the resolution of a vector into three mutually perpendicular components always gives components smaller than the resultant, because the resultant is always the diagonal of the parallelepiped. Thus the resultant just found is the maximum moment the force F can have about any axis in space through point O, and hence the resultant vector must be perpendicular to the plane through the force and O.

Theorem V relates to the moment of momentum instead of to the moment of a force. For a single particle dm of the body we can think of the vector $v\,dm$ as of a force and apply the proof just given. Thus in theorem IV we can replace the word "force" by "momentum of a single particle." If the system consists of many particles there are many small $v\,dm$ vectors. By the procedure discussed on page 106, we now replace the conglomerate of all these vectors by a statically equivalent space cross, one of whose component vectors passes through the intersection of the three mutually perpendicular axes, while the other one does not. Then, by Varignon's generalized moment theorem of page 136, the moment of that last vector about any axis equals the moment of the many $v\,dm$ vectors about that axis.

Therefore theorem V is proved and we recognize from the proof that it is not restricted to rigid bodies, but holds true for swarms of particles or for deformable bodies as well. For the same reason as explained above for a force, the resultant vector has the direction about which the angular momentum of the body is an absolute maximum, and therefore can be regarded as *the* angular-momentum vector of the body. It can always be resolved into three mutually perpendicular components. One particular case of this resolution is when one of these three directions coincides with *the* vector. Then the two other components of the resolution are zero.

The moment vector or the angular-momentum vector can be resolved into components in two or three mutually perpendicular directions only, and not in any number of arbitrary directions, like the angular speeds of theorem III. The physical reason for this becomes clear when we try to resolve the angular-momentum vector into components along two different directions that almost coincide with the vector itself, including angles $d\varphi$ with it. The two components then are about half as large as the original resultant, which is obviously incorrect, because the angular momenta of a system about two axes close to one another ought to differ very little from each other. On the other hand, rotations of 5 rpm and 10 rpm about two axes practically coinciding *do* add up to approximately 15 rpm about the resultant axis.

Theorem VI is an extension of the properties of moments and products of inertia from two- to three-dimensional bodies. A good proof of it usually occupies an entire chapter in the more advanced treatises on dynamics, and therefore will not be given here. As an illustrative example consider a solid rectangular block of side dimensions a, b, and c, (Fig. 280), of which we have seen on page 225 that the three principal moments of inertia are

$$I_{max} = \frac{m}{12}\,(a^2 + b^2),$$

$$I_{min} = \frac{m}{12}\,(b^2 + c^2),$$

$$I_3 = \frac{m}{12}\,(a^2 + c^2)$$

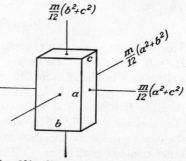

From symmetry we can deduce that the products of inertia about these principal axes are zero (Fig. 198, page 223). The-

FIG. 280. Shows the three principal axes of inertia of a parallelepiped.

orem VI now states that a rigid body of any complicated, unsymmetrical shape has the same properties; in other words, that for any such body it is possible to construct a rectangular parallelepiped (Fig. 280) that has exactly the same moments and products of inertia as the original body, about axes in all possible directions.

Theorem VII is illustrated in Fig. 281. We have seen on page 277

that if a body rotates about an axis O with speed ω_0, its angular
momentum is $I_0\omega_0$. Theorem VII concerns itself with possible additions to this amount caused by the simultaneous rotation ω_1 about
another axis. In Fig. 281 the velocity of a particle dm at P as a

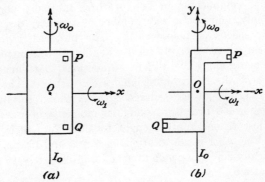

FIG. 281. In (b) the y axis is not a principal axis of inertia; then a rotation about
another axis, like ω_1 about the x axis, contributes to the angular momentum about the
y axis. This is not the case in (a), where the y axis is a principal axis.

result of ω_1 is $\omega_1 y_1$, coming perpendicularly out of the paper. The
momentum is $\omega_1 y\ dm$, and the angular momentum about the y or
I_0 axis is $\omega_1 yx\ dm$. Integrated over the body, the increment in angular momentum about the y or I_0 axis caused by rotation ω_1 about the
x axis is

$$\int \omega_1 xy\ dm = \omega_1 I_{xy}$$

This addition is zero only if the y axis is a principal axis, so that
$I_{xy} = 0$, or if the y axis is *the* axis of rotation, so that $\omega_1 = 0$. Physically we see in Fig. 281a that the contributions to the angular momentum due to ω_1 about I_0 by the elements P and Q cancel each other
(because I_0 is an axis of symmetry and hence a principal axis). On
the other hand in Fig. 281b, where the y axis is not a principal axis,
the contributions by the elements P and Q have the same sign and
hence reinforce each other.

Theorem VIII is a direct consequence of theorems III and V combined, and is mentioned because it serves to clarify the physical picture
of the relationship between the angular velocity vector ω and the
angular-momentum vector $\mathfrak{M}$.

Theorems IX, X, XI, and XII again follow directly from the
previous theorems and require illustration on an example rather than
formal proof. For the example we choose Fig. 282, a cylindrical rotor
of which the shaft center line passes through the center of gravity,

but is offset by angle α from the geometrical center line, which is a principal axis of inertia. This rotor spins with constant angular speed ω about its shaft, in bearings that are solidly anchored to ground. Obviously the angular-speed vector ω is permanently along the shaft center line. In order to find where the $\mathfrak{M}$ vector is, we resolve ω along the two principal directions, which is permitted by Theorem III, so that $\omega_1 = \omega \cos \alpha$, $\omega_2 = \omega \sin \alpha$, and $\omega_3 = 0$. By theorem VII

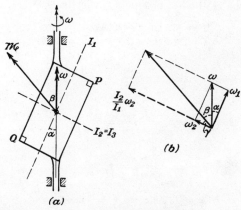

Fig. 282. A rotor turning about a shaft not coinciding with a principal direction of inertia has an $\mathfrak{M}$ vector in a direction different from the shaft ω.

the components of angular momentum are $I_1\omega_1$ and $I_2\omega_2$, or $I_1\omega \cos \alpha$ and $I_2\omega \sin \alpha$. In the example I_2 is substantially larger than I_1; in the figure $I_2 = 5I_1$. Then, by theorem V, we may add these two components vectorially, as is done in Fig. 282b, with the result that

$$\mathfrak{M}_{\text{total}} = \omega \sqrt{(I_1 \cos \alpha)^2 + (I_2 \sin \alpha)^2}$$

The direction of $\mathfrak{M}$ is seen to be entirely different from that of ω, which is what theorem IX affirms.

Now let us apply theorem XI [Eq. (33)] to this situation. From the geometry of Fig. 282 we see that the $\mathfrak{M}$ vector is forced by the bearings to rotate in a conical path. Hence $d\mathfrak{M}/dt$ is not zero and there must be an external moment acting on the body, which must be furnished by the bearings.

This moment is caused, of course, by the centrifugal forces of P and Q, so that the upper bearing pushes the rotor to the left, and the force was calculated on page 216 (Fig. 190). Now we will calculate that force by the method of theorem XI [Eq. (33)].

The end point of the $\mathfrak{M}$ vector is seen to describe a circle in a horizontal plane of radius $\mathfrak{M} \sin \beta$. During time dt the rotor turns through angle $\omega\, dt$ and the end point of $\mathfrak{M}$ displaces a distance $\mathfrak{M}\omega \sin \beta\, dt$. Therefore $M = d\mathfrak{M}/dt$ is a vector of magnitude $\mathfrak{M}\omega \sin \beta$, directed perpendicular to the paper and pointing towards us. This is a counterclockwise couple in the plane of the paper and represents the couple of the two bearing forces on the rotor. We can satisfy ourselves by algebra that this result is the same as that obtained on page 216 by the method of products of inertia.

$$M = \mathfrak{M}\omega \sin \beta = \mathfrak{M}\omega \cos (\alpha + \gamma)$$
$$= \mathfrak{M}\omega[\cos \alpha \cos \gamma - \sin \alpha \sin \gamma]$$

But

$$\tan \gamma = \frac{I_1\omega_1}{I_2\omega_2} = \frac{I_1 \cos \alpha}{I_2 \sin \alpha}$$

Therefore

$$\sin \gamma = \frac{I_1 \cos \alpha}{\sqrt{(I_1 \cos \alpha)^2 + (I_2 \sin \alpha)^2}} = \frac{I_1 \cos \alpha}{\mathfrak{M}/\omega}$$

and

$$\cos \gamma = \frac{I_2 \sin \alpha}{\sqrt{(I_1 \cos \alpha)^2 + (I_2 \sin \alpha)^2}} = \frac{I_2 \sin \alpha}{\mathfrak{M}/\omega}$$

Hence

$$M = \omega^2[I_2 \sin \alpha \cos \alpha - I_1 \sin \alpha \cos \alpha]$$
$$= \omega^2 \frac{I_2 - I_1}{2} \sin 2\alpha = \omega^2 I_{xy}$$

by the equation on the top of page 223.

From this example the truth of theorem XII is immediately recognized. A rigid body rotating permanently about an axis that is *not* a principal one requires bearing forces. Consequently, if the bearings are removed, the body can no longer rotate about that axis. Only when the bearing axis coincides with a principal axis of inertia ($I_{xy} = 0$) can the bearings be removed without interfering with the motion.

Problems 324 and 325.

60. The Principal Theorem of the Gyroscope. By a gyroscope we mean a rigid body that rotates at a very large angular speed Ω about one of its principal axes of inertia, and of which the rotation ω about axes perpendicular to this "gyroaxis" are very slow compared to the main rotation Ω. This definition involves two important simplifications from the general rotating rigid body:

    a. The axis of rotation is a principal axis.

    b. The ratio Ω/ω between angular speeds is "infinitely" large.

As an example consider the motion of a top or of a half dollar that has been set spinning on the table about a vertical diameter. As long as the speed of the top or of the coin is large, it stays upright perfectly and without hesitation; when the spin slows down it starts to "precess" or "gyrate," and just before falling down, the motion becomes extremely complicated. We will call the top or the coin a "gyroscope" only in

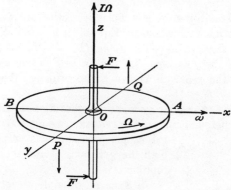

FIG. 283. To derive the principal theorem of the gyroscope from the angular momentum Eq. (33).

the first phase of this process when the motion is quite simple and Ω/ω is very large.

In practically all technical applications the gyroscope is not free, but held in bearings, usually ball bearings, and rotates at as high a speed as is practically possible, of the order of 20,000 rpm. As we have seen in the example of Fig. 282, the bearings will be subjected to forces proportional to $\Omega^2 I_{xy}$, and since Ω is enormously large, these forces become prohibitive even for very small values of I_{xy}, that is, for very small deviations of the axis of rotation from a principal axis of inertia. Therefore gyro rotors for technical work have to be balanced to an extraordinary high degree of precision.

As a consequence of the two simplifications just mentioned, the angular-momentum vector $\mathfrak{M}$ will always be directed along the axis of rotation, and this fact in conjunction with theorem XI [Eq. (33) of page 315] suffices to explain all technically important properties of gyroscopes. Consider as an example a bicycle wheel spinning at high speed Ω about the vertical z axis (Fig. 283), and suppose that this wheel, while spinning fast, is given a very slow rotation ω about the horizontal x axis, by means of two handles, attached to it, so that point P goes down and Q goes up very slowly. We ask what forces have to be exerted on the handles to bring about this slow ω motion.

We reason as follows.   On account of the ω motion the angular-momentum vector $\mathfrak{M} = I\Omega$ turns slowly in the fore-and-aft yz plane. During time dt the angle of rotation of the vector is $\omega\,dt$ and its end point moves forward, in the xy direction, through a distance

$$\mathfrak{M}\omega\,dt = I\Omega\omega\,dt,$$

or at a rate of $I\Omega\omega$ lb-ft-sec/sec.   This then is the rate of change of the angular-momentum vector, and by Eq. (33) it is equal to the moment of the forces acting on the body.   The $d\mathfrak{M}/dt = \mathbf{M}$ vector points in the direction of the y axis, so that the forces $\mathbf{F}$ acting on the system are parallel to the x axis as shown.   **The principal theorem of the gyroscope is**

If a gyroscope of angular momentum $I\Omega$ rotates slowly with speed ω about an axis perpendicular to the gyro axis, a couple $I\Omega\omega$ acts on the

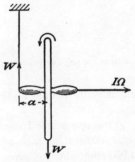

gyroscope about an axis perpendicular to both the gyro axis Ω and the axis of slow rotation ω.

As a second example we consider in Fig. 284 the same bicycle wheel with its axis horizontal, suspended from a string at the end of one handle.   The experiment shows that the wheel does not fall down, as it undoubtedly would if it were not rotating, but moves very slowly in a horizontal plane. The center of gravity does not accelerate appreciably, so that by the theorem of page 313, the resultant force on the wheel must be zero.   The weight W is acting downward;

FIG. 284.   A rotating bicycle wheel suspended from a string precesses slowly around the string.

hence the string force must be W upward and the wheel experiences a moment Wa with its arrow into the paper.   This then is the rate of change of $\mathfrak{M}$, or in other words the end point of the $\mathfrak{M}$ vector moves into the paper by a distance $Wa\,dt$ during time dt.   Physically this means that the wheel axle remains horizontal and the whole wheel rotates about the string in a counterclockwise direction seen from above.   Again we have three mutually perpendicular directions; $\mathfrak{M}$ to the right; $\mathbf{M}$ into the paper and ω upward.   From the principal theorem of the gyroscope we know that

$$I\Omega\omega = M = Wa$$

but we will deduce it once again directly from Eq. (33).   During time dt the end point of $\mathfrak{M}$ moves $Wa\,dt$, but, by geometry, this is also

$\mathfrak{M} \, d\varphi = \mathfrak{M}\omega \, dt = I\Omega\omega \, dt$. Hence

$$Wa = I\Omega\omega \qquad \text{or} \qquad \omega = \frac{Wa}{I\Omega}$$

The slow ω motion is called the *precession*, a term originating in astronomy in connection with the precession of the equinoxes (page 333).

We will now give a second, different, proof of the principal theorem of the gyroscope, which will throw additional light on the physical significance of it. In Fig. 283 we consider a single particle dm of the disk, and calculate its acceleration and consequent inertia force. The components of this inertia force in the xy plane have no moments about the x or y axes and do not interest us, but the z or vertical component is important. The motion of a point dm can be looked upon as the sum of the fast rotation Ω about the z axis and the slow rotation ω about the x axis. Only the latter motion causes vertical velocity of the point dm. When dm is at location P in Fig. 283, its vertical velocity is maximum downward; at Q it is maximum upward; at the intermediate points A and B it is zero. But the rate of change of vertical velocity is maximum at A and B. When passing through A the point dm is coming out of a region of downward speeds and is going into a region of upward speeds, hence its acceleration at A is upward.

There must be upward force $\ddot{z} \, dm$ acting on the particle at A, which is furnished by the surrounding particles of the disk. By action equals reaction, the particle dm at A pushes down on the disk. By a similar reasoning we deduce that a particle at B must push up on the disk, while particles at P and Q are neutral. Integration of these effects over all the particles of the disk leads to the conclusion that these inertia forces exert a clockwise couple on the disk about the

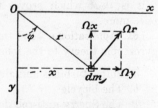

FIG. 285. Proving the principal theorem of the gyroscope by integration of the moments of the inertia forces of the constituent particles.

y axis, which is held in equilibrium by the counterclockwise couple of the two forces F.

To reduce this consideration to a formula, let $\varphi = \Omega t$ be the angle between dm and the y axis (Fig. 285), and let $\psi = \omega t$ be the angle between the disk and the horizontal xy plane.

When we reach the final result of the analysis we will consider $\psi = 0$ (the disk is horizontal), but $\dot{\psi} = \omega$ is never zero, and we cannot assume $\psi = 0$ at the beginning, because if we did that, we could not differentiate. Figure 285 shows the velocity components in the plane of the disk. The component Ωy, parallel to the x axis, has no vertical component; but the speed Ωx points upward at the small angle ψ. Furthermore point dm is subjected to a rotation ω about the x axis. Thus its velocity upward is

$$v_z = \dot{z} = \Omega x \psi - \omega y$$

Differentiating, we find

$$\ddot{z} = \Omega \dot{x} \psi + \Omega x \dot{\psi} - \omega \dot{y}$$

as our final result. Now we are ready to consider the disk horizontal, $\psi = 0$, so that the first term vanishes. In the other two terms we have

$$\dot{\psi} = \omega, \qquad \dot{y} = (\overline{r \cos \varphi})^{\cdot} = -r \sin \varphi \, \dot{\varphi} = -x \dot{\varphi} = -x \Omega$$

so that

$$\ddot{z} = 2 \Omega \omega x$$

The inertia torques about the x and y axes are

$$M_x = \int y \ddot{z} \, dm = 2 \Omega \omega \int y x \, dm = 2 \Omega \omega I_{xy} = 0$$
$$M_y = \int x \ddot{z} \, dm = 2 \Omega \omega \int x^2 \, dm = 2 \Omega \omega I_{\text{diametral}}$$

But by page 226 the diametral moment of inertia is half the polar moment of inertia about the z axis, so that the torque about the y axis is $\Omega \omega I$, which proves the principal theorem of the gyroscope once more.

61. Applications. In this article we discuss applications of gyroscopic theory to the following cases:

a. Ships' turbines and aircraft propellers
b. The bicycle
c. The Sperry anti-roll device for ships
d. The artificial horizon for aircraft
e. The precession of the equinoxes

a. Ships' Turbines and Aircraft Propellers. The rotors of steam turbines for ship drives or the propellers of airplanes are objects of considerable moment of inertia rotating at high speed about a principal axis. When the ship or airplane turns during a maneuver, or rolls or pitches, the direction of the axis of rotation of the turbine or propeller is forcibly altered with the ship's motion and hence a gyroscopic torque appears, which is furnished by extra bearing reaction forces. As an example, we calculate those bearing forces for an especially severe case: an auxiliary steam-turbine rotor consisting of a single Laval disk is mounted with its shaft athwartships in a destroyer. The turbine disk runs at 10,000 rpm; its radius of gyration $k = 8$ in. [see Eq. (12), page 224]; its bearings are 8 in. apart, and the destroyer in very bad weather rolls 45 deg each way from the vertical with a period of 10 sec per complete cycle.

Assuming harmonic oscillation (page 159), we have for the angle of roll, φ, of the ship

$$\varphi = \varphi_0 \sin \frac{2\pi t}{T} \quad \text{and} \quad \dot{\varphi} = \frac{2\pi}{T} \varphi_0 \cos \omega t$$

The maximum angular velocity of the ship about the roll axis thus is

$$\omega = \dot{\varphi}_{\max} = \frac{2\pi}{T} \varphi_0 = \frac{2\pi}{10} \frac{\pi}{4} = 0.5 \text{ radian/sec}$$

The angular velocity Ω of the turbine-gyroscope is

$$\Omega = \frac{\text{rpm}}{60} 2\pi = \frac{\pi}{30} 10,000 = 1,050 \text{ radians/sec}$$

If we denote the mass of the rotor by m lb in.$^{-1}$ sec^2, its moment of inertia is $mk^2 = 64m$ lb-in.-sec^2. The "gyroscopic couple" is

$$M = I\Omega\omega = 64m \frac{1,050}{2} = 33,500m \text{ lb-in.}$$

and the "gyroscopic force" F_{gyro} on a bearing is this amount divided by the bearing distance of 8 in.

$$F_{\text{gyro}} = 4,200m \text{ lb}$$

The static force F_{st} on each bearing, which is the force for a non-rolling ship, is

$$F_{st} = \frac{1}{2}mg = 193m \text{ lb}$$

We see that the gyroscopic bearing force is more than 20 times the static force. The various directions are shown in Fig. 286, where the ω vector, which is the rolling-velocity vector of the ship, points forward. This means that the starboard or right-hand side of the ship is going down in the rolling motion. The turbine rotates clockwise when looking toward starboard. Under these circumstances the reader should satisfy himself that the forces F_{gyro} and F_{st} *from* the bearings *on* the shaft have the directions indicated in the figure. As an exercise the reader should sketch a figure like Fig. 286 for the case of an airplane carrying an engine and propeller when the plane is flying horizontally, making a sharp turn to the right.

b. The Bicycle. We know from experience that a bicycle moving on a straight path is stable in the upright position, although it rests on the ground in two points only. In order to explain this stability, we must show that if the bicycle is inclined at a small angle with respect to the vertical plane, forces come into play that drive it back into the vertical. The explanation is that when the rider feels that he is falling toward the right, he turns his front wheel slightly towards the right.

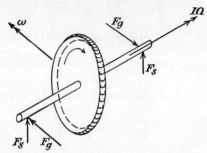

Fig. 286.   A small, auxiliary turbine disk mounted athwartship in a rolling destroyer.

thus guiding the bicycle into a curved path. The centrifugal force set up by this motion, acting on the center of gravity, pushes the bicycle back into the vertical position. Thus the ordinary stability of the device has nothing to do with a gyroscope.

However, we also know by experience that it is possible to ride the bicycle without our hands on the handle bars. In this case the gyroscopic action of the front wheel causes it to move in the same direction as the rider would ordinarily force it to move. The angular-momentum vector $\mathfrak{M}$ of the front wheel points to the left. If the bicycle falls towards the right, the end point of the $\mathfrak{M}$ vector moves upward. Thus the moment vector $\mathbf{M} = d\mathfrak{M}/dt$ points upward, which is a counterclockwise torque seen from above. This is the torque that must be acting on the wheel if it is to move straight forward, which means that the rider if he had his hands on the handle bar would have to exert a couple pushing the wheel to the left while it was going straight forward. When he does not exert that couple, the front wheel will turn to the right, thus giving the centrifugal force its chance to push the bicycle back to the vertical position. This explains why, for stability without our hands on the handle bars, we require greater forward speed than for operation with the use of our hands. It also explains why the motions of the handle bars are unnoticeably small at reasonable speeds, while for very slow speeds stability can be had only with violent and large turns of the front wheel.

c. The Sperry Anti-roll Device for Ships. The Sperry anti-roll device for ocean-going ships consists of one or more very large gyroscopes rotating about a vertical axis in a horizontal plane, mounted in a frame *A* (Fig. 287). This frame is mounted in bearings so that it can rotate (slowly) about an athwartship axis *BB*. The frame *A* carries

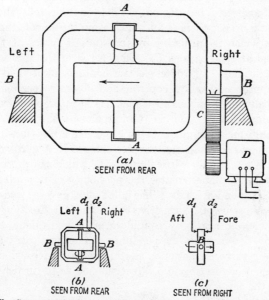

(a)
SEEN FROM REAR

(b)
SEEN FROM REAR

(c)
SEEN FROM RIGHT

Fig. 287. The Sperry anti-roll gyroscope for ships. The large gyro with a vertical axis can be precessed slowly fore and aft by the precession motor *D*, causing a gyroscopic reaction torque on the ship against the direction of the rolling motion.

a segment of spur gearing *C*, meshing with a pinion on the shaft of the "precession motor" *D*.

The angular-momentum vector $\mathfrak{M}$ of the gyroscope is normally vertical, and when the precession motor rotates the frame about *BB*, the end point of the $\mathfrak{M}$ vector tips in a fore-and-aft direction, up to a maximum of about 30 deg on each side of the vertical. Thus the moment of the forces on the gyro frame is about a fore-and-aft axis, and its reaction is taken in the form of vertical forces on the bearings *B*, one up and one down. These bearings are mounted in the ship, and thus we see that a rotation of the precession motor results in a torque on the ship about the longitudinal or roll axis. For proper operation it is necessary to reverse the direction of rotation of the precession motor periodically, in such a manner that the gyroscopic torque exerted on the ship is always directed against the rolling motion. This is accomplished by means of a small "pilot gyro" (Figs. 287b and c), which is mounted in the ship in the same manner as the full-size gyro, but which carries no gear segment and has no precession motor. Instead of this there are two stops, also functioning as electric contacts, preventing the pilot frame A from rotating about BB, except for a small angle of a few degrees. When the ship rolls, the $\mathfrak{M}$ vector of the pilot gyro experiences an athwartship increment, the necessary couple for which is furnished by fore-and-aft forces at the electric contact and at the bearings B. Thus a roll of the ship in alternate directions produces alternate electric contacts at d_1 and d_2, which through appropriate relays control the full-size precession motor D.

Practically the same arrangement has been proposed for a monorail car, which is a railroad car running along a single rail. Without special stabilizing equipment such a car is like a bicycle with clamped handle bars and is obviously unstable sidewise. A gyroscope aboard the car with a normally vertical spin axis, which can precess by ± 30 deg in a vertical plane, creates the possibility of providing a torque about the fore-and-aft axis. The pilot that governs the precession motor now must be made sensitive to the vertical direction and therefore must contain a gravity pendulum or its equivalent. When the car heels over to one side the pendulum makes an electric contact, or closes an air valve, or actuates another relay device, which sets the precession motor rotating in a direction such as to furnish a righting torque on the car.

d. The Artificial Horizon for Aircraft. The Sperry gyro horizon or artificial horizon is an instrument used in aircraft to indicate the horizon when it is not visible, as at night or in clouds. Such an instrument is useful, because the pilot cannot orient himself with respect to the horizon by his sense of balance or by observing a pendulum hanging before his eyes in the airplane. When the plane is flying continuously in a horizontal circle at the proper angle of bank, such a pendulum will hang in a direction perpendicular to the floor of the airplane, and not vertically with respect to the horizon.

The instrument is shown schematically in Fig. 288. It is essentially a free gyroscope of which the gyro disk A rotates at high speed about a vertical axis aa; the inner gimbal ring can rotate about an athwartship axis bb and the outer gimbal ring can rotate about the fore-and-aft axis *cc*, which is tied to the airplane. The inner gimbal ring carries a mechanism *B*, to be explained presently, and counterweights *C*, placing its center of gravity in the *bb* axis, so that the gyro itself and both its gimbal rings are completely balanced. In the instrument the inner gimbal ring is attached to an indicator, which is painted to represent an artificial horizon. When the ship is in level flight, a locking mechanism enables the pilot to lock the gyro disk in a horizontal position, and the artificial horizon appears level. The gyro can then be unlocked, and if the airplane then banks or climbs,

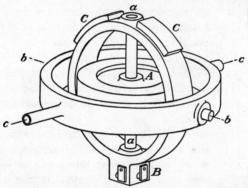

Fig. 288. The gyro horizon for aircraft is a gyroscope mounted in gimbal rings to be free about all three axes.

the gyro disk remains horizontal, so that the artificial horizon appears inclined with respect to the airplane. Only bearing friction will throw the instrument off level after some time. If in cloudy weather the pilot gets occasional glimpses of the horizon, he can lock the instrument in place when the plane is level and unlock it again. But if the horizon remains invisible for a long time, this cannot be done and for this contingency the attachment *B* is provided. The rotor *A* is driven by an air stream, which enters the apparatus through the hollow shaft *cc*, then passes via the outer gimbal ring to the inner one through the hollow shaft *bb*, and from there via *aa* to the rotor wheel, which has some primitive turbine buckets cut on its periphery. The air stream, after having done its work, is discharged at *B* through four horizontal openings 90 deg apart. The detail of this is shown in Fig. 289. The four openings are half shut off by small pendulums. When the shaft *aa* departs from the vertical by rotating about *cc* through a small angle, pendulum *p* will open its hole wider and pendulum *q* will close its hole. The reaction of the escaping air stream thus exerts a couple about axis *bb*, which, by the principal theorem, will turn the gyro about axis *cc*. The reader should reason for himself that the gyro should have its 𝔐 vector pointing downward in order that this reaction torque may make the axis *aa* line up again with the pendulums *p* and *q*. The other pair of pendulums, *r* and *s*, comes into play when the axis *aa* deviates from the vertical by a small angle about *bb*. Thus the axis *aa* tends to seek the "vertical" as determined by the small pendulums.

FIG. 289. Detail of the mechanism *B* of Fig. 288, showing four small pendulums half closing off four air-exit passages.

These pendulums, of course, do not indicate the true vertical except when the plane is flying in a straight path at constant speed. The usefulness of the instrument lies in the fact that the gyro has great inertia and ignores quick changes of the pendulum positions, following only their average inclinations over a considerable time.

A simplified construction of the gyro horizon dispenses with the device *B* and depends for its operation altogether on occasional resetting by the pilot.

e. The Precession of the Equinoxes. The precession of the equinoxes is a phenomenon in astronomy, which was observed thousands of years ago, and was explained by Newton on the basis of gyroscope theory deduced from his own three laws. Casual observation shows that the axis of rotation of the earth practically stands still in space, pointing toward the star Polaris in the sky. More careful measurements at intervals of many years or centuries disclose that the axis of rotation does not stand still but describes a cone, with respect to the fixed stars, which has an apex opening of about 46 deg (the angular distance between the two "tropics" on earth), once around in 26,000 years.

For Newton's explanation of this effect we turn to Fig. 290, showing the earth in midsummer (midwinter in Australia). The earth is attracted by the sun at the right, and the attractive force is held in equilibrium by the centrifugal force of the earth's orbital motion about the sun. The attractive force follows the law $F = C_1/r^2$, and the centrifugal force is $F = C_2 r$ (page 165). When these two forces balance exactly for a particle at the center of the earth C, they do not balance at points at different distances from the sum like A or B. For points in the half earth away from the sun (at night), the centrifugal force is the larger of the two; for points in daylight the attractive force is larger. Thus there is a field of small surplus forces (Fig. 291), being the difference between these two; directed towards the sun at B and away from the sun at A. (A similar condition holds for the moon's attraction and explains why ocean tides are high *twice* in

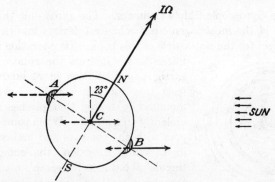

FIG. 290. Centrifugal force *increases* with the center distance r; Newton attraction *decreases* with r; thus, if these two forces balance for a particle at the center of the earth, they do not balance at other points.

24 hours.)   If the earth were an exact sphere, these small forces would have no resultant at all and would form a system in equilibrium.   But the earth is not a sphere; it is an ellipsoid, flattened at the poles.   Thus there is an "equatorial protuberance," and the small forces acting on this protuberance (Fig. 290) have a re-sultant couple, directed perpendicular to the paper.   This couple, by Eq. (33), is the rate of change of the $\mathfrak{M}$ vector, the end point of which will precess toward the reader out of the paper.

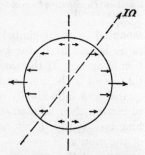

FIG. 291.   The field of small forces being the difference between the gravitational attractive forces and the centrifugal forces.

Six months later the Australians have summer and the sun is to the left of Fig. 290. The reader should sketch the forces for him-self and see that the resultant couple has the same direction as before.   At the equinoxes, in March and September, there is no result-ant torque.   Newton performed the nec-essary integrations of this effect in detail. The period of precession, 26,000 years, was known to him, but the amount of the protuberance, or the difference between the polar and equatorial radii of the earth was not known in his day, so that the calculations could not be checked.   In the eighteenth century, follow-ing Newton's work, the lengths of a degree of arc on the earth near the equator and near the poles were measured accurately by several governmental expeditions, verifying the predicted result completely. *Problems* 326 *to* 331.

62. The Gyroscopic Ship's Compass. The gyroscopic ship's compass is one of the most ingenious mechanical devices known and is a fitting subject for the last article of this book. Its operation is almost

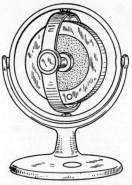

miraculous. It feels the rotation of the earth, one revolution in 24 hours, and it feels this rotation correctly, even when mounted in a ship that pitches and rolls violently in a storm. When someone tampers with the compass and makes it point in a wrong direction, the compass distinguishes between the earth's rotation and the ship's motions, corrects its own error, and after a few hours points to the true north again. Obviously a mechanism that can do all of this cannot be explained in a few words, and in order to understand its operation we consider some preliminary stages before coming to the actual construction of Fig. 300.

Fig. 292. A gyroscopic disk mounted in three well-balanced gimbal rings preserves its direction in space and therefore is a primitive form of compass.

First we imagine a gyroscopic disk mounted in three gimbal rings (Fig. 292). If the base of the apparatus is held fixed, the gyro axis can be made to point in any direction in space. If the three gimbal axes all intersect in one point and that point is the center of gravity of the gyro rotor as well as of each gimbal ring individually, and if the gyro rotor axis is a principal axis of inertia, then no moments act on the gyro rotor. If it is set spinning fast, it will therefore preserve its direction in space, independent of the motion of the base. In a sense this device could be used as a compass: if the gyro axis is pointed toward Polaris, it will stay there. However, there are three things wrong with this kind of compass:

a. It points north, but it is not horizontal, except at the equator.

b. Friction in the gimbal bearings causes small torques, which will gradually throw it off.

c. It is not self-correcting.

If the compass of Fig. 292 were mounted at the equator and pointed north, *i.e.*, parallel to the axis of rotation of the earth, it would stay in that position indefinitely, except for bearing friction. Now let us consider Fig. 293, which represents the earth as seen by an observer in space far above the North Pole. The earth rotates from west to east once every 24 hours. If the compass of Fig. 292 is mounted in position 1 at the equator, pointing east and horizontal, it will find itself 6 hours later in position 2, pointing vertically up; after another 6 hours, in position 3 pointing west, and finally in position 4, pointing down. After this is understood, we imagine our compass in Fig. 293 started at position 1, not pointing east as shown, but practically north with a small easterly

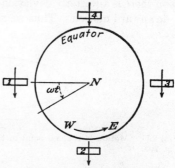

Fig. 293. The earth seen from far above the North Pole.

error. Then at 2 we have a small upward error, at 3 a small westerly error, and at 4 a small downward error.

Next we proceed to the construction of Fig. 294, which is closer to the actual compass than Fig. 292. The gyroscopic disk is still mounted in the same three gimbal rings, but now a pendulous mass has been added. When the gyro axis is horizontal this pendulous mass has no influence, but as soon as the gyro axis has an upward deviation, the

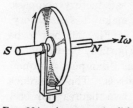

Fig. 294 A gyroscopic disk carrying a gravity pendulum is the second step in the explanation of the marine compass.

pendulum exerts a torque, which drives the end point of the $\mathfrak{M}$ vector to the west. Similarly when the $\mathfrak{M}$ vector points north and is dipped down slightly, the pendulum torque drives the top of the $\mathfrak{M}$ vector toward the east.

Now imagine that a sheet of paper is held perpendicular to the $\mathfrak{M}$ or gyro axis at a distance of several feet north of the compass, and that we mark on this paper the point where the gyro axis intersects it.

The result is Fig. 295, showing a coordinate system (east-west; up and down) with an origin O, and the coordinate axes marked in degrees. If the compass (Fig. 294), still at the equator, is started with an easterly deviation of 4 deg from the true-north horizontal position, the gyro axis intersects the paper at point A. The rotation of the earth, by Fig. 293, introduces an upward deviation, so that the axis moves to point B of Fig. 295. Then the pendulum torque takes hold and drives it to the west. We thus reach point C, where the tangent to the curve is horizontal, because there is no longer an easterly deviation at that point. The pendulum torque continues its westward push, so that soon there is a westerly deviation, which by the earth's rotation causes a downward motion. Thus we arrive at D, then at E and back to A.

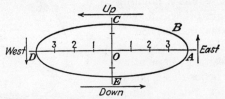

FIG. 295.   Motion of the end point of the angular momentum vector of the compass Fig. 294.

The period of this elliptic motion depends on the intensity of the pendulum torque, and in an actual compass it is so adjusted that one full cycle takes 84 minutes, which is the period $T = 2\pi \sqrt{R/g}$ of a simple pendulum with a length equal to the radius of the earth. The practical reason for this particular relation is too complicated for explanation in this text.

Thus we see that the compass of Fig. 294, when placed at the equator with a deviation, will move around the desired true-north horizontal

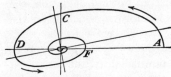

FIG. 296.   Modification of the motion diagram Fig. 295 by the introduction of damping.

position, but will never reach it: it is not yet a satisfactory compass. In order to make it satisfactory, the diagram of Fig. 295 has to be modified to Fig. 296, by the introduction of a "damping torque" into the compass. The difference between the two figures can best be seen at point C. In the first figure the curve there is parallel to the east-west axis; in the second figure it is not. In the first figure a deviation causes a reaction pushing the compass in a direction 90 deg from the deviation. In the second figure this is still so, but in addition there is a reaction diminishing the original deviation directly. At point C in Fig. 296 the deviation is upward only, but the curve at C has a westward as well as a downward slope. In this manner the ellipse of Fig. 295 is transformed into the spiral of Fig. 296. The latter diagram describes a satisfactory compass, because a deviation gradually corrects itself.

In order to introduce the damping torque into the mechanism, Fig. 294 has to be modified radically into Fig. 297 in which the gyro rotor

A is mounted in a casing B, which is held in a gimbal ring C by means of two hinges. The ring C is suspended from a piece D by means of a flexible wire or very thin rod. This part D is known as the *phantom element*. It in turn carries the pendulum E, which thus is not carried by the gyro rotor directly. The pendulum E is coupled to the rotor casing B by means of a pin F, which is drawn in eccentrically, but which for the moment may be considered to be just in the vertical center line.

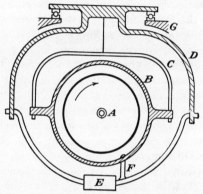

Fig. 297.   Third step in the explanation of the gyroscopic marine compass.

The phantom element rests on a piece G through a ball bearing. The piece G is carried through a complete set of gimbal rings by the ship.

The connection between E and B through the pin F is such that the pin is rigidly built into E but rides in a long slot of B, such that the force between F and B can only be perpendicular to the paper. If for any reason the assembly ABC tends to swing to the right in the plane of the paper, this can take place without any restraint on account of the slot. If on the other hand ABC tends to swing out of and into the paper with respect to EF, the pin couples the two motions.

Then there is another thing yet. If the assembly ABC turns (about a vertical axis) with respect to D, twisting the supporting wire, an electric motor is started which turns the phantom D on its ball bearings on G so as to set it in line with ABC again. Thus the phantom always follows the rotations of ABC about the vertical axis and no twist in the wire can be permanent. This turning is regulated by means of a pair of electric contacts and is so sensitive that the following action is started as soon as the twist between C and D is $\frac{1}{4}$ deg.

It is easy to see that the compass with the *central* pin F acts exactly like Fig. 294. When it is deviated east, the earth tips the point of the $\mathfrak{M}$ vector of A upward. Then the casing B pushes against the pendulum pin F, and a torque is created on A about a horizontal axis passing through the two bearings of B. This torque has exactly the same sense as the gravity torque of Fig. 294. The apparatus, although much more complicated, accomplishes exactly the same thing as Fig. 294.

But it is now capable of introducing a *damping* as specified in Fig. 296, simply by setting the pin F somewhat to the right shown in Fig. 297. When that is done, the force between F and B (which is perpendicular to the paper) not only gives a moment about the horizontal axis but also about the *vertical* center line AE. In the case of Fig. 297 the main $\mathfrak{M}$ vector points into the paper away from the reader. When the point of that vector is deflected upward, the casing B pushes the pin F into the paper and consequently F pushes B out of the paper. This force on B gives a couple about the vertical center line represented by a vector pointing downward. Thus an *upward* deviation of the compass causes a *downward* precession as desired.

The pendulum E of Fig. 297 can be made of an effective length of 6 in. or so and thus has a natural frequency of several cycles per second. If the ship rolls (with a period of 5–25 sec., the larger value for larger ships), it is clear that the pendulum E will follow the roll and react on the compass. Whether this is serious or not we do not discuss; let it suffice that under certain conditions it may cause wrong readings. Therefore it is desirable to have a pendulum E that will *not* follow the roll of the ship. This can be done by making the natural period of the pendulum very long, say 5 minutes. For that it is necessary to make E extremely long, about 250,000 ft. Evidently this is mechanically impossible, but first it should be explained that if such a pendulum existed, it would not respond to the roll of the ship with a period of from 5 to 25 sec.

Consider a pendulum of 3-ft length (with a period of about 1 sec). Apply to its bob a force alternating at the rate of say 30 times per second. The bob is so inert that it cannot move fast enough for this force; it will practically stand still. The conditions with a 250,000-ft pendulum (with a period of 5 minutes) acted upon by a force of a period of 10 sec or so are similar. Such a pendulum is known as a *ballistic* pendulum. The name "ballistic" is frequently applied to apparatus that is so sluggish that it cannot respond to quick impulses (for example, the ballistic galvanometer).

Now a pendulum of 250,000-ft length cannot be made. Therefore we abandon the pendulum and go to something else that will accomplish the same purpose. In Fig. 298 is shown a bicycle wheel pivoted about its center C. It carries a glass vessel attached to it in such a manner that its center of gravity is located in C. The vessel is partly filled with water such that the center of gravity of the water also coincides with C. Evidently the wheel is now in equilibrium. Let us tilt it to the left. Then water will flow from the vessel R to the vessel L, and the center of gravity of the water mass is now to the left of C. There is a moment tending to tip the wheel still farther to the left. The wheel is *unstable*. If a solid pendulum is turned a small angle from its equilibrium position, a couple is set up tending to restore the pendulum to its original position of equilibrium.

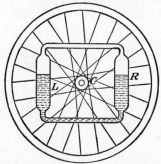

Fig. 298. A bicycle wheel with communicating vessels partly filled with a liquid tends to be unstable, like a ship with shifting cargo. It is a model of the "mercury ballistic."

There is a *positive restoring couple*. In the case of the wheel of Fig. 298 there is a *negative restoring couple*. If a mass were attached to the bottom of the rim of the wheel, that would give rise to a positive restoring couple, which may or may not be larger than the negative couple of the water for the same angular displacement. Then the combination is stable in the first case and unstable in the latter case.

Now we modify Fig. 297, by eliminating the weight of the pendulum E, and replacing it by the vessels of Fig. 298. Two vessels filled with mercury are solidly attached to the pendulum ring E, one in front of the paper and the other behind it. These two reservoirs are connected to each other by a tube of very small diameter, running perpendicular to the paper. This is done in such a manner that the center of gravity of the ring E plus the empty vessels lies in the horizontal bearing axis of E. Moreover the center of gravity of the mercury (when in the equilibrium position) also lies in the bearing axis. On account of the small diameter of the connecting tube, the mercury flows very slowly from one vessel to the other when tilted. In fact during a rolling period of the ship (approximately 10 sec) the transfer of mercury between the two vessels is insignificantly small. In that case the ballistic acts as a solid body, and no restoring torque is exerted, because the pendulum E is in indifferent equilibrium. For very much slower motions however (like the rotation of the earth) the mercury can flow freely from one vessel to the other, and in that case produces a negative restoring torque, which will precess the gyroscope to east or west in the same manner as the pendulum of Fig. 294 or 297. Only the direction of the precession is reversed, so that the $\mathfrak{M}$ vector now must be made to point south instead of north.

Figures 299 and 300 are reproduced from a pamphlet of the Sperry Gyroscope Company and represent the two compasses both

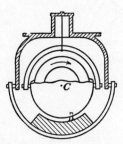

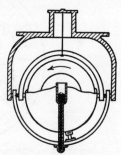

FIG. 299. The gyrocompass with a phys-ical pendulum.

FIG. 300. The final form of the Sperry gyrocompass, incorporating a mercury bal-listic instead of a pendulum.

seen from the south. In the pendulum compass the gyro rotation is clockwise giving an $\mathfrak{M}$ vector into the paper toward the north, whereas in the mercury ballistic compass the rotation is counterclockwise giving an $\mathfrak{M}$ arrow towards the reader, towards the south.

This explains the principal features of the gyroscopic marine com-pass when operating at the equator. The behavior of the compass at latitudes different from the equator is similar, but its detailed explanation would lead us too far. The reader is referred to a book entitled "The Theory of the Gyroscopic Compass,"[1] written by A. L. Rawlings, the inventor of the mercury ballistic of Figs. 298 and 300.

Problems 332 *to* 334.

[1] The Macmillan Company, New York, 1944.

PROBLEMS

1. In the figure, let the horizontal distance between the two pulley centers be $2l$, let the pulleys be at equal height, of negligible diameter, and let the height of the knot A under this horizontal line be h. Let further $W_1 = W_2$, while W_3 is different. Derive a formula showing the relation between the ratio h/l and the ratio W_3/W_1, and plot a graph of this relation.

2. If $W_1 = 3$ lb, $W_2 = 4$ lb, and $W_3 = 5$ lb, plot the position of the string graphically, and find the angle P_1AP_2.

3. If W_1 is made 10 lb and we want to make $\angle P_2AR = 30°$ and $\angle P_2AP_1 = 90°$, what are the weights W_2 and W_3?

4. Make $W_1 = 20$ lb, $W_2 = 10$ lb, and $W_3 = 25$ lb. By graphical construction and using a simple protractor, find the angles of the string.

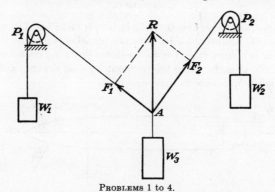

PROBLEMS 1 to 4.

5. The four strings of a violin are strung over the bridge B at an angle of about 20 deg. The pressure of the bridge on the front panel of the fiddle is about 20 lb.

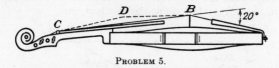

PROBLEM 5.

a. Assuming all strings to have the same tension, what is that tensile force?

b. The free length BC of the string is about 12.5 in. What is the force required to pull up one string $\frac{1}{4}$ in. at a point D midway between B and C, assuming that the force in the string is not changed by it?

341

6. A square plate of 1-ft side is subjected to five forces, applying at the corners, the mid-points, and quarter-length point of the sides, as shown. Determine the location and magnitude of the resultant.

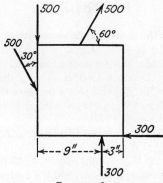

PROBLEM 6.

7. A twin-screw ocean liner arriving in port is being eased to dockside by two tugs pushing against the side, while one propeller is pushing forward and the other one is pulling aft. The forces shown in the diagram are expressed in tons of 2,000 lb each. Find the resultant force on the ship and the point where this resultant intersects the center line.

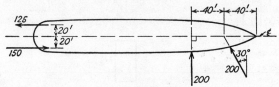

PROBLEM 7.

8. A heavy barge in a canal is being pulled by two towlines, powered by horses, and is pushed from the aft end by a man handling a long pole. One horse is stronger than the other; all forces are at 30 deg with respect to the center line of the barge. Determine the resultant force.

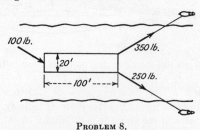

PROBLEM 8.

9. A weight W of 100 lb. is supported by a system of wires as shown. Find the tension T_1, T_2, T_3, and T_4 in the wires of the system.

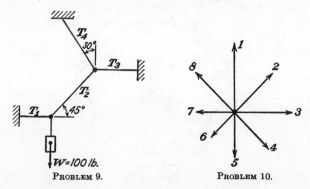

PROBLEM 9.   PROBLEM 10.

10. In the design of an eight-cylinder Diesel engine it is necessary to evaluate the resultants of a star of vectors of different magnitudes but equiangularly spaced. Sometimes these vectors are negative, *i.e.*, pointing in a direction opposite to that shown in the diagram. Find the resultant for the two following cases:

No.	1	2	3	4	5	6	7	8
Case *a*	10	9	8	7	5	4	3	2
Case *b*	10	4	2	−2	−4	0	5	2

11. A boom AB, hinged at the bottom, is 10 ft long. At B a rope pulls at 30 deg with a force $P = 1,000$ lb.

   a. What is the moment of force P about the point A?

   b. What is the moment of the weight force W about A?

   c. If these two moments are alike, what is the weight W?

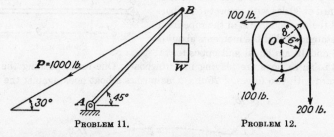

PROBLEM 11.   PROBLEM 12.

12. Two wheels, one of 8-in. radius and another of 6-in. radius, are rigidly attached to each other concentrically. There are three forces acting on it, all tangentially as shown.

   a. What is the moment of the three forces about the center O?

   b. What is that moment about point A?

13. An old-fashioned overshot waterwheel carries on its periphery wooden buckets, which are filled with water on the descending side and are

PROBLEM 13.

empty on the ascending side.   Let the total number of buckets be 16 and the weight of the water in the buckets be as follows:

No.	1	2	3	4	5	6	7	8	etc.
Weight (lb)	0	50	75	75	75	50	25	0	etc.

Further let these weight forces be acting at a radius of 8 ft from the axle and at the angles:

No.	2	3	4
Angles (deg)	10	$32\frac{1}{2}$	55,

and so on, each subsequent weight being 22.5 deg further along the periphery. Calculate the moment exerted by the water weight forces about the axle.

14. A weightless horizontal bar is pivoted at one end O and carries a load $W = 1,000$ lb at quarter length. The bar is held in place by a rope at the end, inclined at 30 deg with respect to the bar. Find the tensile force P in the rope from the statement that the moments about O of P and W are equal and opposite.

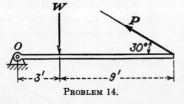

PROBLEM 14.

15. Three boys are playing in a rowboat.   One rows, pushing the boat forward with two forces of 20 lb.   One pushes a boat hook against the shore

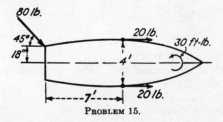

PROBLEM 15.

with a force of 80 lb, and the third one push-pulls against a dock exerting a torque of 30 ft-lb. Find the magnitude, location, and direction of the resultant force.

16. On an elevator hoisting drum of 2-ft diameter is acting a tangential force $P = 1,200$ lb (the weight of the elevator cab) and a couple $M = 1,200$ ft-lb (the torque in the shaft of the electric drive motor). What is the magnitude and the location of the resultant force acting on that drum?

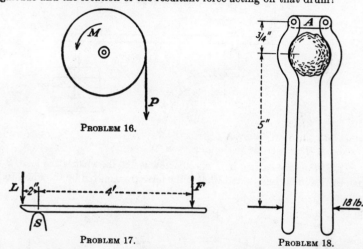

PROBLEM 16.

PROBLEM 17.

PROBLEM 18.

17. A crowbar finds a solid support at S; a load L is to be lifted by a man's push at F. If that force F is 100 lb, what load can be lifted for the dimensions shown?

18. A nutcracker is pinched together with a pair of forces of 18 lb each.

a. What is the force exerted on the nut?

b. What is the force in the link A between the two levers of the nutcracker?

19. A scale consists of a horizontal bar, carrying a weighing pan at one end, and a weight W_2 capable of sliding along the long end of the bar. With an empty pan ($W_1 = 0$) and the weight W_2 removed, the bar is in horizontal equilibrium.

a. When weighing an unknown quantity W_1, what is the relation between x and W_1?

b. For $W_2 = 1$ lb, and $a = 1$ in., $b = 12$ in., what is the maximum capacity of the scale, and how is the long arm to be marked?

PROBLEM 19.

These scales were extensively used by the Romans; they were excavated at Pompeii and are on display in many museums.

20. A safety valve on a steam boiler has a diameter d and a steam pressure p. The horizontal arm itself weighs w lb, concentrated at distance c from the valve.

a. What is the value of W required?

b. Substitute the numerical values: $p = 200$ lb/in^2., $d = 2$ in., $a = 12$ in., $b = 3$ in., $c = 5$ in.

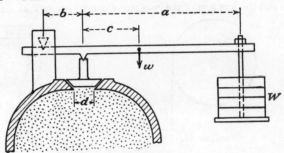

PROBLEM 20.

21. If the load carried in a wheelbarrow is 120 lb, what is the load P the man carries in his hands, and what is the force transmitted by the wheel to the ground?

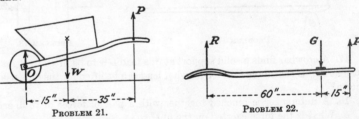

PROBLEM 21.   PROBLEM 22.

22. An oar in a rowboat, when being pulled, does not really stand still; its end slips through the water somewhat. However, the motion through the water is very slow, and to all intents and purposes the end of the oar finds a point of solid support in the water. If the oarsman pulls with a force of 50 lb, what is the force exerted by the water on the oar tip? And what is the force exerted by the oarlock on the boat?

23. What is the force P required to keep the double lever system in equilibrium under a load of 300 lb? Neglect the weight of the levers themselves.

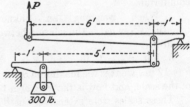

PROBLEM 23.

24. A machine element in common use is the bell crank, so named because our grandparents' doorbells were operated by it.

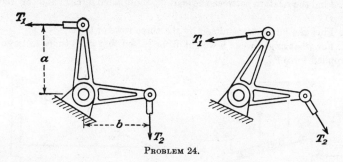

PROBLEM 24.

a. If the two arms are perpendicular to each other and the pulls T_1 and T_2 are perpendicular to their arms, what is the relation between T_1, T_2, a, and b for equilibrium?

b. In case these angles are all different from each other, state which dimensions of the figure should replace a and b, if the answer to the previous question is to apply here without change.

25. A ship's ladder is supported at the top by a hinge H and at the bottom by a rope with tension T pulling at 30 deg with respect to the vertical. The weight of the ladder is 1,000 lb and is considered to be concentrated at the center. A man weighing 200 lb stands at one-fifth distance from the bottom. The ladder itself is inclined at 45 deg. Calculate the pull in the rope and the force at the hinge.

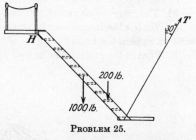

PROBLEM 25.

26. A jeep is being used to pull out stumps. Find the vertical load on each axle when $P = 1,500$ lb, and the weight of car and driver is 2,500 lb.

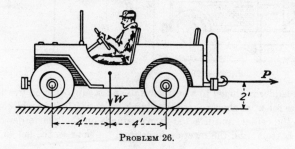

PROBLEM 26.

27. The figure shows how a weight can be raised by an ideal frictionless wedge.

   a. Find the relation between the load W, the applied force F, and the wedge angle α.

   b. Find the forces transmitted by the three rows of balls.

   c. For what angle α can a weight W be raised that is 10 times as large as the applied force F?

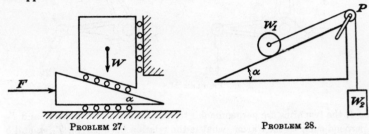

PROBLEM 27.   PROBLEM 28.

28. The weight W_1 is supported on a frictionless inclined plane α by a flexible cable, which passes over the frictionless pulley P and is fastened to the hanging weight W_2.

   a. Find the general relation between W_1, W_2, and α.

   b. For $\alpha = 30°$ and $W_1 = 50$ lb, what is W_2?

29. The crane truck shown weighs 7,000 lb.

   a. What is the largest purely vertical load L_1 the crane can safely handle without tipping the truck?

   b. What is the largest 30-deg load L_2?

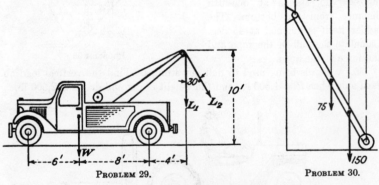

PROBLEM 29.   PROBLEM 30.

30. A ladder is hinged to a vertical wall and is supported on the floor by a frictionless roller, as shown. If the ladder's weight of 75 lb can be considered to act at the center, and if a 150-lb man stands on the quarter-length point from the bottom, find the reaction force from the floor on the ladder.

31. A toggle joint is an element of mechanism whereby a large force P can be exerted by a much smaller force Q. Analyze this system and find the relation connecting α, P, and Q, if friction can be neglected.

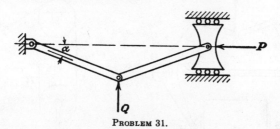

PROBLEM 31.

32. A heavy uniform rope, 20 ft long and weighing 20 lb, is slung over a smooth pulley 1 ft in diameter and carries the weights $W_1 = 10$ lb and $W_2 = 20$ lb. Find the equilibrium position of the system in terms of the distance x.

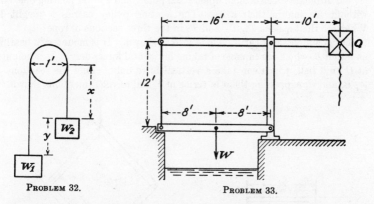

PROBLEM 32.   PROBLEM 33.

33. The Dutch canal bridge carries a counterweight Q of 3,000 lb. The weights of all the members can be neglected in comparison with the weight of the bridge deck itself, which is 4,000 lb.

a. Find the necessary pull in the counterweight rope in order to just begin lifting the bridge.

b. What is the pull in the rope when the bridge deck is up 30 deg?
Neglect all friction.

34. The lifting tong shown is suspended from a chain attached at D. The pivot blocks G and H grip the load W, supporting it by means of friction. The bell cranks CAG and EBH pivot about the pins A and B on the crossbar. Assuming no friction in the pivots and neglecting the weight of the mechanism, find the force in the crossbar AB and the gripping force on the load.

The bellcranks are symmetrical about their centers; the lines CG and EH are vertical, and the dimensions are $CG = 40$ in., $CD = DE = 18$ in., $DF = 3$ in.

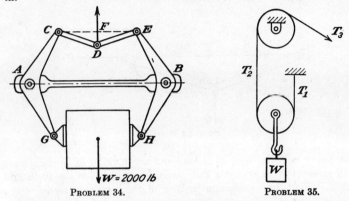

PROBLEM 34.   PROBLEM 35.

35. A tackle consists of an upper fixed pulley and a lower floating one, both without friction in their axles. The floating pulley carries a weight W. What are the tensions T_1, T_2, and T_3 in the three sections of rope?

36. A simple hoist carries a load P lb as shown. It is mounted in bearings at A and B, which are capable of taking horizontal forces only. It is mounted at C on a ball, which can take a vertical force only. Find the reactions at A, B, and C, expressing them in terms of P and of the dimensions shown.

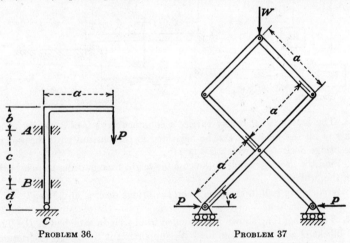

PROBLEM 36.   PROBLEM 37

37. The diagram shows a simplified lazy-tongs linkage, of which the members are supposed to be weightless and pinned together by frictionless joints. Analyze the system and determine the ratio P/W.

38. Two 20-ft beams are laid across three supports A, B, and C as shown. What is the largest load X that can be applied to the left beam without upset-

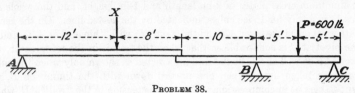

PROBLEM 38.

ting the system when the right beam is supporting $P = 600$ lb? Neglect the weight of the beams.

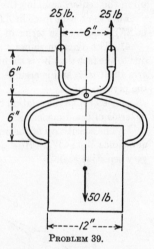

39. The iceman cometh, carrying a 50-lb cake with ice tongs, having 25-lb vertical loads in each of his hands.

a. Calculate the normal force with which the points of the tongs dig into the ice.

b. What force is transmitted by the pin connecting the two arms of the tongs?

PROBLEM 39.

40. The iceman is getting out of date. For us moderns ice grows in little cubes. The sketch shows schematically a lever system used in one of the standard home refrigerators for cracking ice cubes out of the container. The

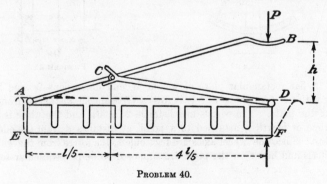

PROBLEM 40.

piece $ADEF$ is an aluminum plate about 1 in. high, 8 in. long, and of negligible thickness. It has seven slits cut in it almost to the bottom. In these slits fit aluminum cross plates of 2-in. length and 1-in. height, and the whole sits loosely in a tray for 16 ice cubes, indicated by the dotted line. On the center plate is attached the lever AB and the strut CD, with "hinges" at A, C, and D. The pivot at C is collapsible so that when the tray is in the refrigerator, the levers AB as well as CD lie flat and are erected as shown only when the cubes are to be taken out. When one presses down with the thumb on B, the strut CD is put in compression. This causes tension in the "rod" AD, which is the top line of the center plate $ADEF$. Since EF cannot stretch, but AD can stretch (on account of the slits), the entire plate $ADEF$ bends and humps up in the center, cracking the ice cubes loose.

Find the tensile force in AD (and hence the equally large compressive force in EF), expressed in terms of P, l, and h.

41. The diagram shows a scale, which is in balance in the position shown due to the fact that W_1 is slightly heavier than W_2. The weight of the balance arm itself is w and is assumed concentrated at a point G at distance a below the pivot point O.

a. Determine the relation between the angle α and the difference $W_1 - W_2$ in terms of the dimensions.

b. The sensitivity of the scale is great when the angle α is large for a small unbalance $W_1 - W_2$. How do the dimensions have to be chosen to make a very sensitive scale?

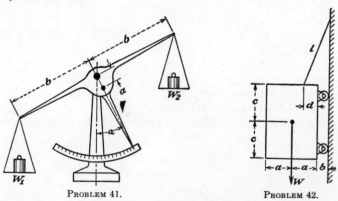

PROBLEM 41.   PROBLEM 42.

42. A flat rectangular plate of weight W and dimensions $2a$ and $2c$ is resting with two frictionless wheels against a vertical wall, distance b from the edge of the plate. The wheels are distance $2e$ apart, and the plate is held by a rope of length l, attached to a point at distance d from the edge of the plate. Since the wheel axles are frictionless, the forces from the wall on the wheels and hence on the plate are supposed to be perpendicular to the wall.

a. Find the rope tension and the two wheel reaction forces.

b. For what relation between the dimensions will the lower wheel leave the wall?

43. What horizontal force P is necessary to pull an 18-in., 300-lb lawn roller over a 1-in. plank?

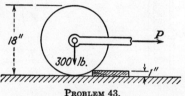

PROBLEM 43.

44. A bascule bridge consists of a tower AB, a bridge deck pivoted to the tower at A, two link bars BC and CD, pivoted together at the ends, carrying a counterweight Q. Let the weight of the bridge deck be W, concentrated at a point as shown; and let the weights of the link bars themselves be negligible. The counterweight is a large block of concrete having a specific weight of 2.5.

a. What volume of concrete at Q is required to keep $W =$ (short) 10 tons in equilibrium?

b. Answer the same question for the case that the bridge deck is tipped up 30 deg.

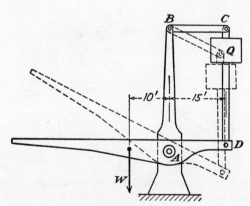

PROBLEM 44.

45. A set of meshing spur gears of radii r_1 and r_2 is mounted in bearings B_1 and B_2, which are bolted to the ground by means of a frame. A moment or torque T_1 is applied to the pinion shaft, and the system is held at rest by another torque T_2.

Calculate T_2, the force between the pinion tooth and the gear tooth, the forces between the shafts and the bearings B_1 and B_2, and finally the moment transmitted by the frame to the ground.

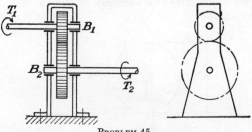

PROBLEM 45.

46. The weighing platform of the Toledo scale ("No Springs; Honest Weight for a Penny; Your Money Back if You Guess Right") is supported at four points A by two levers L_1 and L_2, which are pivoted about horizontal axes

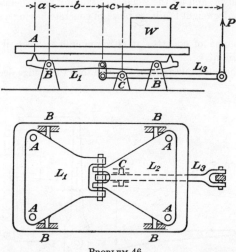

PROBLEM 46.

BB. The levers L_1 and L_2 (roughly triangular in shape) are pivoted together in their center and transmit their combined force through a short vertical link to a third lever L_3, pivoted at C.

a. Study the system and explain how the combination of levers acts as an "equalizer" in the sense of Figs. 26 and 27 (page 30), so that the force P is independent of the location of the load W on the platform.

b. Determine the relation between W and P, assuming that W includes the weight of the platform itself.

47. The sketch shows the pendulum system of a Toledo scale, and the bottom force P in this sketch connects to the force P of Problem 46. The force P connects to the center of an "equalizer lever," and the two equal half forces are transmitted by flexible steel bands a, which wrap around the periphery of two circular segments b, having their centers at C. The centers C carry smaller circular segments d, on which flexible steel bands e are wrapped,

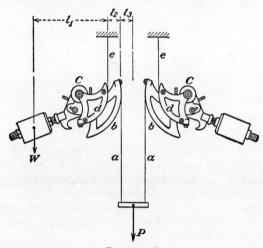

PROBLEM 47.

carrying the load up to ground, *i.e.*, to the top of the housing of the scale indicator. The centers C are not fixed; they are "floating" and held in position only by the forces in the steel bands a and e. The pendulums W are integral with the two segments and C.

a. Deduce a relation between P, W, l_1, l_2, and l_3, expressing the equilibrium of the floating pendulum lever.

b. Examine the system and recognize that it is drawn in the position of the maximum possible load P. Sketch the position of the pendulums for a load P half as large.

48. Find the reactions at the supports of the two-bar linkage (or hinged arch) under the influence of three forces.

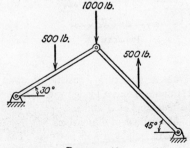

PROBLEM 48.

49. A two-bar linkage is loaded with three equal forces of 500 lb, one in the center of one of the bars, the other two at the third-length points of the other bar. Determine the reactions on the ground supports and also the force transmitted by the center hinge.

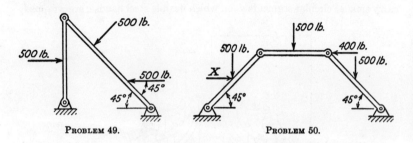

PROBLEM 49.   PROBLEM 50.

50. A three-bar linkage consisting of three bars of equal length is loaded by five forces, of which four are given and the fifth one X is unknown. Determine the value of X for equilibrium and also the ground support reactions and the forces transmitted by the intermediate hinges.

51. Two of the bars of the three-bar linkage shown are subjected to forces at their mid-points. What is the force X required for equilibrium, and what are the ground support reactions?

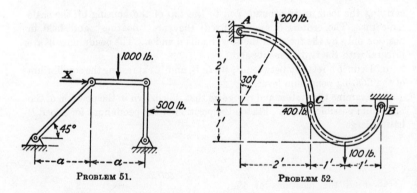

PROBLEM 51.   PROBLEM 52.

52. A structure consists of two circularly curved beams hinged at A, C, and B, and is loaded with forces of 100, 200, and 400 lb as shown. Find the reactions at A and B.

53. A door of dimensions 3 ft by 6 feet is hung on two hinges, 4 ft apart and 1 ft from top and bottom. The door weighs 30 lb, and its weight is

uniformly distributed over its area. Determine the forces exerted by the
door on the hinges.

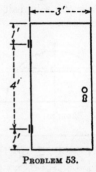

PROBLEM 53.

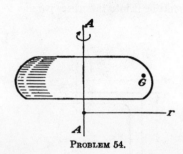

PROBLEM 54.

54. A curve, drawn in the plane of the paper, is revolved around an axis
AA, also lying in the paper, thus forming a surface of revolution. Prove that
the area of this surface equals the product of the length of the curve and the
periphery $2\pi r$ of the circle described by the center of gravity of the curve.
This is the first theorem of *Pappus*, a Greek mathematician who lived in
Alexandria, Egypt, about A.D. 400.

55. An area A in the plane of the paper is revolved about an axis also
lying in the plane of the paper, thus generating a toruslike volume. Prove
that the volume is equal to the product of the area and the periphery $2\pi r$ of
the circle described by the center of gravity of the area. This is the second
theorem of Pappus.

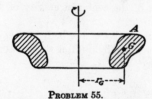

PROBLEM 55.

56. A uniform flat plate is made up of a square and a 45-deg triangle with
dimensions a and $2a$ as shown. Find the center of gravity, expressing it in
terms of the x and y coordinates shown.

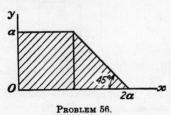

PROBLEM 56.

57. Find the location of the center of gravity of a piece of angle iron, **6 ft** long and of a cross section as shown.

58. Determine the location of the center of gravity of a solid half sphere by cutting it into thin slices parallel to the meridian plane.

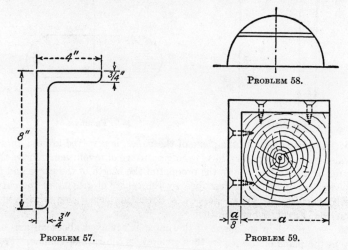

PROBLEM 58.

PROBLEM 57.

PROBLEM 59.

59. A composite beam consists of a square wooden section a of specific gravity 0.8, to which is screwed an angle iron of wall thickness $a/8$ and of specific gravity 7.8. Find the location of the center of gravity.

60. Find the location of the center of gravity of a flat plate consisting of a triangle and a half circle.

61. Find the center of gravity of the body of revolution, consisting of a cone and a half sphere, generated by rotating the plate of the previous problem about its axis of symmetry.

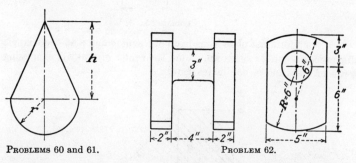

PROBLEMS 60 and 61.

PROBLEM 62.

62. A crank consists of two cheeks and a crankpin with dimensions as shown. Approximate the cheeks by rectangular parallelepipeds by laying it out full scale on a sheet of paper and sketching in the straight line (see Fig. 43, on page 41).

a. Determine the location of the center of gravity.

b. Calculate the error in that location caused by a shift of 0.01 in. in the location of the straight line replacing the bottom circular arcs, and also for a similar shift for the upper circular arc.

63. A block has the shape of a truncated (cut off) four-sided pyramid. The base has sides a, the top has sides $a/3$, and the height of the block is $2h/3$.

a. Find the height of the center of gravity above the base.

b. Generalize the answer to the case of truncated cones of any arbitrary base shape.

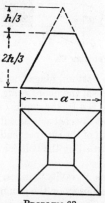

PROBLEM 63.

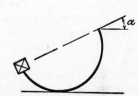

PROBLEMS 64 and 65.

64. A semicylindrical trough of uniform wall thickness weighs w_1 lb per linear foot in the direction perpendicular to the paper. At one edge it carries a linearly distributed load of w_2 lb per ft. Determine the relation between the ratio w_1/w_2 and the angle of inclination α when the trough rests on a horizontal plane.

65. A semispherical shell of uniform wall thickness and total weight W_1 carries a concentrated weight W_2 at one point of its periphery. Find the relation between the ratio W_1/W_2 and the angle α when resting on a horizontal plane.

66. Verify Pappus's first theorem (Problem 54) in connection with the relation between the area of a sphere and the center of gravity of a heavy semicircular arc.

67. *a.* From the known formulae for the volume of a sphere and the area of a circle, derive the expression for the location of the center of gravity of a semicircular flat plate by means of Pappus's second theorem (Problem 55).

b. Verify Pappus's theorem in connection with the volume of a cone and the center of gravity of a triangle.

68. A tank contains water 10 ft deep at its deepest point. The dimension perpendicular to the paper is also 10 ft so that the free water surface is a 10-ft square. In the bottom it contains a 45-deg trap door, hinged at the top and laid against a lip at the bottom. Assuming no leakage and no friction, and water weighing 62.4 lb/cu ft, find the forces at the hinge and lip.

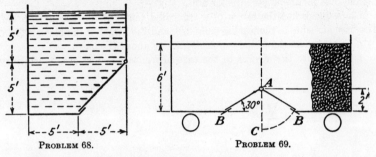

PROBLEM 68.   PROBLEM 69.

69. A coal dump car has two trap doors in the bottom, hinged at A and held at B by bolts. When the bolts B are withdrawn, the doors swing down to C and the coal falls out. The length of the door is 4 ft; the depth of coal, 6 ft. As an engineering approximation, assume that the coal acts as a fluid of specific gravity twice that of water, and let the width of the car (perpendicular to the sketch) be 6 ft. Calculate the total force carried by the bolt (or bolts) at B.

70. Concrete dams are sometimes designed as "gravity" dams. The assumption is made that the block of concrete rests loosely on the ground and that the water pressure acts on the back side of it. The water is not supposed

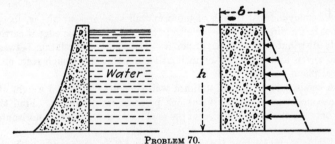

PROBLEM 70.

to penetrate under the dam, so that the pressure on the ground under the dam is due to its weight. Assuming that the specific gravity of concrete is $2\frac{1}{2}$ times that of water, and considering the simplified case of a dam of constant width b, calculate the necessary width b in order that the dam not tip over about its downstream edge.

71. As an idealization a little closer to the truth, consider a dam of triangular cross section loaded with water all the way up to the top. Further assume that the ground under the dam is porous and that water slowly leaks under it, giving a water pressure falling off linearly under the dam from the full pressure on the upstream side to zero on the downstream side.

a. Calculate the minimum value for b/h for which the dam can stand without tipping over on its downstream edge.

b. Sketch a dam cross section of more reasonable shape, and assuming leakage at the bottom, sketch the three forces acting on the dam and express the condition for equilibrium in a sentence.

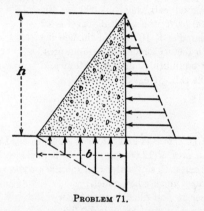

PROBLEM 71.

72. A culvert of diameter d is built under a roadbed and is covered with earth to a depth h as shown.

a. Assuming that the earth acts as a fluid of specific gravity $\gamma = 2\frac{1}{2}\gamma_{\text{water}}$, calculate the total vertical force on the culvert. (The pressure varies with the depth, and the vertical component of this pressure varies with the angular location on the semicircle. Hence this is a problem in integration.)

b. Substitute numbers as follows: $d = 6$ ft; $h = 6$ ft.

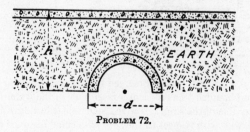

PROBLEM 72.

73. In the example of Fig. 46 (page 44) let $l = 12$ ft, $w_1 = 20$ lb/ft, $P = 1,000$ lb, and $k = 100$ lb/ft/in. At what distance b from the right-hand end can the load P be placed so that the left-hand end of the beam does not lose contact with the ground?

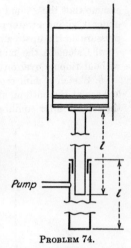

74. A hydraulic elevator is operated by a pump furnishing oil under pressure to a well, into which a plunger fits with some clearance. The elevator cab is shown almost in its highest possible position; the plunger and the well are somewhat longer than the possible vertical travel of the elevator. The loaded cab weighs 8,000 lb; the plunger weighs 100 lb/ft; the vertical travel is 30 ft; the plunger diameter is 10 in., and the specific gravity of the oil is 0.9. Calculate the pump pressure required for equilibrium in the bottom and top positions of the cab. Neglect friction.

PROBLEM 74.

75. A ship of 10,000 tons ("long" tons of 2,240 lb) dead weight has a waterline cross-sectional area of 12,000 sq ft. When it passes from a fresh-water harbor into the ocean with salt water of a specific gravity of 1.02, does it sink in deeper or less deep and by how much?

76. A container is half-filled with mercury and with water above the mercury. A rectangular steel block is floating in it. What percentage of its total height will be in the water? The specific weight of mercury is 13.6; of steel, 7.8; and of water, by definition, 1.0.

PROBLEM 76.

PROBLEM 77.

77. A ship, idealized in its submerged part as a rectangular parallelepiped of depth d and breadth b has its center of gravity in the water line and in the vertical center line of the ship. Due to a load redistribution inside, the point G is shifted sidewise with respect to the ship by a small distance δ. As a result of this the ship will incline by the small angle α. Derive the formula relating α to δ and the dimensions b and d.

78. A cylindrical container of cross section A is filled with water to a height H above the bottom. In it floats a cylindrical block of ice of cross section a (necessarily smaller than A) and height h, of which $0.9h$ is immersed and $0.1h$ protrudes above the water.

The ice is allowed to melt completely. What is the rise (or fall) of the water level *H*?

79. A true perpetual-motion machine. A rectangular box carries in it a circular cylinder pivoted at both ends, so that it can rotate inside the box. The box is divided into halves by a vertical partition. It is filled on one side with mercury, on the other side with water; the joint between the partition and the cylinder is leakproof as well as without friction. The inventor states that the buoyancy of the mercury acting on the center of gravity of the submerged half cylinder, being greater than the water buoyancy on the other side, causes a large unbalanced torque on the cylinder, which can be used to raise weights outside the box by means of a pulley.

a. Explain what is wrong.

b. Figure the force on the pivot.

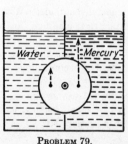

PROBLEM 79.

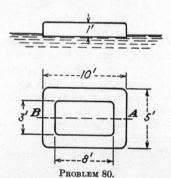

PROBLEM 80.

80. The figure shows a balsa-wood float, such as was standard equipment aboard ship during the war for the purpose of lifesaving in case of shipwreck. The float has the dimensions shown and weighs 270.4 lb or 10.4 lb per sq ft area. It floats in fresh water of 62.4 lb/cu ft.

a. What weight can be placed at *A* so that the end *B* is just dry?

b. If 300 lb is placed at *A*, how far will *B* be above the surface?

Hint: Of the 10-ft length let *x* ft be wet and $(10 - x)$ ft dry. Solve the cubic equation for *x* by trial and error.

81. A truss is loaded by two forces *P* as shown. Find the forces in the bars 1, 2, 3, and 4.

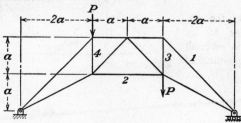

PROBLEM 81.

82. A railroad flatcar, roughly speaking, is a long beam, supported on wheels at its ends, carrying a load between the wheels. In order to increase the carrying capacity of the car, it is often reinforced by tension bars below it. The figure shows this in idealized form. The upper bars are supposed to be the bottom of the flatcar, and for the purpose of this analysis are considered to consist of three pieces, each $l/3$ long, hinged at their junction points. The loading is as shown. Note that the truss so formed is not a rigid one. It

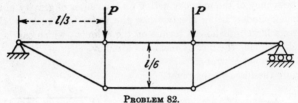

PROBLEM 82.

can be made rigid by adding a diagonal in the central rectangle. With this diagonal absent, the truss can support only symmetrical loadings (for which the force in the diagonal bar, if existing, would be zero). In an actual freight car, unsymmetrical loadings are carried by virtue of the fact that the upper member is not hinged at the two intermediate points but is continuous. Find the bar forces in the idealized system sketched.

83. A roof with a 30-deg slope is supported by a truss as shown. The joints on top are 3 ft apart. The truss is loaded by equal vertical loads of 1,000 lb at each joint. Determine the bar forces in section AA.

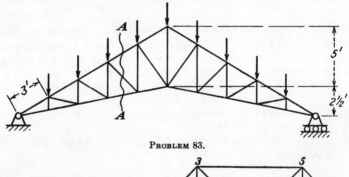

PROBLEM 83.

84. Complete the discussion of page 55 by calculating the forces in all the bars by the method of sections.

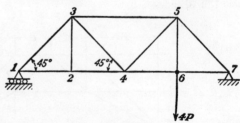

PROBLEM 84.

85. Complete the discussion of page 56 by calculating the reactions and the forces in all the bars by the method of sections.

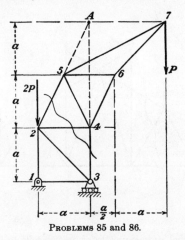

PROBLEMS 85 and 86.

86. Construct the Maxwell-Cremona diagram for the crane truss of page 55, and check the results against those of the previous problem.

87. A saw-tooth factory roof truss consists of 11 bars. The three top stringer bars are of equal lengths and are perpendicular to the three equidistant and parallel uprights. The longest of these three uprights, *i.e.*, the left one, has the same length as one of the three horizontal bottom stringers. The four loads are vertical, and each is equal to P. Construct the Maxwell-Cremona diagram and state in which bar the maximum force appears.

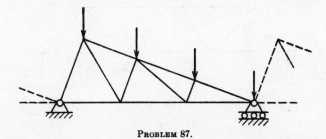

PROBLEM 87.

88. The roof truss shown has its nine upper joints at equal distances apart horizontally. The upper stringers, the four bars in the skylight, and the few bars near the supports are all at 30 deg with respect to the horizontal. The truss is loaded by nine equal and equidistant loads P.

Start the construction of a Maxwell-Cremona diagram from the left support and carry it through 5 joints. Then calculate the forces in section AA by the method of sections and check the results against the Maxwell-Cremona answers.

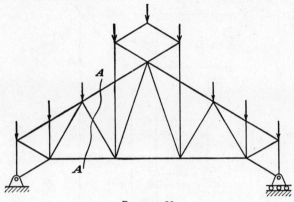

PROBLEM 88.

89. A crane in a coal storage yard runs on two rails 12 ft apart. The load of 20,000 lb is at a 6-ft overhang. The height of the truss is 3 ft, all angles being 90 and 45 deg. There are two counterweights of 15,000 lb each to prevent the left wheel from lifting off its rail. All bars have the same cross section. Calculate that cross section in square inches, if the stress is not to exceed 10,000 lb/sq in. anywhere in the structure.

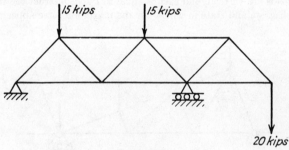

15 kips *15 kips*

20 kips

PROBLEM 89.

90. A wall hoist truss has a horizontal bottom stringer of 12 units of length, subdivided into four bars of 3 units each. The four vertical bars have lengths 1, 2, 3, and 4 units, respectively, and all other dimensions follow.

a. The truss is loaded by only a vertical end load P (disregard the Q's shown). Find all the bar loads by the method of joints. Reason much and calculate little.

b. The truss is loaded by the three Q's only, the end load being zero. Construct the Maxwell-Cremona diagram.

c. The truss is loaded with both P and Q's, where $P = 3Q$. In which bar does the maximum force appear?

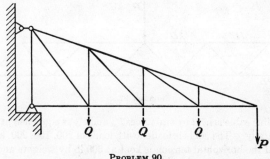

PROBLEM 90.

91. A symmetrical railroad bridge, half of which is shown in the sketch, consists of three trusses hinged together at A and A', supported on piers at B, B' and on shore at C, C'. It is loaded by four equal loads P at the joints shown, representing a single locomotive passing over the bridge. No other loads apply to the left side of the structure, which is not shown in the sketch. The seven sections of the big trusses are all alike; $AB = 3l$; $BC = 4l$, while $AA' = 3l$. The height of the small center truss is l, and the perpendicular distance from the support B to bar 2 is $2l$. Find the forces in bars 1 and 2.

Hint: A truss is a rigid structure. When several such "trusses" are joined together, they are first treated as rigid bodies, and the various forces between them and the ground are determined. Then each truss is handled separately by the usual methods.

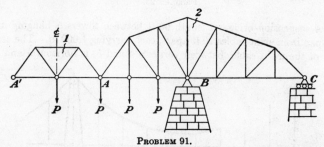

PROBLEM 91.

92. A simple bridge truss is made up of bars with angles of 45 and 90 deg. The height is h; the horizontal bars consequently are of length $2h$, and the 45-deg bars are of length $h\sqrt{2}$. The loading is a central vertical one of P.

a. By the method of sections find the forces in all the bars.

b. Generalize the problem to a long truss containing N upper bars instead of the four shown. Find the force in the nth upper bar from the left, the nth lower bar from the left, the nth right-slanting diagonal bar, and the nth left-slanting one. (Do it only for $n \leq N/2$.)

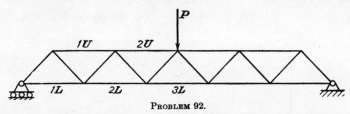

PROBLEM 92.

93. A flexible cable is held by two forces P_A and P_B at equal height and 10 ft apart horizontally. The cable is loaded with loads of 200, 100, 200, and 100 lb as shown, and the horizontal tension is kept at 500 lb by weights and pulleys. Find the deflections of the cable at the four load points by accurate graphical construction.

Hint: Determine the vertical reactions at the end pegs before choosing the pole of the force diagram.

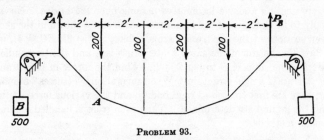

PROBLEM 93.

94. A suspension bridge of length 90 ft between towers is hanging in its center span from eight cables 10 ft apart, each carrying 1,000 lb. The center portion of the main cable, between verticals 4 and 5, is horizontal and 6 ft

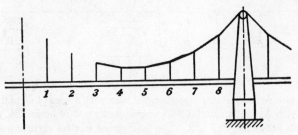

PROBLEM 94.

above the bridge deck; the top of the tower is 36 ft above the bridge deck. Calculate the total tension in the section of the main cable adjacent to the tower. Also calculate the slope of that section.

95. A cable is stretched horizontally with a tensile force P over two pulleys. Then two 60-deg forces, also equal to P, are applied at C and D, retaining their 60-deg. direction when the cable sags. At A and B sufficient vertical forces are applied to keep A and B always on the original horizontal. We have $AC = CD = DB = l/3$. Find the sag and the tension of the center section.

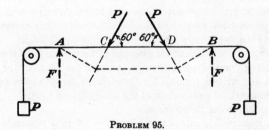

PROBLEM 95.

96. A string is attached to a wall at A. It carries three weights on definite points on the *string* so that $AC = CD = DE = l$. (Note the difference between this case and that of Problem 93.) The string then passes through a point B located at $3l$ horizontally to the right of A and l vertically below it, which is secured by an appropriate vertical force at B. The horizontal tension in the string is 300 lb. Find the shape assumed by the string.

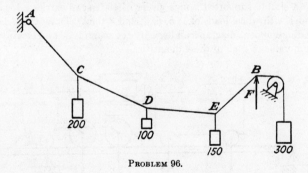

PROBLEM 96.

Hint: This problem cannot be solved directly because the horizontal locations of C, D, and E are not known at the start. Start by *assuming* a vertical position of the pole in the force diagram and construct the shape of the string beginning at A, which will then end up either above or below B. Repeat the construction with an appropriately chosen second pole, and for the third trial interpolate (or if necessary, extrapolate) between the two first trials.

97. A transmission line consists of copper wire (weighing 0.333 lb/cu in.) of 0.5 sq in. cross section. The span between towers is 400 ft. The line is going uphill, and the top of one tower is 20 ft higher than the top of the next one. The lowest point of the span is 20 ft below one tower top and 40 ft below the other tower top. Calculate the horizontal tension of the line and the location of the low point.

98. Calculate the sag in a transmission line with spans of 500 ft, tower tops at equal heights, with a horizontal tension of 10,000 lb, and a weight of 2 lb/ft.

99. A flexible cable weighing 1 lb/ft is stretched with a horizontal pull of 1,000 lb between two points 100 ft apart and at equal height. In addition to its own weight, the cable carries a central concentrated load of 50 lb.

a. Assuming "small" sags, describe in words the shape of the two halves of the cable and write a formula for the slope or slopes at the center point, expressing the vertical equilibrium of the concentrated load.

b. Calculate the sag in the center.

100. An inextensible cable of length $2L = 200$ ft is suspended from two points at equal height and distance $2l = 100$ ft apart. How far will this cable sag below the level of the supports?

Hint: This problem involves two properties of the catenary: its sag, for which a formula is given in the text, and its length, for which the formula is to be derived. The two equations thus obtained in terms of the two unknowns y and H are "transcendental" and cannot be solved exactly; the numerical solution is determined by consulting a table of hyperbolic functions, to be found, for instance, in Marks' "Handbook."

101. A simply supported bridge girder of 40-ft span carries a locomotive and tender with wheel loads of 1, 5, 5, 3, and 3 tons. Draw the shear-force and bending-moment diagrams and specify the magnitude and location of the maximum value of each of these quantities.

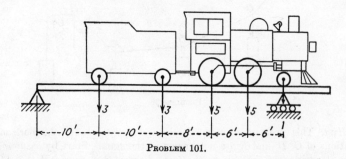

PROBLEM 101.

102. A hoist consists of a 16-ft beam pivoted at one end and supported by a 30-deg cable at the other end, as shown. The hoisting unit is a two-wheel truck, and carries a central load of 10 tons. Construct the shear-force and bending-moment diagrams of the horizontal portion of the beam.

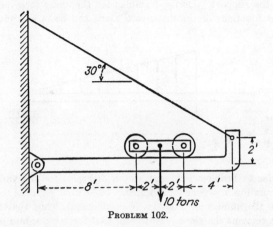

PROBLEM 102.

103. A beam of length $3a$, supported at one end and at a point $2a$, is loaded by a concentrated load P at mid-span and by a distributed load of total magnitude P on the overhang. Draw the shear and bending-moment diagrams true to scale and indicate the scale in the drawing. Specify the location and magnitude of the maximum values of the shear force and bending moment.

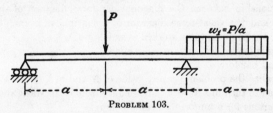

PROBLEM 103.

104. An attic floor in a factory building of 40 ft span carries distributed loads as indicated in the sketch. Draw the shear-force and bending-moment diagrams.

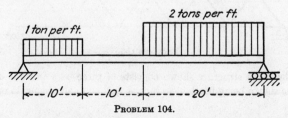

PROBLEM 104.

105. A simply supported beam of span l carries a distributed load that varies linearly from a maximum value of w lb/ft at the quarter-length point to zero at the support. Derive formulas for the shear force and bending moments as functions of the distance x along the beam and plot them in diagrams.

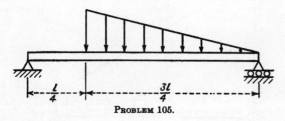

PROBLEM 105.

106. Refer to Problem 92b, and sketch diagrams for the various forces along the truss for $N = 8$ as an example. Note what happens to these diagrams when the number of triangles N is increased, while the total length of the truss remains the same. Note also that for large values of N and n the compressive force nP in the upper girder becomes practically equal to the tensile force in the bottom girder. Verify that for the limiting case $N \to \infty$, the forces in the upper and lower girder form a bending moment

$$nPh = Pnh = Px/2,$$

as for a rigid beam, in the manner of Fig. 72 (page 69). Now deduce the similar relationship between the diagonal bar force diagram of the truss of Problem 92 and the shear-force diagram of a continuous beam. After this explanation, the problem to be worked is as follows:

The truss is loaded with five equal loads Q.

a. Find all bar forces.

b. Generalize the problem for any value of N and n.

c. Proceed to the limit $N \to \infty$, and deduce from it the shear and bending-moment diagrams for a uniformly loaded beam.

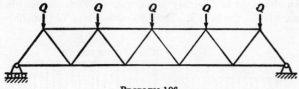

PROBLEM 106.

107. The plane structure shown consists of three bars and one cable, all connections shown being frictionless hinges, and all angles being 45 or 90 deg.

The length $BC = a$; $AB = 2a$; the hinge A is located vertically above D, and all other dimensions follow.

a. Calculate all hinge forces everywhere, and show in separate sketches the forces on the cable and on the three bars individually.

b. State the value and the location of the maximum bending moment in all three bars.

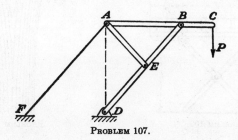

PROBLEM 107.

108. A pulley of diameter $l/3$ has to be mounted at the end of a horizontal strut of length l; the strut has to be properly braced, and a weight W has to be supported by a cable, of which the other end must be horizontal. Two designers, one of them named Rube Goldberg, tackle the job and produce the designs sketched. Calculate the bending moments in the three bars shown.

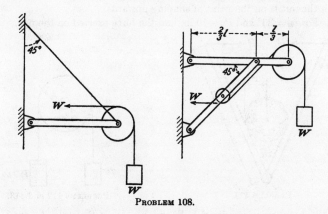

PROBLEM 108.

109. A table top is supported by crossed legs which are hinged together in the center. There is no friction on the ground or in any of the hinges. The table is uniformly loaded with w lb/ft, and the dimensions l are expressed in feet. Find the bending-moment diagrams of the top and of one of the legs.

What and where is the maximum bending moment in the legs?   What is the
distribution of compressive force in the legs?

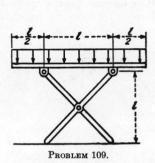

PROBLEM 109.

PROBLEM 110.

110. A wedge is used to split a log.   It is common experience that such a
wedge is self-locking, but it is conceivable that a well-greased, blunt wedge
might be squeezed out by the log.   Find the relation between f and α for which
this will take place.

111. A nutcracker with arm lengths $l = 5$ in. is used to crack a 1-in.-diam-
eter nut.

a. Find the relation between the friction coefficient and the angle α for
which the nut is on the point of slipping upwards.

b. For $\alpha = 30°$ and $P = 10$ lb, find the force exerted on the nut.

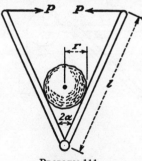

PROBLEM 111.

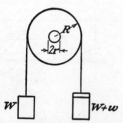

PROBLEMS 112 and 113.

112. A pulley carries weights W and $W + w$ at the ends of a string slung
over it.   Find the maximum value of w for which the pulley remains in equilibrium.   The pulley diameter is R, the axle diameter is r, the coefficient of
friction in the axle is f.   Assume that the pulley load is transmitted to the
axle on a small area on the top of the axle and that the clearance is so large
that the pulley wheel does not touch the axle on the sides.

113. Assuming that the friction in the axle is small, calculate the ratio
between the friction coefficients at the pulley periphery (f_{pulley}) and at the axle

(f_{axle}) so that slipping will take place simultaneously at the axle and at the periphery. By "small" is meant that an e function can be expanded into a Taylor series and only the first significant term retained. The axle clearance is the same as in the previous problem.

114. A 200-lb packing case, 2 ft high and 3 ft long, has its center of gravity at mid-height but at one-third length horizontally. The friction on the ground is 60 per cent.

a. If pulled to the right by a force P at the top of the case, will it tip or slide, and at what value of P will this happen?

b. For what coefficient of friction on the ground will the case be on the border line between sliding and tipping?

c. Answer questions *a* and *b* when the case is pulled to the left instead of to the right.

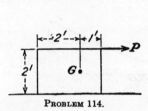

PROBLEM 114.

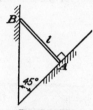

PROBLEM 115.

115. A uniform ladder of length l rests at A on a 45-deg incline and at B against a vertical wall. The ladder itself is also at 45 deg so that it is perpendicular to the incline at A.

a. In the absence of friction, which way will the ladder start sliding at A, up or down?

b. Assuming no friction at B, what coefficient of friction at A is necessary to prevent slipping?

116. A round vertical shaft of diameter $2r$ rests on a horizontal plane with a total force W. When the shaft is turned, the friction at the bottom support causes a resisting torque. Calculate this friction torque, assuming that the force W is distributed uniformly over the area πr^2, and calling the friction coefficient f.

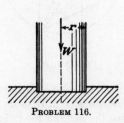

PROBLEM 116.

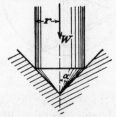

PROBLEM 117.

117. A vertical shaft, sustaining a weight W, has a cone-shaped bottom end and rests in a cone-shaped bearing. Assuming that the normal pressure is the same over the entire bearing surface, calculate the friction torque in terms of W, R, f, and α.

118. A rope of length l and weight $W = w_1 l$ is to be dragged along a rough floor with a coefficient of friction f. It can be lifted at the pulling end to a height h above ground. In the calculation, the difference in length between the curved portion of the rope and its horizontal projection is to be neglected.

a. What horizontal pull H is required to just drag the rope, and what is the distance x in this condition?

b. What is the vertical component of the pull at the pulling end in that condition?

c. Calculate the total pull (the resultant of the horizontal and vertical components) for the case that $l = 100$ ft, $w_1 = 2$ lb/ft, $f = 0.50$, and $h = 0$ ft and 3 ft.

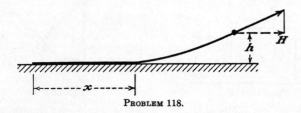

PROBLEM 118.

119. A bar of length $R\sqrt{2}$ rests in a rough semicircular trough of radius R. Find the relation between the angle φ and the friction coefficient f for which the bar is on the point of slipping down. Solve this problem by setting up the equations of equilibrium.

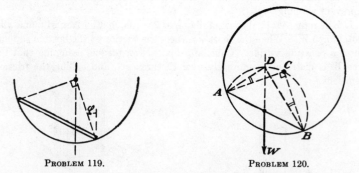

PROBLEM 119.   PROBLEM 120.

120. If in Problem 119 a semicircle ABC is described on the bar as diameter, and if the weight force W intersects this semicircle in point D, prove that AD and BD represent the contact forces between the bar and the cylinder. From this property deduce the answer to Problem 119.

121. A uniform bar of length l rests with one end on a horizontal floor and with another point on a half cylinder of radius r. The coefficient of friction on the ground and on the cylinder are both equal to the same value f. Derive the relation between the angle α and the coefficient of friction for impending slip.

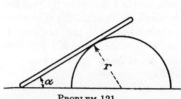

PROBLEM 121.

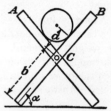

PROBLEM 122.

122. For the purpose of supporting a log to be sawed, a pair of sawhorses is made up of two-by-fours A and B joined by a 1-in. broomstick C. Idealize the problem by neglecting the thickness of the two-by-fours and assuming no friction between the log and the sawhorses or in the broomstick hinge. The only friction is between the legs and the horizontal ground.

a. Find the necessary friction coefficient on the ground to prevent slipping under the weight of the log, expressed in terms of a, b, and α.

b. Calculate f numerically for the case that $b = 2a$ and $\alpha = 45°$, and express your opinion on the technical competence of the designer of this device.

c. How would you improve the sawhorse?

123. A rope carrying a weight of 300 lb passes with ample clearance through a piece of pipe bent into a quarter circle. The pull at the other end of the rope is P, and the pipe is rigidly attached to a foundation. Assuming 25 per cent friction between the rope and the pipe, calculate the range of values of the pull P for which the weight neither goes up nor down. Also calculate the force (magnitude and direction) transmitted by the pipe to its foundation.

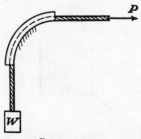

PROBLEM 123.

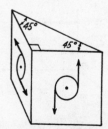

PROBLEM 124.

124. A box, 3 ft high with two square sides 3 ft by 3 ft and a 45-deg face, is supported on balls on the floor where it can roll freely in all directions. On each of the three upright faces there is a 1-ft-diameter pulley, solidly attached to the face in its center, and two ropes are attached to each pulley. On the front face the ropes are pulled with 100 lb up and down; on the side face, with 100 lb fore and aft, as shown. The ropes on the back-side 45-deg face are pulled horizontally. What forces must be in them to keep the box in equilibrium?

125. A long flexible steel wire is solidly attached to 1-ft circular disks at the ends, each carrying two handles, a black one and a white one. All of this has negligible weight. A man has one disk in his hands, and he pushes down on the black handle and up on the white one, with equal forces F. Another man holds the other disk in his two hands. What forces must he exert for equilibrium? Answer this for cases a, b, and c.

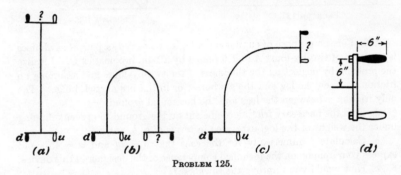

PROBLEM 125.

126. The figure shows a cube of side a. Three forces F are acting on it as shown.

a. By shifting the point of application of all three forces to O, find the resulting "screw" in direction and magnitude.

b. Verify this result by shifting the A and B force to point C, finding the resulting force and couple and from it again the "screw."

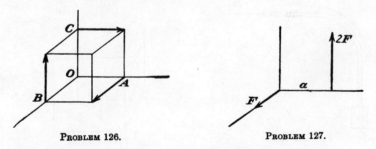

PROBLEM 126.   PROBLEM 127.

127. Two forces F and $2F$ are at right angles and at distance a apart. Find the resulting "screw."

128. Two forces F_1 and $F_2 = F_1/\sqrt{2}$ are distance a apart and at 45 deg with respect to one another. Find the resulting screw.

Hint: Is there a way of using the known solution of Problem 127 for this case?

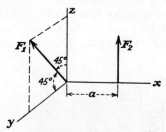

PROBLEM 128.

129. A bar of length $a\sqrt{2}$ is pivoted by a frictionless ball joint at A. The upper end of the bar rests against a rough, vertical wall at distance a from the joint. Find the relation between the coefficient of friction, f, on the vertical wall and the angle α of the vertical projection of the bar.

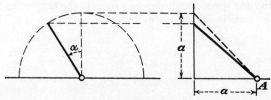

PROBLEM 129.

130. The sketch shows a perspective and two projections of a space structure consisting of three bars, OC, AC, and BC, hinged to the ground at their

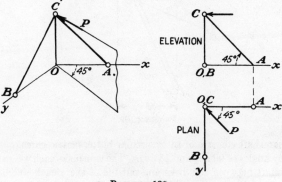

PROBLEM 130.

bottoms and hinged together at the top. At point C a force P is acting as shown, parallel to the xy plane and at 45 deg with respect to both the x and y axes. All angles between the bars and the ground are 45 and 90 deg. Find the three bar forces.

131. A turbine rotor weighing 15,000 lb is supported on two bearings 15 ft apart. In the planes I and II there are forces acting as shown, both projections as seen from A to the right. Find the bearing reactions.

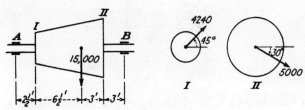

PROBLEM 131.

132. A pole AD is attached to a vertical wall by a ball joint at D and two horizontal guys at A and C. It carries a vertical load of 1,000 lb at B. Find the forces in the guy wires, the three components of the supporting forces at D, and the maximum bending moment in the bar.

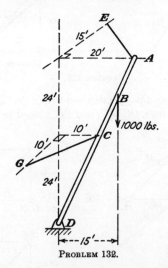

PROBLEM 132.

133. A camshaft consists of three circular but eccentric cams, offset from each other by angles of 90 deg and 45 deg. The cam disks each weigh 2 lb; the sections of shaft between the disks and outside of them each weigh 1 lb, so that the total weight of the assembly is 10 lb. The eccentricity of the points

C is 1 in.; the diameter of the discs is 4 in.   Determine the location of the center of gravity.

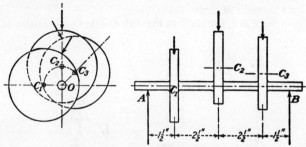

PROBLEMS 133 and 139.

134. A bar of length $2r$ and of weight W, with its center of gravity in the center, is suspended from two vertical strings of length l.   Two horizontal forces P at distance $r/2$ apart are made to act on the bar, and these forces remain perpendicular to the bar after the bar has turned.

a. Show by simple reasoning that the bar will rotate in its plane about its center.

b. Calculate the angle of rotation α as a function of P, W, r, and l.

c. What forces P are required to rotate the bar 90 deg and 180 deg?

d. The case when the center of gravity of the bar is no longer in the center, but a distance a from it, is more complicated.   Assuming small forces P, find the point of the bar about which it rotates through a small angle, and derive this angle as a function of P, W, r, l, and a.

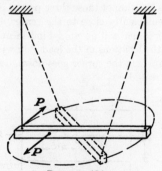

PROBLEM 134.

135. A solid circular disk of radius r and weight W with its center of gravity in the center O is suspended from three vertical strings of length l attached to points A, B, and C of the periphery, 120 deg apart.   Calculate the displacement of the disk in each of the four following cases:

a. As a result of a force P in the plane of the disk in the direction AO.

b. By a force P in the plane of the disk in the direction perpendicular to AO.

c. By a moment M in the disk.

d. By a force P in the plane of the disk directed tangentially at point A.

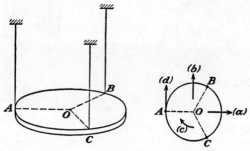

PROBLEM 135.

136. A stiff flat rectangular plate is placed on an elastic foundation giving a vertical reaction force of $k\delta \, dA$ lb on an element of area dA pushed into the ground through a distance δ. A load P is placed on it off center at location c,d.

a. Find the deflection of the plate $\delta = f(x,y)$.

b. Find the condition for c and d so that one corner of the plate just has zero deflection, the entire plate being pushed down into the ground.

Hints: a. Read Sec. 11, and note particularly Figs. 46 and 47 (page 44). Resolve the load into a central load and into two couples, tending to turn the plate about the x and y axes, and solve for these three cases separately. The total deflection is the sum of those three partial deflections.

b. If the load is sufficiently close to the center, the entire plate will go down; otherwise, one corner will go up. Find in one quadrant the line or curve that separates the locations of the load P in which the opposite corner goes up from those in which the corner goes down. By symmetry the other three quadrants are treated.

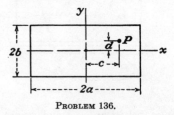

PROBLEM 136.

137. In a machine there was a square plate, which was loaded at its four corners, two pushing down and two pushing up. As a result the plate warped as indicated in the sketch, and the warping was considered greater than permissible, so that it became necessary to stiffen the plate. This could be done with a minimum of extra weight by erecting on it a space structure of four thin bars meeting at a point above the center of the plate. The situation is idealized in the nine-bar five-joint space frame shown, in which $AB = CD$. Point E is above the center of the square, and angle AEC of the vertical diagonal plane is 90 deg.

a. Indicate how this space frame can be constructed by the repeated tetrahedron method of page 121.

b. Calculate the forces in all the bars.

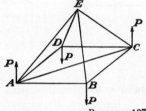

PROBLEM 137.

138. A tetrahedron is a six-bar four-joint space frame, the simplest possible one. In problem 137 we have three more bars and one more joint. In this problem we go one step further by adding again three bars and a joint. The space frame of this problem can be derived from the previous one by distorting it, *i.e.*, pulling point E sidewise to bring it above point B, and incidentally making $BE = AB = AD$. Then we add point F, tying it to E, D, and A. The loading we keep as before. Find the forces in all bars.

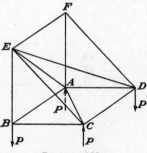

PROBLEM 138.

139. Referring to Problem 133, let the cams each carry a load, applying on their peripheries at the point where they intersect the vertical plane OC_2. These three forces all have the same vertical component $P = 40$ lb, and enough horizontal component to make the resultant pressure perpendicular to the cam.

Moreover, let each cam carry a "centrifugal force" of 20 lb, directed radially out from O through C. Let the cam be supported on bearings A and B, 8 in. apart.

Determine the bending-moment diagram in the horizontal plane OC_1 only.

140. A two-throw crank with the cranks 90 deg offset consists of nine pieces of straight shafting, all of length a. It is held at A by a ball joint (worth three supports), at B by a bearing (worth two supports), and at C by a double-hinged strut (worth one support). It is loaded by a single force P.

a. Draw the bending-moment diagrams for the horizontal and vertical planes, and the twisting-moment diagram, for the center piece of main shafting only.

b. Determine the magnitude and location of the maximum bending moment, both in the horizontal and in the vertical planes, and of the maximum twisting moment.

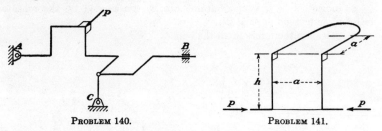

PROBLEM 140.   PROBLEM 141.

141. A steam pipe line contains a temperature-expansion bend. The center line of the pipe is shown in the sketch; it has vertical stretches of length h, followed by horizontal stretches of length a, joined together by a horizontal semicircle of diameter a. When heated, the pipe tends to expand, but this expansion is impeded by the structure of which the pipe forms a part. As a result, forces P appear as shown.

Draw the bending-moment diagrams in the vertical and horizontal planes and the twisting-moment diagram. State where and how large are the maximum total bending moment and the maximum twisting moment.

142. A wheel of 4-ft diameter rolls without slipping over a horizontal floor through a distance of 1 ft to the right. On it are acting a force $F = 100$ lb to the right and a counterclockwise moment of 200 ft-lb.

a. What is the work done by the force F?

b. What is the work done by the moment M?

c. What is the work done by the force and moment combined?

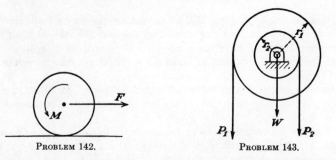

PROBLEM 142.   PROBLEM 143.

143. A solid wheel of weight $W = 100$ lb has two pulley grooves cut into it at radii $r_1 = 8$ in. and $r_2 = 4$ in. The ropes shown pull with forces $P_1 = 100$

lb, $P_2 = 150$ lb. The wheel is allowed to rotate through a full revolution in a clockwise direction. What is the work done by all three forces combined?

144. The cart of a roller coaster in an amusement park weighs 300 lb. The part of the track we are considering has the shape of a quarter sine wave of height $h = 30$ ft and $l = 100$ ft. While the cart is going down, two forces are acting on it: the weight W and the normal force N.

a. What is the work done by the weight force on the cart when it goes down from the top to the bottom?

b. What is the work done by the normal force N?

c. What is the work done by all the forces (W and N) acting on the cart?

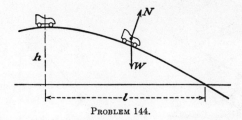

PROBLEM 144.

145. A platform scale is supported at A and B by means of a system of levers, which are all pivoted at their ends without friction. The dimensions are given by a, b, c, and d, while the distance e is variable. The adjustment weight w is placed at an appropriate location e so as to counterbalance the load W to be weighed.

a. Find a relation between the dimensions a, b, c, and d that must be satisfied if the reading e is to be independent of the location of W on the platform.

b. With the above relation satisfied, what is e?

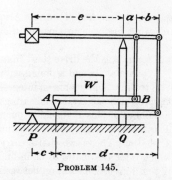

PROBLEM 145.

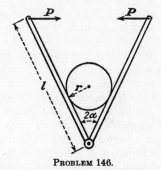

PROBLEM 146.

146. A trough, hinged at the bottom, carries a cylinder of radius r and weight W. What forces P are necessary to keep the system in equilibrium in the absence of friction?

147. The figure shows a system like that of Fig. 131 (page 148), except that here the angles are 30 deg. Determine the forces X in terms of P by the method of work.

Determine the reactions between the bars at their hinge points 1, 2, 3, etc. (by the method of equilibrium of successive bodies).

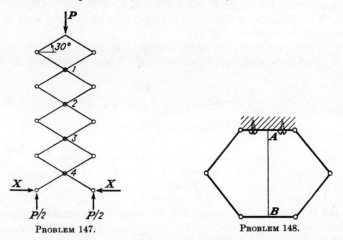

PROBLEM 147.   PROBLEM 148.

148. Six bars of equal lengths and equal weights are hinged together at their ends so as to form a hexagon. One of the bars is screwed to a ceiling, and the hexagon hangs down in a vertical plane. A weightless string AB connects the center points of the upper and lower bars. If l_b is the length of one bar, l_s the length of the string, and W_b the weight of one bar, find the tensile force in the string.

Hint: Cut the string, call its force X, and use the method of work. Also solve by equilibrium of the individual bars; what happens to this solution for the extreme cases $l_s = 0$ and $l_s = 2l_b$?

149. In the differential drive in the rear of a car, A is the drive shaft from the engine, B and C are the wheel shafts. A carries at its end a pinion with n_A teeth on it, meshing with a gear D with n_D teeth. This gear is an integral part of a frame, shown in thin lines in the sketch, which can rotate on bearings around the shafts B and C, and it carries a number of pins E (usually four), on which planet gears can freely turn. These planet gears mesh with the gears mounted on the ends of the B and C shafts. These two shafts carry protuberances at the center of the figure, which force the shafts to be in line with each other, but which oppose no restriction to relative rotation between them. Not shown on the drawing is a housing surrounding everything, with bearings around A, B, and C, through which the three shafts protrude. This housing is filled with grease and furnishes a point of attachment to the car body (through the springs).

A torque of moment M_A is applied to the non-rotating drive shaft, while B, C, and the outside housing are held by appropriate forces and moments to keep the entire system at rest. Find the moments on the shafts B and C and the moment on the housing.

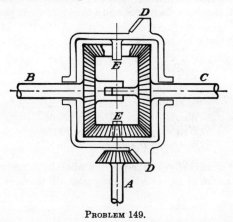

PROBLEM 149.

150. A *vélotaxi*, or French wartime equivalent of the Oriental rickshaw, is a bicycle used for hauling behind it a light carriage with one or more passengers. Calculate the ratio between the load Q and the pedal push P necessary for starting, in terms of the following quantities:

a = moment arm of pedal
r_1 = radius of the large or pedal sprocket
r_2 = radius of the small or rear-wheel sprocket
R = radius of the rear wheel

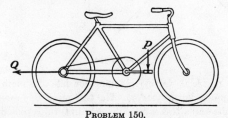

PROBLEM 150.

151. A grab bucket for hoisting coal hangs from a crane at point A. The crane can lift and lower point A, but for the problem consider A fixed in space. Then by pulling up on the "closing rope" at P, the buckets can be closed. At C there is a drum with a portion of large diameter over which the closing rope is wrapped several times, and a portion of smaller diameter over which a chain is wrapped several times, with the other end attached at A. If the rope P is pulled, the drum rotates, and the chain is wound up the

small diameter part of it, thus bringing C closer to A.   One bucket then moves
as a two-bar link mechanism with the bars AB and BC, where A is fixed and
C moves along a vertical line.

   a. Set up an equation expressing the pull P in terms of the bucket weights,
the drum weight, the drum diameters, and a ratio ϵ/δ, where you define ϵ and δ
appropriately.

   b. Substitute the following values:

$$AB = 48 \text{ in.}, \qquad BC = 36 \text{ in.}, \qquad BG = 18 \text{ in.}$$
$$\angle CBG = 90°, \qquad \angle DCB = 45°$$

   Large-drum diameter = 18 in.
   Small-drum diameter = 4½ in.
   Weight of bars AB is negligible
   Weight of each bucket = 100 lb
   Weight of the drum = 50 lb

   c. What is P numerically for the position when BC is horizontal?
   d. For what angles of BC with the horizontal will the force P be zero?

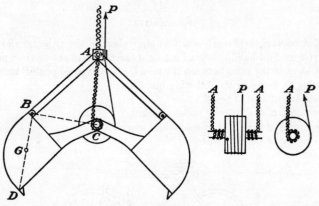

PROBLEM 151.

   152. A *Riehle* materials-testing machine consists of a main frame A supported on an elaborate lever system.   The test piece T is mounted between
the top of the frame and a test table B, which, in turn, is pulled down by four
large screws C.   The screws are turned in the bottom pit by a gear-and-pinion
drive and are held by thrust bearings to the ground in the pit.   In this manner, a very large downpull can be exerted on B, sufficient to break the test
piece T.   The rest of the equipment serves only to measure the force in the
test piece.   The right-hand end of the frame A is supported by the lever E,
and the left-hand end by the lever D, which is only partly drawn, but which
extends all the way to F, where it attaches to the lever E in a pivot joint.
This is done by giving E a V shape, with the point at F, the two arms at the

table A. The lever D passes between the two arms of the V to F. The force in the test piece is measured by adjusting the position of the moving weight H until the end of the lever at G plays between the two stops. The counterweight J is so dimensioned that when H is set at zero, the arm G is floating when the test-piece pull is zero, *i.e.*, the weight of the frame A plus the weight of the levers and the weight J together balance the arm G and the weight H at zero.

a. What relation between a, b, c, and d must exist in order that the frame force A be equally distributed among the two levers D and E?

b. Find the relation between the test-piece pull T, the weight H, and the various dimensions shown.

c. In which of the levers and at which location does the maximum bending moment occur?

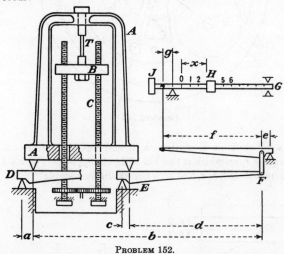

PROBLEM 152.

153. The speed-reduction gear between an airplane engine and the propeller usually is of the planetary type. The engine is directly coupled to a central gear A, which meshes with a number of planet gears B. The planet gears mesh with an internal gear C, which does not rotate, but which is rigidly mounted to the engine frame and hence to the airplane. The planet-gear shafts fit in bearings in a spider D, which has a central bearing fitting around the shaft of the center gear and engine A. It is to this spider that the airplane propeller is attached. Thus the engine shaft A is free to rotate inside the spider bearing D.

The sketch shows the case of five planet gears, all of the same diameter as the central gear A.

a. If a torque moment M_A is applied to the engine shaft, what torque must act on the propeller D and on the engine frame C for equilibrium?

b. Answer the same question for the more general case that the diameters of *A* and *B* are not the same but are d_A and d_B.

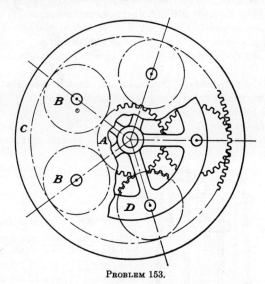

PROBLEM 153.

154. A solid block of dimensions *h*, *l*, and weight *W* rests on a rough cylinder of radius *r*, on which it can roll without slipping. For what relation between the quantities *l*, *h*, *r*, and *W* is the equilibrium stable?

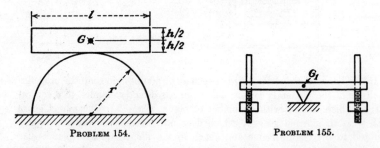

PROBLEM 154.   PROBLEM 155.

155. The assembly consisting of the horizontal bar, knife-edge, and two vertical crossbars (but without the weights) weighs 10 lb, and its center of gravity G_1 is 3 in. above the knife-edge. The two weights shown are 3 lb each and can be screwed up and down, always moving together, so that they are at the same distance from the main horizontal bar. Describe accurately the positions of the weights for which the equilibrium is stable. (In the good old days the stores used to sell small Chinese wood carvings utilizing this principle.)

156. A uniformly solid half cylinder of radius r_1 rests, with the curved side down, on top of a fixed half cylinder of radius r_2. The surfaces are sufficiently rough to prevent slipping. Under what circumstances is the equilibrium stable?

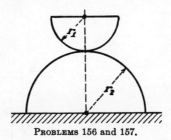

PROBLEMS 156 and 157.

157. *a.* The same as Problem 156 when the half cylinders are replaced by uniformly solid half spheres.

b. Which answer applies when one of the objects is a half sphere and the other one is a half cylinder?

158. In Problem 93 the deflections of a flexible cable were calculated under the influence of four loads that could move along vertical lines only. It is possible to calculate the position of the center of gravity of the *six* weights involved.

a. Using the theorem of work, make a statement in one sentence about a property of that center of gravity, and in that sentence consider the two cases of an extensible and an inextensible cable.

b. Calculate numerically the position of the center of gravity for the three following cases: (1) The cable is in the position of equilibrium of Problem 93. (2) The point A is 1 ft higher than before, and consequently point B is in a different position, but all other points are in the same position as before. (3) Point A is 1 ft lower.

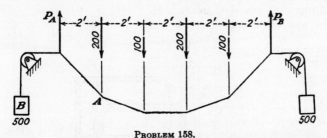

PROBLEM 158.

159. Four rectangular plates are hinged together along their sides so as to form a square or diamond-shaped parallelepiped. A spring of stiffness k connects two opposing hinges such that in the square shape shown, the spring tension force is P. This force is held in equilibrium by two externally applied forces P along the other diagonal. Find under what circumstances this equilibrium is stable or unstable.

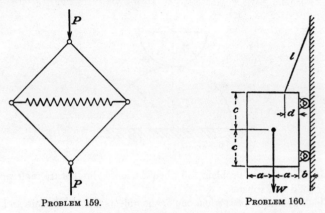

PROBLEM 159.   PROBLEM 160.

160. Return to Problem 42, where both wheels rest against the wall. Consider the possibility of the plate and rope leaving the plane perpendicular to the wall and rotating about the vertical line connecting the two wheels until the plate may go so far as to lie flat against the wall. Assume that the wheels do not slip sidewise. For what relations between the dimensions is the equilibrium in the mid-position stable, unstable, or indifferent?

161. A rotating crank carries at its end a block that can slide in a guide. The guide is attached to a piston and piston rod and is itself constrained to move in a direction perpendicular to its own center line AA. The apparatus is known as a "Scotch crank."

  a. Calling the crank radius r, the constant angular speed of the crank ω, and the angle of the crank with respect to the vertical $\alpha = \omega t$, find the acceleration of the piston.

  b. Reduce the answer to numbers for $r = 12$ in., the crank rotating at 200 rpm, and $\omega = 45$ deg.

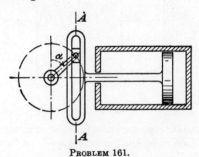

PROBLEM 161.

162. In Leonardo da Vinci's experiment (page 159), we suppose that the vertical boards are 3 ft long. The leaking faucet is at such height, and the rate of dropping is so regulated, that when a drop is just coming out of the faucet, the next or second drop is just at the top of the boards, and the sixth drop is at the bottom of the boards, while drops 3, 4, and 5 are between the boards. Calculate the number of drops leaving the faucet per second, and sketch the distance between drops.

163. A local train has a normal speed of 60 mph between stations. When putting on brakes, it can decelerate at a rate of 2 mph/sec, and when starting from a station it can accelerate at a rate of 1 mph/sec.

a. Calculate the time lost in making a stop of 1 minute at a station.

b. Sketch curves for velocity versus time and displacement versus time for the entire period.

164. In Problem 163, calculate the minimum distance between two stations for which the train just can reach its 60 mph speed before putting on the brakes again.

165. The motion of a piston of a steam engine is described by the equation

$$x = a[\cos \omega t + \tfrac{1}{16}(\cos 2\omega t - 1)]$$

Find the expressions for the velocity and acceleration and calculate the value of the maximum velocity occurring during this motion.

166. A point moves along a straight line with a constant third derivative $\dddot{x} = 10$ in./sec^3. It starts off at time $t = 0$ with an initial acceleration $\ddot{x}_0 = 10$ in./sec^2 and an initial velocity $\dot{x}_0 = 5$ in./sec. How far does the point travel

a. During the first second?

b. During the first 5 seconds?

167. A wheel has an initial speed of 500 rpm and is being slowed down at a rate of 2 rpm/sec. How many revolutions does the wheel make before it comes to a stop?

168. A rocket is shot vertically straight up, and, once in flight, is subjected to a downward acceleration $g = 32.2$ ft/sec^2.

a. Calculate the initial velocity required to shoot a rocket to a height of 100 miles.

b. Calculate the time elapsed between the moment of launching and the moment of return to earth of this rocket.

169. A wheel with a peripheral radius of 1 ft carries a string wrapped around it. A weight on the string goes down with a velocity of 5 ft/sec and an acceleration of $\tfrac{1}{2} g = 16.1$ ft/sec^2. Calculate

a. The magnitude and the direction of the acceleration of a point on the periphery of the wheel.

b. The same for a point midway between the center and the periphery of the wheel.

PROBLEM 169.

170. A mass at the end of a 3-ft string is swung around a steady point in a vertical plane. The mass can pass through the overhead position A in the manner described only when the centripetal acceleration exceeds g. Calculate the velocity v required to insure a centripetal acceleration $1.20g$.

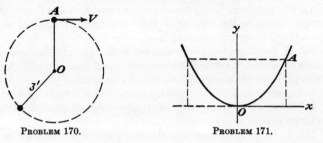

PROBLEM 170.   PROBLEM 171.

171. A point moves with a constant velocity of 1 ft/sec along a parabola, defined by the apex in the origin O and the point A with $x = 1$ ft and $y = 1$ ft. Calculate the acceleration of the point when it passes through the apex O.

172. The sketch shows a "crown wheel," an ancient piece of mechanism that has acquired new interest during the last 15 years as an element in the large computing machines that are called "mechanical brains" by the newspapers. With a crown wheel it is possible to integrate a function mechanically.

  a. Prove that if the crown wheel is rotated at constant angular speed ω_1, and if the rider wheel is moved along its own axis according to a given relation $r = f(t)$, and if the rider wheel does not slip, then the angle through which the rider rotates is proportional to the integral $\int f(t)\, dt$. What is the proportionality constant?

  b. Through how many revolutions does the rider wheel rotate in 5 sec if $r_0 = 1$ in., ω_1 is found from 8 rps, and r increases from 2 to 4 in. during 5 sec with a constant velocity $\dot{r}$?

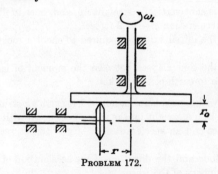

PROBLEM 172.

173. Another piece of mechanism that has been applied to calculating machines is the differential gear, familiar from the rear end of automobiles

(Problem 149).   It is used as a device for adding and subtracting continuous functions.

Prove that if axis C is forcibly rotated so that its angle $\varphi_C = f_C(t)$ is a given function of time, and if axis B is rotated as $\varphi_B = f_B(t)$, then the third or central axis rotates as

$$\varphi_A = \text{constant } [f_B(t) + f_C(t)]$$

Determine the proportionality constant from the dimensions of the device.

174. A third device that has been employed in computers is "Peaucellier's inversor," consisting of six rigid bars, two of length L, four of length l, linked together as shown.   The point P is guided to move along a straight line through O, and the point Q necessarily moves along the same line OPQ.   The apparatus is called an "inversor," because with $OP = r_P$ and $OQ = r_Q$, the relation $r_P r_Q = L^2 - l^2$ exists for all positions of the points P and Q, so that r_P and r_Q are inverses of each other.

a. Prove the property $r_P r_Q = L^2 - l^2$.

b. From it find the ratio of the velocities of points P and Q, and in particular, find the velocity of Q if $L = 8$ in., $l = 4$ in., $OP = 5$ in., and $v_P = 4$ in./sec.

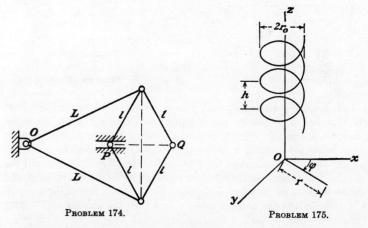

PROBLEM 174.   PROBLEM 175.

175. A point moves at constant velocity v_0 along a spiral curve, which is described in the simplest manner in semipolar coordinates, r, φ, z, as follows:

$$r = r_0$$
$$z = \frac{h}{2\pi}\,\varphi$$

in which r_0, the radius, and h, the pitch, are constants.   Calculate the components of acceleration in the three component directions, and find the resultant acceleration.

176. A point moves in a plane along an Archimedean spiral at constant angular speed ω_0, described by

$$r = r_0 + v_0 t, \qquad \varphi = \omega_0 t$$

a. Calculate the total velocity at time t, its r and φ components, and its x and y components.

b. Calculate the x and y components of acceleration at time t, and the total acceleration at time t.

c. Substitute numbers for point A, where $\varphi = 135°$, the angular speed is 1 rps, $r_0 = 10$ in., and $v_0 = 1$ in./sec.

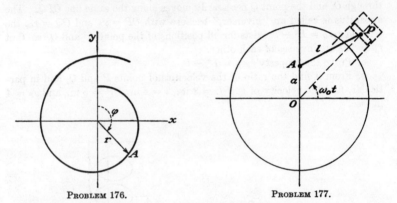

PROBLEM 176. PROBLEM 177.

177. The Gnome-Rhone aircraft engine, widely used during the First World War, had a stationary, non-rotating crank $OA = a$, while the seven cylinders, equiangularly spaced, rotated at constant speed. The propeller was attached to the rotating cylinders, either directly or through a gear reduction. The sketch shows one of the seven cylinders. The cylinders and the radius OP rotate at uniform speed ω_0; the piston P can slide in the cylinder, and the connecting rod AP of length l consequently rotates at non-uniform speed.

a. Calculate first the angular speed ω, and then the angular acceleration $\dot{\omega}$ of the connecting rod AP.

b. Substitute numbers: $a = 4$ in., $l = 8$ in. OP rotates at 2,000 rpm. Plot ω and $\dot{\omega}$ versus the angle $\varphi = \omega_0 t$ for a full revolution.

178. The two extremities of a rod of length l are made to slide along two perpendicular lines, the bottom block A with a constant horizontal velocity v_0.

a. Determine the path described by the mid-point B.

b. Find the velocity of mid-point B, and also the velocity of the upper block C, as functions of the time t. The bar is in the vertical position along the y axis at time $t = 0$.

c. Find the acceleration of points C and B.

d. Substitute the numbers $l = 2$ ft, $v_0 = 1$ ft/sec, and plot the various velocities and accelerations versus time for 2 sec, indicating the vertical scales in proper units.

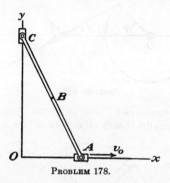

PROBLEM 178.

179. The main drive rod of a locomotive connects the block sliding in the guide to the crankpin of the front driving wheel. Let the center line of the guide be tangent to the circle constituting the path of the crankpin with respect to the wheel center. The system thus is a crank mechanism differing from Fig. 148 of the text only in that the piston-rod center line is tangent to the crank circle instead of passing through its center. Let the locomotive move forward at constant speed V, let the wheel radius be R, the crank radius r, and the connecting-rod length l.

a. Derive an expression for the displacement of the slide with respect to the wheel center and also with respect to the rail.

b. Find from it expressions for the velocity and acceleration of the slide.

c. Substitute the numbers $V = 60$ mph, $R = 2\frac{1}{2}$ ft, $r = 1\frac{1}{2}$ ft, and $l = 6$ ft, and calculate the maximum acceleration of the guide.

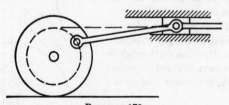

PROBLEM 179.

180. A crank $OB = r$ is rotating at a constant angular velocity $\dot{\varphi}$. It drives a rod, which passes through a rocking slide, pivoted at a fixed point A located at a distance $OA = 2r$. Let the angle of the rod BAC be ψ. Derive a formula for the ratio of the angular speeds $\dot{\psi}/\dot{\varphi}$ in terms of the angle φ only (the angle ψ must not appear in the answer). Check this formula for the four points where the result can be seen almost immediately, namely, for $\varphi = 0$ and 180° and the two locations where $\dot\psi = 0$.

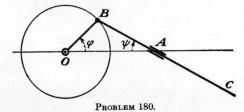

PROBLEM 180.

181. A mechanism known as a "Geneva motion" consists of two flat disks of equal diameter, set parallel to each other, face to face, a short distance apart.

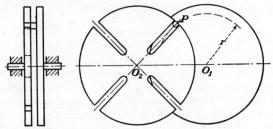

PROBLEM 181.   The Geneva mechanism.

One of the disks O_1 (the driving one) carries a pin P, perpendicularly attached to itself and at distance r from its center. The other disk has four slots as shown. The distance between the two disk centers O_1 and O_2 is $r\sqrt{2}$, which means that their 45-deg radii are perpendicular to each other, as shown in the sketch. The driving disk O_1 rotates at constant speed ω, and once per revolution the pin engages a slot and turns the other disk O_2 through 90 deg. During the next 270 deg of motion of the driving disk, the driven one stands still.

a. Let $\angle O_2 O_1 P = \varphi$ and $\angle O_1 O_2 P = \psi$. Find a simple relation between φ and ψ, not necessarily explicit in ψ, and plot φ versus ψ.

b. Differentiate and find a relation for $\dot\psi/\dot\varphi$ as a function of φ and ψ. Calculate this numerically and plot $\dot\psi/\dot\varphi$ versus φ.

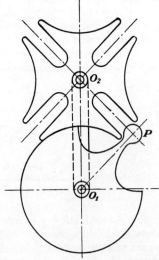

PROBLEM 181.   The Maltese-cross mechanism.

c. Notice that with the dimensions shown, the driven angular speed $\dot{\psi}$ is zero when the pin leaves the wheel O_2. Derive an expression, relating the wheel-center distance $d = O_1O_2$ to the pin radius $r = O_1P$ and to the number n of equidistant slots, that will ensure this desirable characteristic of the Geneva mechanism for a number of slots, n, different from four.

A construction of this mechanism where the two disks are in the same plane requires cutting away some parts of the driven disk. It is then called a "Maltese-cross mechanism."

182. A rod is pivoted at its end O. A block A can slide freely along the rod, and the block A is attached to another block B by a frictionless hinge pin. The block B can slide freely along a slide of which the center line has a perpendicular distance a to the center O. The bar is made to rock between two extreme 45-deg positions P and Q in a harmonic manner (see page 159), expressed by

$$\varphi = \frac{\pi}{4} \sin \omega t$$

a. Calculate the velocity of the block B.

b. What is the numerical value of that velocity for $\varphi = 30$ deg, $a = 10$ in., and $\omega = 6\pi$ radians/sec (which means 3 full back-and-forth oscillations per second)?

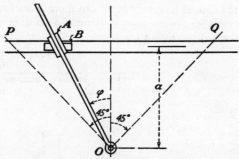

PROBLEM 182.

183. It is said that a parachutist lands on the ground with a speed equal to the speed he would have had if he had jumped from a height of 16 feet. Express that velocity in miles per hour.

184. In Galileo's problem (Fig. 154, page 180), the locus was found of the points reached by a particle in equal times starting from rest for various inclinations of a smooth plane.

Now find the locus of the points where particles starting from rest at various inclinations of a smooth plane reach equal velocities.

185. A particle slides down a 20-deg, rough inclined plane with a coefficient of friction of 10 per cent. Starting from rest, what is the speed reached after 3 ft of travel along the plane? What time has elapsed?

186. A block weighing 5 lb is pulled up along a 30-deg incline with a parallel rope having a tension of 5 lb. The coefficient of friction is 10 per cent. The block starts from rest and the string pulls during 2 sec, when it is suddenly slacked off.

a. How long does it take the block to slide down to the starting point at the bottom again?

b. How long does the incline have to be?

187. A four-engine airplane flies horizontally at uniform speed with a "drag" or air-resistance force equal to 5 per cent of its weight, which force is balanced by the four propeller pulls. One engine goes out, and the pilot leaves the adjustments on the remaining three engines unchanged, and also keeps the ship at the same speed as before by permitting it to glide down. How much altitude does the airplane lose per mile?

188. A particle is thrown up a smooth inclined plane with an initial velocity v_0. It will go up a certain distance and then reverse.

a. Find expressions for the time elapsed and for the velocity at a point on the way down midway between the starting and the topmost positions.

b. Give numbers for the case that $\alpha = 30°$ and $v_0 = 20$ ft/sec.

189. A 5-lb weight hangs from a spring of such stiffness that the spring stretches $\delta = \frac{1}{4}$ in. under the weight.

a. Calculate the stiffness of the spring.

b. Calculate the natural frequency of up and down vibrations.

c. Write a formula for the frequency in terms of δ alone, in which neither the mass m nor the spring constant k appears.

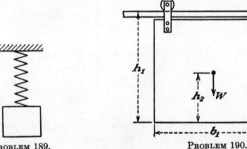

PROBLEM 189.   PROBLEM 190.

190. A sliding garage door of the dimensions shown is pushed by a force P at the bottom edge. Assume no friction anywhere, and assume that the door acts as a particle of weight W, located at height h_2, as shown. For what value of P does one of the wheels lift off, and which wheel is it?

191. An object of 1-cu-ft volume and of 65.0-lb weight is released from rest in salt water of 64.0 lb/cu ft. It sinks down so slowly that the water-resistance force is negligible at the beginning of the motion. How long does it take to sink down 10 ft?

192. An object sinks down in water, and is subjected not only to the gravity force, but also to a water-resistance force that is proportional to the velocity: $F = c\dot{x}$. In the beginning of its fall it will accelerate, but after some time it will acquire a terminal velocity that is constant.

  a. In what units is the constant c expressed?

  b. Find that terminal velocity in terms of whatever variables are necessary.

  c. Set up the equation of motion and integrate it once to obtain an expression for the velocity $\dot{x}$.

  d. If the specific weight of the sinking body is 1.05 and the terminal velocity is 10 ft/sec, calculate the time necessary to reach a speed of 8 ft/sec, starting from rest.

193. Calculate the length of a simple pendulum with a frequency of 50 cycles/minute, *i.e.*, making 50 swings to the right and 50 swings to the left each minute.

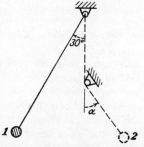

194. A simple pendulum with a 1-lb bob and of 2-ft length is released from rest from a 30-deg position. When it reaches the vertical it strikes a peg 1 ft under the point of suspension. It swings up to an extreme position 2. Calculate the time elapsed between 1 and 2. Calculate the tension in the string at position 2.

PROBLEM 194.

195. A cone with an apex angle of $2\alpha = 60°$ rotates at uniform speed about its vertical axis. On it lies a 1-lb particle hanging from a string. The particle rotates with the cone.

  a. Derive formulae for the tension in the string and for the pressure on the cone.

  b. Plot these numerically against the rpm, indicating proper scales on your graph.

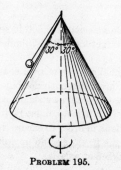

PROBLEM 195.

196. In a semispherical bowl of radius r there is a particle of mass m. In the absence of friction it makes no difference whether the bowl rotates or not; the condition is determined by the rotation ω of the particle. Find the relation between the angle α and ω. For $r = 6$ in., what is the required rpm for $\alpha = 45°$?

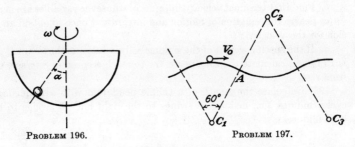

PROBLEM 196.   PROBLEM 197.

197. A portion of the smooth track of a roller coaster is made up of 60-deg arcs of a circle, of radius $r = 30$ ft, lying in a vertical plane as shown. A particle moves along this track having a velocity v_0 at the top, limited by the fact that on reaching the point A the pressure on the track is just zero. Determine the velocities and the pressures at the top and bottom points of the track.

198. A smooth track consists of a 45-deg incline tangent to a 225-deg arc of circle, as shown. A particle is launched with zero starting speed from point A at equal height with the top of the circle. At a point B, the particle will leave the track. Calculate the angle α where this takes place.

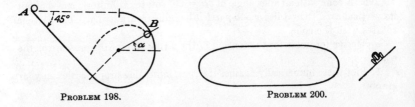

PROBLEM 198.   PROBLEM 200.

199. A gun has a muzzle velocity of 2,500 ft/sec. Calculate the angle of elevation required to hit a target located at 20 miles from the gun at the same level as the gun.

200. A track for automobile races of oval shape consists of two straight pieces and of two arcs that are not semicircles but curves of which the curvature increases gradually from zero at the straight portion towards a maximum at the center of the arc. Such a track must be inclined towards the inside, or must have an angle of bank in order to ensure that the racing vehicles have reactions perpendicular to the track, and do not slip sidewise (outward at too high speeds, inward at insufficient speeds).

PROBLEMS

403

a. Derive a formula relating the angle of bank to the radius of curvature R of the track and the speed of the vehicle v.

b. What is the maximum radius of curvature required at 120-mph speed, if the angle of bank is limited to 45 deg?

201. A flat circular table with upraised edge rotates at uniform speed ω about its axis. On the table lies a particle without friction, which can move along a radius only and is tied to the center by a spring of stiffness k. When the spring is unstretched (at no rotation of the table), the center of the particle is at radius r_1. When the particle reaches the edge, its center radius is r_2.

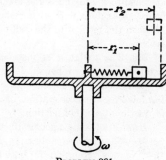

a. Set up an equation for the relation between r and ω for speeds ω in the region $r_1 < r < r_2$.

b. Set up the relation between the speed ω and the pressure against the edge of the table for higher speeds ω.

PROBLEM 201.

c. Let $r_1 = 5$ in., $r_2 = 8$ in., $w = 1$ lb, $k = 50$ lb/in. At what rpm will the particle reach the edge?

202. If in the Scotch crank mechanism of Problem 161, the piston, its rod, and the slotted guide have a total weight w, write

a. An expression for the torque that has to be exerted on the crank to rotate it at a uniform velocity ω.

b. For $w = 20$ lb, $r = 5$ in., and 500 rpm, what is the maximum value of the torque and at what angles α does it occur?

203. A bucket filled with water is swung at the end of a rope of length l, in a circular path in a vertical plane. If m_1 is the mass of the water, considered to be a particle, and m_2 is the mass of the bucket, find the minimum speed in the top of the path for which the water still stays in the bucket.

What is the tension in the string and what is the pressure of the water on the bottom of the bucket as a function of velocity in that top position?

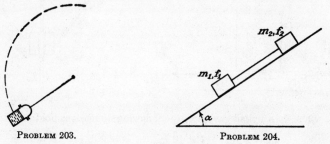

PROBLEM 203.   PROBLEM 204.

204. Two particles of different weight and different surface roughness slide down an inclined plane, *a.* Set up a formula for the downward acceleration and for the tension in the connecting string. What must be true about the two friction coefficients f_1 and f_2 if the string is to have tension?

b. What is the string tension if $w_1 = 2$ lb, $w_2 = 1$ lb, $f_1 = 10$ per cent, $f_2 = 30$ per cent, and $\alpha = 30°$?

205. The system of Problem 204 is being pulled up the plane by a string attached to the top mass and inclined at angle β with respect to the plane (and hence at angle $\alpha + \beta$ with respect to the horizontal). Write the relation between the tension in the pulling string and the upward acceleration of the system.

206. The wedge α of mass m_2 lies on a horizontal plane. On it rests another mass m_1. When m_1 slides down, the mass m_2 goes to the right. Find formulae for the accelerations of m_1 and m_2 in the absence of friction.

Hint: First write a geometric relation between the displacements x_1, y_1, and x_2; then Newton's equations for both bodies.

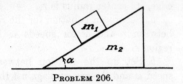

PROBLEM 206.

207. The wedge of the previous problem is pulled horizontally to the left by a string. If the string is pulled sufficiently hard, the mass m_1 will slide up the plane. Derive formulae for the accelerations of m_1 and m_2 as functions of the string tension T. Do this for the case of zero friction only. In particular, what is the tension T required to keep m_1 at rest relative to the wedge m_2?

208. A block resting with large friction on a flat car can be considered as a particle concentrated in its center of gravity. The flat car, on frictionless wheels, is pulled to the right with a force T. Find the force T at which the block will tip over, in terms of such letters as you may find it necessary to use.

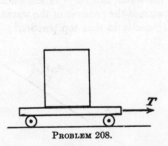

PROBLEM 208.

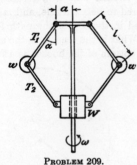

PROBLEM 209.

209. In a flyball governor, the following dimensions hold:

$$a = 2 \text{ in.}, \qquad l = 10 \text{ in.}, \qquad w = 2 \text{ lb}, \qquad W = 10 \text{ lb}$$

and the apparatus turns at 200 rpm. Calculate the angle α.

210. A chain of 4-ft length lies on a smooth horizontal table with 3 ft of its length, while 1 ft hangs down. How long does it take the chain to fall off the table, starting from rest? Use Fig. 165 of page 193 and a table of hyperbolic functions, if one is available.

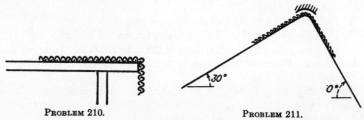

PROBLEM 210.   PROBLEM 211.

211. A flexible chain of total length l is lying partly on a 60-deg smooth inclined plane and partly on a 30-deg smooth plane, joined at the top.

a. Determine how the length l must be divided between the two planes so that there be equilibrium permanently.

b. Derive the equation of motion, starting from rest from a position that differs by a distance a from the position of equilibrium.

c. For $l = 3$ ft, divided into 18 in. on each of the planes, find the time in seconds for the end of the chain to slip over the top, starting from rest.

212. Figure *a* shows a thrust ball bearing in which each ball is supported at three points. The top race rotates while the bottom race stands still. Show that this can take place with pure rolling contact on the three points of the ball without sliding.

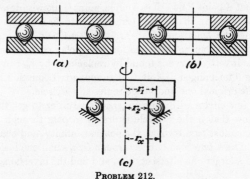

PROBLEM 212.

a. Find the instantaneous axis of rotation of a ball.

In Fig. *b* the balls are supported at four points, in a symmetrical way. Is it possible to rotate the upper race relative to the lower one without slipping on the balls?

b. Determine how the locations of the two upper points of contact have to be modified in order to obtain a proper, non-slipping, four-point bearing.

Figure c shows a combination thrust and axial bearing. Where is the instantaneous axis of rotation of each ball?

213. Consider once more the bar of Problem 178 sliding with its ends along two perpendicular tracks.

a. Determine the location of the velocity pole.

b. Find the pole curve on the x, y plane, *i.e.*, on the bottom sheet.

c. Find the pole curve on the plane of the ladder, *i.e.*, on the upper transparent sheet. Describe the motion in terms of a rolling of the latter pole curve over the first one.

214. This is not a problem in kinematics but one in geometry. You are advised to work it in order to appreciate the mechanisms of the next two problems, which are among the most beautiful ever invented.

The figure shows a circle of radius r_0 and center O, with an interior point P. Through P is drawn the chord perpendicular to the radius OP, and then two tangents are drawn to the circle, leading to the point Q.

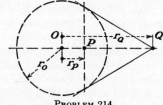

PROBLEM 214.

a. Prove that $r_P r_Q = r_0^2$

Thus, for a given circle, r_Q is r_0^2/r_P, and the point Q is said to be the *inversion* of the point P with the respect to the *circle of inversion*. Now, if point P describes a path or figure, point Q describes another figure, which is called the "inversion" of the figure P with respect to the circle of inversion.

b. Prove that if point P describes a circle, point Q likewise describes a circle, or in other words, prove that circles transform into circles by inversion. Do this by writing the equation of an arbitrary circle in the x,y coordinate system and then transform this equation into polar coordinates r, θ. The equation of the inverted curve is found by replacing r by r_0^2/r in the equation and by leaving θ unchanged. Note that the inverted equation has the same structure as the original one and hence represents a circle.

c. Draw a few circles with their inversions in order to get the feel of the situation. Show that if the P circle is drawn to pass through the point O, then the Q circle must pass through the point at infinity, and hence must be a straight line. Draw two such circles through O, a small one inside the inversion circle, and a larger one intersecting it, and find the inversions of those two circles.

215. The figure shows the mechanism of Problem 174, modified in that point P now is guided by the link O_1P to move on a circle. The dimensions have been so chosen that $\angle POR = 30°$, the long links $L = 5$ units, the short ones $l = 3$ units. Verify that the circle of inversion then must have a radius of 4 units, about O as center, as drawn. The path of point Q is a straight line. The thin line OA shows the extreme position to which the mechanism can move.

Here the distance $OQ = L + l = 8$ units, and the six bars of the inversor are all on top of each other along the direction OA.

a. If point P is given a velocity v_P, necessarily vertical in the position shown, find by graphical construction the velocities of the points R and Q.

b. Suppose we omit the link O_1P and replace it by a link from point Q instead. Where does the ground pivot of that link have to be placed in order that point P describe a straight-line path?

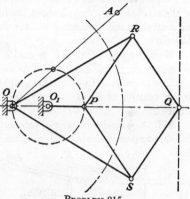

PROBLEM 215.

The device was invented in 1860 by Peaucellier, a lieutenant in the French Navy, about a century after Watt's parallelogram (page 203). During that century many other linkages had been described that could transform a circular into a rectilinear motion approximately, but Peaucellier's was the first exact solution. See the article on Linkages in the "Encyclopaedia Britannica."

216. Another linkage for producing straight-line motion described in the "Encyclopaedia Britannica" is due to Hart (1875). It consists of four bars

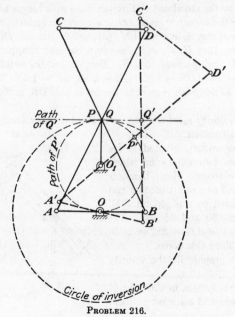

PROBLEM 216.

only—AB and CD of equal length l, and AD and BC of equal length L, pivoted together at their ends as shown. The mid-points P and Q of the long bars are *not* attached to each other. Hart has proved (you don't have to do it) that if the mechanism is moved about the fixed pivot O, the points P and Q remain on a straight line through O and are inversions of each other, as shown in the sketch.

Now the point P of AD is guided on a circle by the link PO_1, and as a consequence, point Q of BC will move on a straight line.

The problem is to determine graphically the velocity of Q in terms of the given velocity of A, in the fully drawn central position of the device. The dimensions are $AB = CD = l = 3$ in., $AD = BC = L = 7$ in.

217. A Walschaert type of locomotive valve gear is shown in the sketch. All dimensions are expressed in terms of the main-crank radius OA, which is

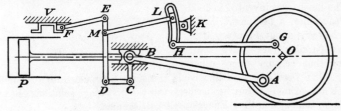

PROBLEM 217.

our unit of length. Then the connecting rod $AB = 4$. The steam valve V is driven partly by the crosshead B, through links with hinges at C, D, E, and F, and partly by the eccentric crank OG, the eccentric rod GH, the link HKL, pivoted at K, and the radius rod LM, pivoted to the bar DE at M. For simplicity, assume that H, K, and L are in a straight line, perpendicular to OB, GH, or CD and parallel to DE. Further let $OG = 0.5$, $HK = 0.5$, $KL = 0.3$, $BC = 0.75$, $DE = 1.80$, $ME = 0.36$, $EF = 1.50$. The angles of OA and OG are 45 deg with respect to the horizontal OB, and EF is parallel to ML.

Assuming a velocity v_P of the piston, determine the velocity of the valve by graphical construction.

218. A 90-deg rocker, pivoted at its corner O, carries two guides in which blocks can slide freely. Each block is pivoted to the end of a rod, the other end of which is pivoted to the ground. All angles shown are 90 and 45 deg. The lengths are $OA = AB = 2OC = 2CD$. The velocity of block C is given.

a. Determine graphically the velocity of block A.

b. Look at this system in the light of Fig. 146 (page 168) and see what you can see.

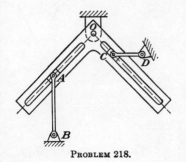

PROBLEM 218.

219. Return to the floating pendulum of the Toledo system (Problem 47).

a. What is the path described by the points *C* when the load *P* varies from the maximum value shown to a minimum value when the weights *W* are down?

b. Prove that the point *C* and the points of tangency of the bands *a* and *e* on their respective circular arcs *b* and *d* are all three lying on a straight line.

c. What is the velocity pole of one of the pendulums?

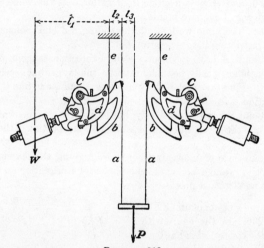

PROBLEM 219.

220. If in Watt's parallelogram (page 203), the bar *OC* is not made equal in length to *DF*, but half as long (*DF* = 2*OC*), and if the anchored hinge *F* is still located vertically below the mid-position of point *A*, find what other changes in the dimensions have to be made in order to obtain decent operation.

221. Prove that for a plane body with zero velocities (starting from rest), the acceleration pole coincides with the velocity pole, by using an argument employing figures like Figs. 178 and 181 (page 206).

222. A linkage of the dimensions shown is moving through the 45-deg position shown, with the bar *AC* turning clockwise at a speed of 1 radian/sec and zero angular acceleration.

a. Find the velocity of *D* and the angular velocity of *BD* in this position.

b. Find the acceleration of *C*.

c. Find the acceleration of *D*.

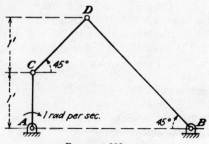

PROBLEM 222.

223. On page 210 it is shown that the location of the acceleration pole of an accelerated rolling disk depends on the angular velocity of the wheel, or rather on the ratio of the accelerations $a\dot{\omega}$ (horizontal, of the wheel center, due to acceleration) and $a\omega^2$ (radially inward, of a point on the periphery, due to angular speed). The text shows that for $a\dot{\omega}/a\omega^2 = 0$, the pole is at the wheel center, and for $a\dot{\omega}/a\omega^2 = \infty$, or for $\omega = 0$, the pole is at the bottom contact point. Prove that the locus of the acceleration pole for values $a\dot{\omega}/a\omega^2$ between $-\infty$, 0, and $+\infty$ is a circle, half as large as the wheel periphery passing through the above two points.

224. For the third time we look at the sliding rod of Problems 178 and 213, this time restricting it to a position $OAC = 45°$.

a. Let the point B have a velocity v_0 in a direction tangent to the bar (verify that this is physically possible), and let the tangential acceleration of point B be zero. Find the accelerations of points A and C and find the location of the acceleration pole.

b. Let the point B have zero velocity, but let it have a tangential acceleration $\dot{v}_0$. Now find the accelerations of A and C and the location of the acceleration pole.

c. Combine the two previous states of motion, giving the center point B a tangential velocity v_0 and a tangential component of acceleration v_0. Find the location of the acceleration pole.

225. Read the story explaining Fig. 181 on page 206. In the figure with this problem, the construction has been repeated in the sequence shown; first point 1, then 2, etc., the quantities 6 and 7 being angles. After the pole $P =$ point 8 has been so found, prove by geometry that angle 8-2-4 equals angle 8-1-3.

Hint: Prove first that triangle 1-2-8 is similar to triangle 5-2-4.

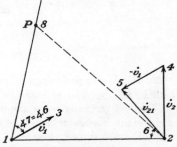

PROBLEM 225.

226. Consider a three-bar linkage as shown in Fig. 176, page 201, but with bars AB and CD parallel and vertical, and with bar BC horizontal. The lengths are: $AB = 1$ ft; $BC = 2$ ft; $CD = 2$ ft; so that the bearing D lies 1 ft lower than A. Bar AB has a constant $\omega = 10$ radians/sec in that position. Determine the acceleration pole of the horizontal bar BC.

227. Return to Peaucellier's linkage of Problem 215. If, in the central position shown, the point P moves with constant velocity through its circular path, determine graphically the accelerations of the points S and Q.

228. Consider the aircraft-engine speed-reduction gear of Problem 153, in which the central or sun gear is rotating at uniform velocity ω, and the outer gear is anchored at rest. Find the velocity pole and the acceleration pole for one of the planet gears.

229. A vertical rotor is suspended from a thrust bearing and supported sidewise by two journal bearings like the generator in a hydroelectric power station. The rotor itself is completely balanced, and when it is rotating, the journal bearings experience no forces. For our analysis we can disregard the mass of the rotor, which from now on is supposed to be weightless. Now we attach two small masses m on the periphery in different places, 180 deg apart angularly, as shown.

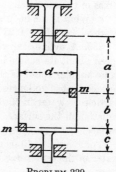

a. Express the bearing forces caused by this in terms of m, of the dimensions, and of the angular speed ω.

b. For $mg = 1$ lb, $d = 6$ ft, $a = 6$ ft, $b = c = 3$ ft, and 500 rpm, what are these bearing forces numerically?

PROBLEM 229.

230. A piston weighing 10 lb is moved harmonically (page 159) through a total stroke of 8 in. (crank radius, 4 in.) at 1,000 rpm. Calculate the force in the piston rod, neglecting friction.

231. A symmetrical locomotive side rod weighs 150 lb, and is attached to crankpins on two driving wheels. The rod is 7 ft long; the two wheel centers are 7 ft apart; the distance from a wheel center to a crankpin is 2 ft, the wheel diameter is 6 ft., and the locomotive is going forward at a constant speed of 60 mph. The "crank" is 45 deg from the vertical, as shown.

a. Determine the forces exerted by the side rod on each crankpin.

b. Let the ends of the rod each weigh 25 lb, and let the remaining 100 lb be uniformly distributed along the length of the rod. Sketch the bending-moment diagram of the rod and find its maximum bending moment.

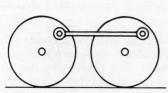

PROBLEM 231.

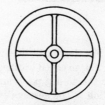

PROBLEM 232.

232. A flywheel consists of a heavy rim and very thin spokes. For simplicity we neglect the effect of the spokes. When the flywheel rotates, each element of it has a centripetal acceleration. Isolate 180 deg of the flywheel, and set up its d'Alembert equilibrium condition, assuming the cross section of the material A to be so small that all mass is concentrated in the circular center line of radius a.

a. Write the condition so that it shows the magnitude of the tensile hoop stress in the material.

b. What is the hoop stress, expressed in pounds per square inch, for a steel flywheel (weighing 0.28 lb/cu in.) of $A = 20$ sq in., $a = 3$ ft, at 500 rpm?

233. A cylindrical vessel partly filled with water is rotated at uniform speed about its center line. After some time all the water rotates with the same angular speed as the vessel, as a rigid body, and it has then acquired a surface that is no longer flat. Set up the condition of equilibrium for a small particle of water in the rotating surface, and from it deduce the differential equation of the surface. Integrate this equation and so find the shape of the surface.

For a vessel of 10-in. diameter, determine the rpm required in order that the water in the center line be 5 in. below the water at the periphery.

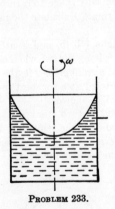

PROBLEM 233.

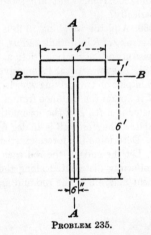

PROBLEM 235.

234. For the fourth time we return to the elliptic motion of Problems 178, 213, and 224, so called because every point of the bar describes an elliptic path. Consider the 45-deg position, and let the center point B have no tangential acceleration. The acceleration pole for that case was found to be the origin O in Problem 224. Calculate the inertia forces on the bar, sketch its bending-moment diagram, and determine the amount and location of the maximum bending moment. Also determine the forces on the two pins at A and C.

235. A flat plate of the dimensions shown has a total weight of 70 lb. Calculate the moments of inertia about the axes AA and BB.

236. A solid steel rotor of the dimensions shown weighs 2,100 lb. Calculate the moments of inertia about the axes AA and BB.

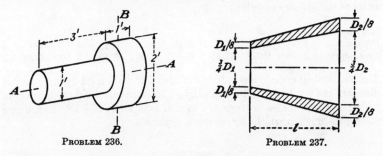

PROBLEM 236. PROBLEM 237.

237. A steam-turbine rotor is idealized into a truncated cone with a wall thickness equal to one-eighth of its outer diameter all along its length. Derive a formula for the moment of inertia about the axis of rotational symmetry, expressed in D_1, D_2, l, and ρ, the mass per unit volume.

238. Consider a Z-shaped flat plate of 14-sq in. area and a total weight of 1.4 lb.

a. Determine the moments and product of inertia about the set of axes through the center of gravity shown in the figure.

From the Mohr-circle diagram, find the direction of the principal axes through O and determine the maximum moment of inertia.

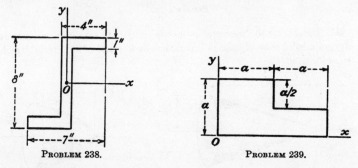

PROBLEM 238. PROBLEM 239.

239. A flat rectangular plate, $2a$ by a, with one quarter removed has a total weight $\frac{3}{4}W$.

a. Find the moments and product of inertia about the x and y axes through the corner O.

b. Find the center of gravity G and the moments and product of inertia about axes through G parallel to the x and y axes.

c. Find the direction of the principal axes through G by means of a Mohr-circle construction.

240. A three-bladed bronze ship's propeller has a diameter of 10 ft. Nine stations are marked out on a blade 6 in. apart radially, and the cross sections at those stations are as follows:

Station No.	1	2	3	4	5	6	7	8	9
Radius, ft	1	1½	2	2½	3	3½	4	4½	5
Area, sq ft	0.40	0.50	0.55	0.58	0.60	0.58	0.50	0.37	0

The hub with the shaft inside it is considered to be a solid cylinder of 2-ft diameter and 3-ft length. Bronze weighs 0.33 lb/cu in. Find the moment of inertia of this propeller in terms of I as well as in terms of WR^2, in each case stating the units in which the answer is expressed.

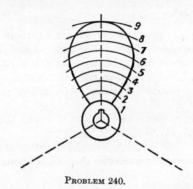

PROBLEM 240.

241. Determine the moment of inertia, about its axis of rotational symmetry, of a solid cone of base radius r_0, height h, made of a material weighing $\gamma = \rho g$ lb/cu in.

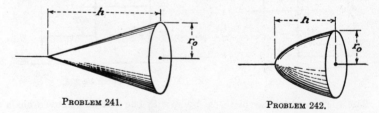

PROBLEM 241.

PROBLEM 242.

242. Find the moment of inertia of a solid paraboloid of revolution, of base radius r_0 and of height h, about its axis of rotational symmetry.

243. A connecting rod of an engine is idealized into a thin circular ring of radius r_1 and weight W_1, a thin bar of length $l - r_1 - r_2$ and weight W_3, and another ring r_2, W_2.

a. Find an expression for the moment of inertia about an axis perpendicular to the paper passing through C_1.

b. Without any further calculations write immediately the moment of inertia about a similar axis through the other center C_2.

c. Determine the center of gravity G and the moment of inertia of an axis through it perpendicular to the paper.

d. Reduce the above formulae to numbers with $W_3 = 3$ lb, $W_2 = 1$ lb, $W_1 = 2$ lb, $r_1 = 2$ in., $r_2 = 1$ in., $l = 8$ in.

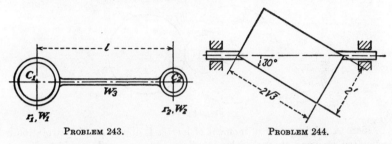

PROBLEM 243.   PROBLEM 244.

244. A steel plate 1 in. thick, 2 ft wide, and $2\sqrt{3}$ (about 3.46) ft long is mounted in bearings to rotate about a diagonal as shown. The weight of 1-in. steel plate is 40 lb/sq ft; the plate rotates at 100 rpm, and the distance between bearings is 5 ft. Calculate the bearing reactions, and in particular state the maximum and minimum values (in time) of the bearing reactions in the vertical direction and also in the horizontal direction.

245. In discussing Atwood's machine on page 191, we considered the pulley to be without inertia. Now we drop that simplifying assumption.

a. Derive formulae for the acceleration of the system as shown, and for the tensions in the two branches of rope.

b. Let $W = 2$ lb, $w = 0.1$ lb, and let the pulley be a uniform disk of 3-lb weight and 6-in. diameter.

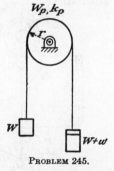

PROBLEM 245.

246. A solid steel horizontal cylinder of 3-ft diameter and 5-ft length is supported in oil-lubricated journal bearings of 8-in. diameter with an equivalent coefficient of friction of 2 per cent. If the rotor is spinning at 1,800 rpm, how long will it take to come to rest as a result of bearing friction alone?

247. A gear train consists of three gears of diameters $d_1 = 2$ in., $d_2 = 4$ in., and $d_3 = 8$ in. having moments of inertia I_1, I_2, and I_3 to be calculated from the fact that all gears are uniform brass disks of ¼-in. thickness. On the shaft of the smallest gear is mounted a steel flywheel I_f of 10-in. diameter and 2-in. thickness. A torque $M_t = 5$ in.-lb is applied to the largest gear.

a. Derive the formula for acceleration of the largest gear in terms of letters only.

b. Substitute numbers to find the numerical answer.

c. If the system were replaced by a single flywheel at the large gear, how large would this "equivalent flywheel" have to be made in order to produce the same acceleration for the same torque?

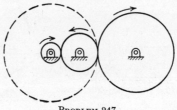

PROBLEM 247.

248. A ship's rudder of moment of inertia I_r about its axis of rotation is operated by hand through a long endless cable wrapped around a capstan r_1, which is on the same shaft with a gear r_2, which meshes with a pinion r_3 on the shaft of the steering wheel r_4. The wheels r_1, r_2, r_3, r_4 have moments of inertia I_1, I_2, I_3, I_4; the weight of the entire cable is W_c, and the weights of the various pulleys and other parts not mentioned specifically are negligible. Calculate the "equivalent moment of inertia" of the steering wheel, *i.e.*, the ratio between the torque applied to the steering wheel and its angular acceleration.

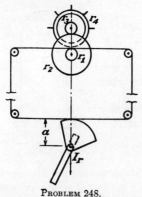

PROBLEM 248.

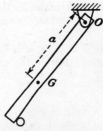

PROBLEM 249.

249. A bar of length l, pivoted at its top O, is held in a 45-deg position by finger support at the bottom end. The bar is not uniform, its center of gravity is at distance a from the pivot, and the moment of inertia about the pivot is I_O.

a. Calculate the pivot reaction and the finger pressure while at rest.

b. Calculate the bearing reaction the instant after the finger has been withdrawn, and the bar starts to swing down.

250. A uniform stick is pivoted at one end and supported at a fairly small angle α. To its end are attached two very light cups, the outer one of which contains a marble. When the bar is allowed to fall to the horizontal position on a felt pad, it is seen that the marble is no longer in the outer cup, but has jumped to the inner one. Explain this experiment.

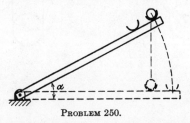

PROBLEM 250.

251. Return to the Maltese-cross mechanism of Problem 181. If the driving disk O_1 has a uniform angular speed $\dot{\varphi}$, with $\ddot{\varphi} = 0$, and the driven disk has a moment of inertia I, find the maximum value of the torque on either disk and the maximum value of the force between the pin and the slot, assuming no friction.

Hint. In differentiating remember that $\dot{\varphi}$ is constant. After finding the answer for the force as a function of φ, plot it and look before proceeding to find the maximum.

252. On page 237, Fig. 214, it was seen that a compound pendulum consisting of a uniform rod of constant length l could be made to swing faster by moving its fulcrum to a proper location along the bar. Now consider a similar, but yet different problem. Let a bar of length l be pivoted at its end.

a. Is it or is it not possible to make this pendulum swing faster by welding on an additional piece of length x to the rod, either at the top end or at the bottom end?

b. Answer the same question for a rod pivoted in its center point.

253. A pulley wheel of radius r carries a uniform chain of length $2a + \pi r$ and of weight μg per unit length. The moment of inertia of the pulley is I_0. Assuming no friction in the pulley axle and large friction on its periphery, derive the equation of motion, starting from rest from an initial position described by $a - x_0$ and $a + x_0$, until the moment that the left overhang becomes zero (see page 193 and Problem 210).

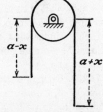

254. In the previous problem, calculate the horizontal and vertical components of the force on the pulley axle during the motion (as functions of the time).

PROBLEMS 253 and 254.

255. A spool-like wheel of total weight 10 lb has an outside diameter of 6 in. It has a thin slit cut into it down to a diameter of 2 in., and with this it can roll without sliding on a thin 45-deg incline. A weightless rope attached to the center point passes over a weightless and frictionless pulley to a hanging weight of 10 lb. Calculate the acceleration of the assembly.

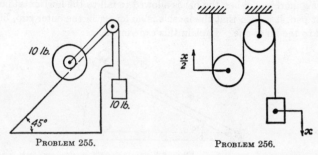

PROBLEM 255.   PROBLEM 256.

256. A system consists of a square weight W and two frictionless pulleys, one supported from the ceiling and one floating as shown. The pulley wheels also weigh W lb; their radius is r and they are uniform disks. Find the three rope tensions T_1, T_2, and T_3, and the acceleration of the square weight.

257. A uniform bar of length $2l$ and weight W is suspended from its center point C by a weightless bar of length l, pivoted at its top A. There is no friction in the top pivot A.

a. Find the frequency of small oscillations and describe the motions of the two bars, assuming that there is no friction in the pivot C.

b. The same question, assuming large friction in C, so that the joint is a rigidly clamped one.

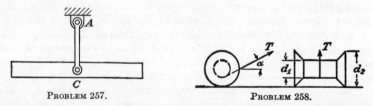

PROBLEM 257.   PROBLEM 258.

258. This is the classical problem of Grandma and the Cat. Grandma sits sewing, drops her spool of thread on the floor, and the cat paws it out of her reach. Under what circumstances can she retrieve the spool by pulling at the thread? Assume sufficient friction on the floor so that the spool will roll without slipping.

259. On pages 210 and 244, and in Problem 223, the accelerated rolling motion of a uniform disk was discussed. Let such a disk be subjected to a horizontal force P at its center (and consequently have a friction force $P/3$ at its bottom contact point). The acceleration $\dot\omega$ is determined by the above, but the angular speed may have any value, depending on the length of time that P has been acting. Write the moment equation about the center of gravity of the wheel and obtain the answer for $\dot\omega$ (independent of ω). Then write the moment equation about the acceleration pole (Problem 223) and show that this leads to the same, correct, answer for $\dot{\omega}$, independent of the ratio $a\dot{\omega}/a\omega^2$, or independent of the location of the acceleration pole in the wheel.

260. A body consists of a uniform bar of length l and mass m. To it is attached at one end a concentrated mass $m/2$. At what point can the bar be struck, in a direction perpendicular to itself, without causing any acceleration of the concentrated end mass?

261. At what height above the table does a billiard ball have to be hit in order that it will not slip on the table, but start in a pure rolling motion?

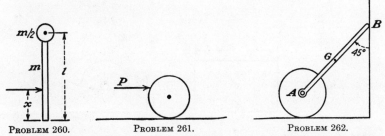

PROBLEM 260.   PROBLEM 261.   PROBLEM 262.

262. A uniform bar of length l and weight w rests against a wall without friction at B and is hinged by a frictionless ball bearing at A to a roller. The roller is a uniform cylinder of weight W and radius r, resting on the ground with sufficient friction to ensure rolling without sliding. The system starts at 45 deg, as shown, from rest (zero velocity).

Let a_x be the (horizontal) acceleration of A in this position. Then

a. Find the acceleration of B by kinematics.

b. Find the acceleration of G and the angular acceleration about G by kinematics.

c. Introduce the reactions of the bar by proper letters and set up Newton's equations for the bar.

d. From the above, solve only for the horizontal reaction at A.

e. By Newton's laws find the relation between this reaction and the horizontal acceleration a_x of the rolling cylinder, and solve for a_x.

263. A uniformly solid cylinder of radius r can roll without sliding in a track of radius R.

a. Draw the forces acting on the cylinder in an arbitrary position φ, and write Newton's equations for that position. Eliminate the two unknown forces between these equations and arrive at a single equation in terms of φ and its derivatives.

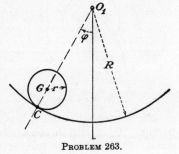

PROBLEM 263.

b. Compare this result with Eq. (9*d*) (page 185) and find the frequency of oscillation of the cylinder for small angles φ.

c. What would be the frequency if the friction were zero and the cylinder could slide freely?

264. A rotor with journals of 6-in. diameter is placed on two tracks with a radius of curvature of 10 ft. By experiment the period of a full back-and-forth rolling motion is found to be 18.0 sec. The rotor is then placed on scales and found to weigh 10,000 lb. Determine from these two measurements the moment of inertia of the rotor, using the result of the previous problem.

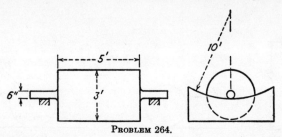

PROBLEM 264.

265. From the theory on page 249, deduce the linear acceleration of the bottom point *A* of a frictionless uniform sliding ladder in the position $\varphi = 60°$.

266. A uniform cylinder of radius *r* and weight W_1 can roll without sliding on a horizontal plane. On its center *C* it carries a frictionless ball bearing forming the pivot of a compound pendulum of weight W_2 with a distance $CG = a$.

a. Draw all the forces acting on the cylinder as well as on the pendulum and write all Newton's equations. Eliminate between these all forces, so that only two equations remain, in terms of the quantities x_C and θ.

267. A particle is sliding down a rough inclined plane of angle α and friction *f*. It starts with a velocity v_0. Calculate the velocity v_1 it reaches after it has moved a distance *s* along the plane.

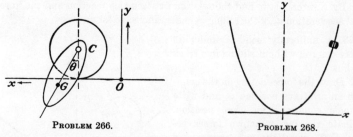

PROBLEM 266.　　　　PROBLEM 268.

268. A bead is strung on a thin, stiff wire, along which it can slide without friction. The wire is shaped according to the parabola $y = 2x^2$ in a vertical plane. The bead starts without initial velocity at the position $x = 1$ ft,

$y = 2$ ft. Calculate the horizontal speed of the bead when it passes through the bottom point $x = 0$, $y = 0$ of the parabola.

269. A weightless spring of stiffness $k = 40$ lb/in. is mounted vertically on a solid horizontal floor. It is deflected ½ in. from its non-stressed position by a weight of 1 lb pushed down by hand. When the hand pressure is suddenly removed, the weight will be thrown upward by the spring.

a. How far will the bottom of the weight rise above the unstressed top of the spring?

b. What is the speed of the weight at the moment that it leaves the spring?

270. The figure shows a rear-axle bicycle coaster brake of the familiar "New Departure" construction. It consists of a non-rotating central axle 1, to which two other parts 2 and 3 are securely fastened. The hub shell 4 can rotate on the ball bearings and carries the wheel spokes attached to the flanges shown. The chain sprocket, part 5, which can rotate relative to the axle through a small angle, is securely fastened to part 6, which is made with a coarse large-pitch screw thread into which meshes part 7. When forward drive is applied at the bicycle pedals and hence to the sprocket 5, part 7 is screwed by part 6 to the right and makes a self-gripping friction lock in the inside conical hub shell, part 4, thus driving the wheel. When torque at 6 is reversed, part 7 is screwed to the left and engages the teeth 8, of a non-rotating part 8, causing pressure on the disks 9. There are 11 stationary and 10 revolving brake disks 9, which can slide freely along the axle or the hub, but cannot rotate with respect to them.

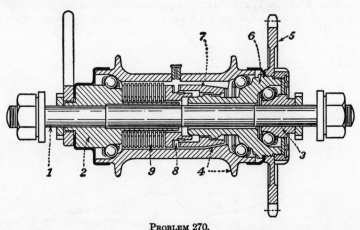

PROBLEM 270.

a. Calculate the braking force on the bicycle, assuming no slip between the rear tire and the ground, based on the following dimensions: The push of the boy's right foot is 40 lb, equal to 50 per cent of his weight. The kinematic chain from his foot to part 7 is such that for 1-in. displacement of his foot, the part 7 moves $\frac{1}{15}$ in. to the left relative to the axle. The screw thread 6 and 7 is 75 per cent efficient as a torque-force converter. There are 20 surfaces of disk contact; the average disk radius is $\frac{1}{2}$ in., and the coefficient of friction is 15 per cent. The outside diameter of the tire is 26 in.

b. Assuming the load on the rear tire to be 80 lb, and the friction between tire and road to be 100 per cent, what braking force from the foot on the pedal will cause the rear tire to skid on the road?

271. An English bicycle usually has friction brakes on the inside of both wheel rims, operated by grips on the handle bars, each hand servicing the brake on a wheel. The brake shoe is a small block of rubber, pulled upward against the rim by a lever having an advantage of about 10 to 1 with respect to the hand. When the bicycle and its rider weigh 200 lb, how many feet of distance does it require to come to a stop from an initial speed of 10 mph, when the gripping force of each hand is 10 lb and the coefficient of friction is 0.3? Neglect the small difference in diameter between the inside of the rims and the outside of the tires.

272. A two-basin tidal power plant such as was discussed for Passama-quoddy Bay, Maine, in the thirties, consists of an upper basin U and a lower basin L, of, say, 40 sq miles area each, both dammed off from the open tide-water T. The powerhouse P is located on a peninsula between the two basins. Assume that the outside tides are 20 ft, sinusoidally varying in time with a period of $12\frac{1}{2}$ hours. By opening and closing gates in the dam, the water level in the upper basin is kept as high as possible, always at least as high as the open tide T, and the lower basin is kept at as low a level as possible. The powerhouse is in continuous operation, the water flowing through it always

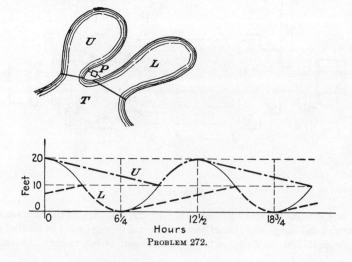

PROBLEM 272.

from U to L, in such a manner that the level of U decreases and that of L increases at the rate of 10 ft in $\frac{3}{4} \times 12\frac{1}{2}$ hours, as indicated in the sketch. Calculate the average available horsepower of the plant, using partly a graphical process on the figure.

273. A windmill driving an electric generator can be built to extract 30 per cent of the kinetic energy of the wind passing through the propeller disk and to convert it into electric energy. The largest windmill built so far (1941 in Rutland, Vermont) had a propeller-disk diameter of 175 ft. Calculate the kilowatt output of the generator, for a wind velocity of 30 mph, using the fact that air is 800 times as light as fresh water.

274. A four-cylinder automobile engine of the usual four-cycle type has a bore of $3\frac{1}{2}$ in. and a stroke of 4 in. Its indicator diagram is shown in the sketch, the suction and exhaust quarter cycles being represented by the horizontal line doubled up. The upper line of the diagram is the firing quarter cycle, and the other line is the compression quarter cycle. The area of the diagram is the same as that of a rectangle on the 4-in. base with a height of 75 lb/sq in. Calculate the indicated horsepower of this engine at 2,000 rpm.

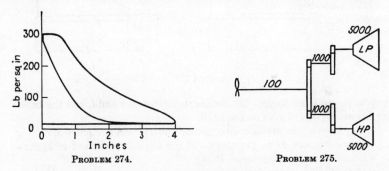

PROBLEM 274.   PROBLEM 275.

275. A steam-turbine ship drive consists of a high-pressure (HP) and a low-pressure (LP) turbine, each rotating at 5,000 rpm and each developing 10,000 hp. They drive the propeller shaft at 100 rpm through double reduction gearing with an intermediate speed of 1,000 rpm.

a. Calculate the torques in the turbine shaft, the intermediate shaft, and the propeller shaft.

b. Calculate the tooth pressure forces on the various gears. The pitch radius of the high-speed pinions is 3 in., that of the low-speed pinions is 6 in.

276. A very heavy train of coal cars has to be pulled up a 2 per cent slope at slow speed. The train resistance is such that without brakes or locomotives the train would just start rolling down a slope of 0.002, and this train resistance is independent of speed for the slow speeds considered here. When the locomotive starts from rest on this slope, its drawbar pull is 1.10 times (10 per cent greater than) the draw-bar pull for pulling up at constant speed.

a. How much time will it take to bring the train speed up to 10 mph and how far will it go in that time?

b. For a train weighing 10,000 (short) tons calculate the required drawbar pull, the required horsepower rating of the locomotive(s), and the necessary weight carried by the driving wheels, if the coefficient of friction under those drivers is 25 per cent.

277. The sketch shows two constructions of a band brake, fitted through 270 deg around a drum of radius r, rotating at N rpm, and operated by a hand force P at the end of a lever. The coefficient of friction is f. Derive the expression for the horsepower dissipated in each case.[1]

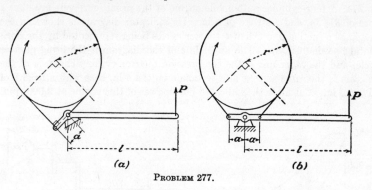

(a)   *(b)*

PROBLEM 277.

278. A crank mechanism has the usual dimensions r and l, and the angles are denoted by φ and ψ as on page 170.

a. Let a force P be acting on the piston, which is held in equilibrium by an appropriate moment M at the crank. If the angle φ is allowed to increase by $d\varphi$, and if as a consequence P is allowed to displace through a distance dx, the work $P\,dx$ of the piston equals the work $M\,d\varphi = Tr\,d\varphi$ of the crank (see page 140). Prove it by expressing M in terms of P, and dx in terms of φ and ψ.

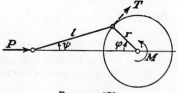

b. In the crank mechanism let the moment of inertia of the crank be I_c;

PROBLEM 278.

let the connecting rod consist of masses m_p and m_c, concentrated at the piston and crankpin, and an additional mass m_r, distributed uniformly along the rod. Further let the system start from rest, from the position $\varphi = 0$, and proceed through half a turn to $\varphi = 180°$, under the influence of a constant external piston force P. What is the speed $\dot\varphi$ at the end of that half revolution?

---

[1] For several other types of brakes and clutches, see Marks' "Mechanical Engineers' Handbook," 4th ed., pp. 942–947.

279. A precursor to the now-familiar helicopter was the autogiro, an aircraft sustained in the air by a large-diameter horizontal rotor of the same general appearance as a helicopter. Whereas the rotor of a helicopter is power-driven, the rotor of the autogiro is freely spinning without any power. Before it can take off from the ground, the rotor has to be set rotating by a starter, and in the latest model the starting rpm of the rotor was made about twice as large as the rpm in the air. While bringing the rotor up to speed on the ground, the angle of attack of the rotor blades was kept at zero, so that no lift occurred. Then, at take-off, the pilot would suddenly increase the angle of attack on the blades, which would lift the aircraft off the ground. The kinetic energy of the rotor would diminish in the process.

Consider an autogiro with a rotor constituting 25 per cent of the total weight, and consisting of four blades that are to be considered uniform radial bars. Just before take-off the peripheral speed of the rotor blades is 400 ft/sec, and in ordinary flight it is 200 ft/sec. Assume that the lifting process is 50 per cent efficient, *i.e.*, that half of the loss in energy of the rotor is converted into lifting work. How high will the aircraft be shot up?

280. A double-block brake is installed on an elevator-hoist motor, and is actuated at P by the magnetic pull of a solenoid.

a. Derive an expression for the number of revolutions required to stop the rotor of moment inertia I from an initial rpm N.

b. Substitute numbers: $r = 5$ in., $a = 9$ in., $b = 3$ in., $c = 9$ in., $P = 100$ lb, $gI = 50$ lb-ft^2, $N = 1,800$ rpm, $f = 0.30d = 7$ in.

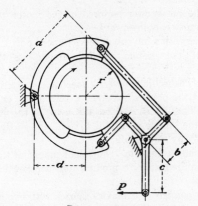

PROBLEM 280.

281. An electric induction motor is being calibrated by means of an electric dynamometer. The shaft of the motor to be tested is coupled to the rotor A of the dynamometer. The dynamometer stator B is again mounted in ball bearings. It carries an arm counterweighted at C, so that its center of gravity is in the center of the dynamometer A. The torque of the motor is

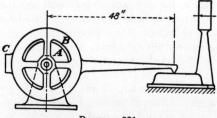

PROBLEM 281.

transmitted from A to B magnetically through the airgap and finally carried on the scale. The measurements made are (*a*) the motor input: 50.0 kw, (*b*) the motor speed: 1,750 rpm, (*c*) the scale load: 47.0 lb.

Calculate the efficiency of the motor under test.

282. An express passenger elevator in a tall office building consists of an empty cab of 3,000 lb (calculated to weigh 5,000 lb when full of passengers), a counterweight of 4,000 lb, connected by a cable slung over a 3-ft-diameter sheave. The cable and sheave weights are to be neglected. The sheave is driven through a 20:1 reduction gear by an electric motor, which can be considered as a solid steel cylinder of 1-ft diameter and 1-ft length. The service speed is 900 ft/min and the acceleration or deceleration is $g/10$.

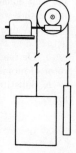

a. Calculate the horsepower required of the motor when accelerating the empty cab downward.

b. When running with a full cab at service speed how is the kinetic energy of the system divided among its three components?

PROBLEM 282.

283. Return to Peaucellier's inversor mechanism of Problem 215. In the position shown, let the point P be given a velocity v_0, and let the mass per unit length of all bars be m_1 and let the mass of the joints be negligible.

a. Determine the kinetic energy of the system, partly by graphical construction and partly by calculation.

b. From the result *a* find the "equivalent mass" of the point P, *i.e.*, the (vertical) force required at P to produce unit acceleration at that point in the same direction.

284. A carriage consists of two axles A, each with two equal wheels solidly attached to it, a carriage body B, and two connecting rods C, one on each side. It can roll without slipping on its level rails, but there is no friction anywhere else. Let φ be the angle of the crank radius with respect to the vertical; let m_A, m_B, m_C be the masses of all the wheels, the body, and both connecting rods, respectively. Let I_A be the moment of inertia of one axle with two wheels about its own center, and let the crank radius be r_c and the wheel radius r_w.

a. When the system starts from rest from a position φ, calculate the speed with which it passes through the bottom position $\varphi = 0$.

b. Reduce the previous answer to small angles φ and then find the frequency of oscillations about the position $\varphi = 0$.

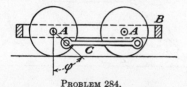

PROBLEM 284.

285. A cylinder, of which the center of gravity G is at distance a from the geometric center C, lies on a rough inclined plane α.

a. Derive a condition between a, r, and α expressing that the cylinder will always roll down the plane, and that it never can roll upward or be in equilibrium.

b. In the on-the-fence case a, describe the position of equilibrium of the cylinder and find whether that equilibrium is stable or unstable.

c. Let $a = r/4$, $\alpha = 30°$, $I_G = mr^2$, and let the cylinder start from rest from a position where CG is perpendicular to the incline. Calculate the velocity of point C for four subsequent positions, after rotations of the cylinder of 90, 180, 270, and 360 deg.

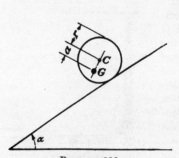

PROBLEM 285.

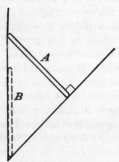

PROBLEM 286.

286. Consider once more the case of Problem 115, this time without any friction on either wall. If the uniform bar of total weight W and length l starts without initial velocity from the position A shown, describe the state of its velocities just before it slams into the vertical wall, position B.

287. A three-bar linkage of three equal lengths l has a base distance of $l(1 + \sqrt{2})$. It is shown in three positions—I, symmetrical upright, II with one bar on the ground, and III with two bars along the same line. Let the weight of each bar be w_b, uniformly distributed along its length; let further w_j be the concentrated weight at each joint. If the system is allowed to fall from rest, under the influence of gravity, from position I via III to II, calculate the velocity of joint A in positions II and III.

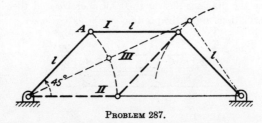

288. Consider the grab bucket described in Problem 151. If with an empty bucket in the closed position the closing rope breaks, the bucket will swing open. Assuming absence of friction, express the downward velocity of point C, when C falls down to the same level as B, in terms of such masses and moments of inertia as you may have to assign letters to.

289. Two light carriages, one weighing W lb, the other weighing $2W$ lb (including the passengers), are at rest on a level track close together. Their passengers push the two carriages apart, without touching ground, so that they acquire velocities in opposite directions. Assuming that the coefficient of rolling friction between the carriages and the track is the same for both, find the ratio of the distances from the common starting point to the positions where the two carriages come to rest again.

290. A 90-deg pipe bend lies in a horizontal plane and is pivoted at a point O about a vertical axis. The point O is the intersection of the center lines of the incoming and outgoing streams of water. If the cross section of the stream is 1 sq in. and the velocity is 50 ft/sec, calculate the direction and magnitude of the pivot force at O. Consider only the component of that force in the horizontal plane, disregarding the vertical component caused by the dead weight of the pipe and the water.

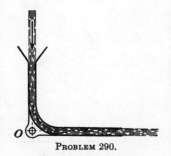

PROBLEM 290.

291. A small .22-caliber pistol weighs 1 lb, and it fires bullets weighing $\frac{1}{100}$ lb with a powder charge in each shell weighing $\frac{1}{400}$ lb. The powder has a chemical energy of one million ft-lb per lb of powder, and the firing process is 25 per cent efficient (25 per cent of the chemical energy is transformed into kinetic energy; 75 per cent goes into heat).

a. Neglecting the momentum of the powder gases, calculate the muzzle velocity of a bullet.

b. With a gun-barrel length of 4 in., and assuming constant acceleration of the bullet while in the barrel, what is that acceleration?

292. A bullet weighing 1 oz is shot with a velocity of 2,000 ft/sec at a
1-in.-thick piece of soft wood weighing 2 lb.
The bullet travels clear through it and comes
out at the other end with a speed of 500 ft/sec.
Assume that the support under the block of
wood does not exert a horizontal force on it.

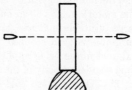
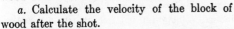

a. Calculate the velocity of the block of
wood after the shot.

b. What percentages of the total original

PROBLEM 292.

energy are in the bullet, in the block, and dissipated into heat after the shot?

293. A 30-lb shell is fired straight ahead from a gun mounted centrally in a
B-25 airplane weighing 35,000 lb. The muzzle velocity is 2,400 ft/sec, and
the length of the gun barrel is 15 ft. Assuming constant acceleration of the
shell in the barrel and assuming that the gun barrel is solidly mounted so that
the entire airplane as a rigid body catches the recoil, calculate

a. The force on the plane.

b. The change in speed of the plane.

294. A weightless rope carrying two unequal weights W_1 and W_2 is slung
over a pulley of weight W_3, radius r, and radius of gyration k. At time $t = 0$,
the smaller weight, W_1 is going up and the larger one, W_2, is going down with a
velocity v_0.

a. Calculate the time t at which these velocities are doubled to $2v_0$.

b. If you write the linear momentum equation in the vertical direction with
an actuating force $W_2 - W_1$, the answer so obtained for question *a* is incorrect.
Explain why.

295. Return to Problem 256, and the discussion on page 265.

a. Calculate the angular speed of the floating pulley 2 sec after starting
from rest, for the case that the radius $r = 6$ in.

b. Through how many revolutions does the floating pulley rotate during
the first 2 sec?

PROBLEM 294.

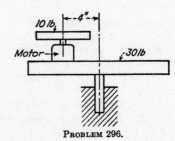

PROBLEM 296.

296. A small turntable is eccentrically mounted on a large one by means of
a vertical-shaft motor, as shown. The eccentric distance is 4 in. The weight

of the small turntable, together with the rotating parts of the motor, is 10 lb, and the radius of gyration is 2 in. The large turntable (including the stator of the motor) weighs 30 lb and has a radius of gyration of 5 in. Initially the large turntable is at rest, and the motor is running at 1,750 rpm. The small turntable is then stopped by means of a magnetically operated brake, no external forces being applied to the system during this operation. Find the final angular velocity of the system. The bearings are all frictionless.

297. One of those "old-fashioned" 50,000-ton battleships with a "battleship admiral" on the bridge fires its nine 16-in. guns broadside simultaneously in a horizontal direction. What is the state of velocity of the ship immediately after the firing? Assume the ship to be a symmetrical rectangular box of 50,000 (short) tons weight, and of a radius of gyration of 20 ft. Let the guns be 30 ft above the center of gravity; let each of the nine shells weigh 555 lb, and let the muzzle velocity be 3,000 ft/sec.

PROBLEM 297.

298. Calculate the diameter and the indicated horsepower of a Pelton water turbine operating at 1,800 rpm under a 2,000-ft head with a single jet 1 sq in. in cross section.

299. A projectile or rocket that can cross an ocean must have an initial velocity that is almost sufficient to give it unlimited reach on earth by having it circle the earth indefinitely like a planet close to the earth's surface. Calculate the speed required for a projectile to circle the earth at 100 miles above the earth's surface indefinitely. The radius of the earth is 4,000 miles, and the gravitational attraction constant g is inversely proportional to the square of the distance from the earth's center.

300. A projectile is shot up vertically at initial speed v_0. Neglecting air friction near the earth, calculate the v_0 required to get out of reach of the earth altogether (infinitely far away), and also the v_0 required to get 80 earth radii away, which is about the distance to the moon.

301. Referring to page 287 and to Problem 299, calculate the ratio of fuel weight to structure weight of a rocket capable of circling the earth indefinitely.

302. A block of 5-lb weight on a 30-deg inclined plane starts sliding down from rest.

a. How far does it go during the first second assuming no friction?

b. How far does it go during the first second when the friction is 10 per cent.

303. A rope slung over a pulley with a completely frictionless bearing carries on one side a dead weight W and on the other side a man of the same weight. The system is initially at rest, and the man is at the same height as the dead weight. Then the man starts climbing up the rope.

   a. Discuss carefully what happens to the accelerations, forces, and displacements in the system, and in particular, deduce what will happen to the dead weight.

   b. What would be the influence of a little friction in the pulley axle?

   c. In some books it is stated that the man is a monkey. Does that make any difference?

   d. Discuss what happens when the man is replaced by a yo-yo wheel, which rolls down its string.

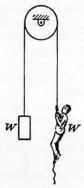

PROBLEM 303.

304. In the billiard-ball problem of page 283, calculate the loss of energy due to sliding friction during the time t_1 that slipping occurs. Do this in two ways: first, by calculating the energies at time $t = 0$ and $t = t_1$ with the results of page 283, and second, by calculating the path of slip and finding the work of the friction force directly. In the latter calculation be careful to consider the details of the slipping motion, and remember that when the ball rolls no relative slip takes place.

305. A pile of 30-ft length, weighing 40 lb/ft, and hence of total weight 1,200 lb, is being driven into the ground by a ram of 2,400-lb weight. With the last stroke the ram is dropped from a height of 16 ft, and when it falls on the pile, the coefficient of restitution $e = 0.5$. With this last stroke the pile is observed to penetrate 1 in. deeper into the ground. Assuming that the retarding force from the ground during that 1-in. motion is constant, and assuming that from now on the pile can carry that same force without sinking in any further, calculate the carrying capacity of the pile.

306. Consider the problem of elastic, oblique impact between two smooth spheres r_1, m_1 and r_2, m_2. Let the sphere m_2 stand still and let m_1 strike it, such that the direction of its incoming speed V_1 includes an angle α with the line connecting the two centers and the point of contact at the instant of impact. Assume the friction between the spheres to be zero, so that they can slip over each other during the impact without tangential force between the balls.

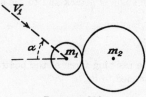

PROBLEM 306.

Set up linear momentum equations in two perpendicular directions and solve for the velocities of the centers of m_1 and m_2 after the impact.

307. Show that a marble on the level floor of a railroad car, on level track, that is being accelerated by a locomotive has a motion relative to the car as if the marble were on an inclined floor in a car standing still. Describe the apparent inclination of the floor in terms of the acceleration.

308. A simple pendulum of length l is at rest, suspended from the ceiling of a vehicle. The vehicle is suddenly given a horizontal acceleration a, which persists indefinitely from that time on.

a. Find the differential equation of motion of the pendulum relative to the vehicle.

b. Assuming that a is small with respect to g, so that the pendulum angle is small, find the solution to the differential equation, and describe the motion.

309. A man stands in a subway train without being able to reach a strap to hang on. As long as the acceleration of the car is zero, the man can stand upright without trouble, his center of gravity being located vertically above his center of support. Even when the car has a constant acceleration $\ddot{s}$, he can stand quietly, but no longer upright.

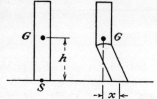

PROBLEM 309.

a. Find the relation between the horizontal car acceleration $\ddot{s}$, the height h of the man's center of gravity G, and the horizontal distance x between his feet and G.

b. Find the relation between the constant "jerk" $\dddot{s}$ and the necessary velocity $\dot{x}$ of the man's feet.

c. If $\dddot{s}$ equals $0.25\, g/0.1$ sec. and h equals 4 ft, what is the required velocity $\dot{x}$?

310. Reread the story of page 31 about sailing against the wind, especially the concluding remark. In this problem we consider a boat sailing with a speed V_b in a direction perpendicular to the wind velocity V_w, the relative wind speed including an angle $\alpha = \tan^{-1} V_w/V_b$ with respect to the boat center line. The sail is set at angle β with respect to the boat center line. The wind force F_w on the sail is assumed to be perpendicular to it, proportional to the angle of attack and to the square to the relative wind velocity.

$$F_w = C_w(\alpha - \beta)V_r^2$$

The resisting force of the boat against forward motion through the water is assumed to be

$$R_\text{boat} = C_b V_b^2$$

a. Set up the equation of equilibrium and prove that

$$\frac{C_b}{C_w} = \sin \beta \left(\tan^{-1} \frac{V_w}{V_b} - \beta \right) \left(1 + \frac{V_w^2}{V_b^2} \right)$$

For a given boat and sail C_w and C_b are known constants, and if the wind velocity is given, there are two unknowns in this equation, V_b and β, and they can be plotted against each other for the purpose of finding the sail angle β for which the boat speed V_b is a maximum. This can be done by numerical computation only, because the equation is too complicated. But we can jump ahead and suspect that β will be a small angle, and also that the maximum boat speed will be considerably greater than the wind speed, so that α likewise is a small angle.

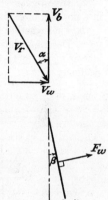

b. Simplify result *a* for small angles α and β, setting sines and tangents equal to the angles themselves, and neglecting higher powers of the angles.

The constants C_w and C_b depend on the construction. We strive to make C_w large by having a large sail, and attempt to make C_b small by having a smooth hull. For a good sailboat $C_w = 30C_b$.

c. Solve the equation of part *a* for $C_w = 30C_b$ and from it find the (approximate) values for the maximum boat velocity and for the best sail angle.

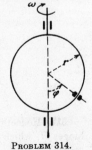

PROBLEM 310.

311. At the equator a projectile is shot vertically straight up. It reaches a height h and falls down again. Find the distance y, between the point where it lands and the point where it was shot up, as a function of the height h. Calculate this distance numerically for $h = 90$ miles.

312. Calculate the east or west deviation of a shell fired (*a*) to the north and (*b*) to the south from a point at 45° northern latitude with a gun elevation of 45 deg and a muzzle velocity of 2,000 ft/sec. Neglect air resistance.

313. A stretch of the Mississippi river at latitude 40° during flood condition is flowing to the south with a speed of 10 mph, and it is then 1.5 miles wide. Coriolis claims that the water surface is not level. Calculate the difference in level between the two banks of the river and state which one is higher.

314. A particle of mass m has the shape of a bead fitting without friction around a thin but stiff wire, bent in the shape of a circle of radius r. This circle is forcibly rotated with constant angular speed ω about a vertical diameter.

a. Write the equation of motion of the bead relative to the wire, expressing its position in terms of an angle φ.

b. State what the Coriolis force is and how it influences the motion.

PROBLEM 314.

315. A turntable carries on it a contrivance consisting of two chemical test tubes each about 6 in. long and mounted at angles α with respect to the vertical. Both tubes are filled with water, closed off with tight corks at the top. One of the tubes has a steel ball in it and the other one has an air bubble. The assembly is rotated at angular speed ω about the vertical center line.

a. Investigate whether, for a given ω, the steel ball or air bubble may be in a position of equilibrium, r.

b. Find whether this equilibrium is stable or unstable.

c. Describe what will happen if the turntable starts from rest, is given an increasing ω up to a maximum, and then is allowed to slow down again gradually.

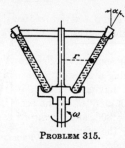

PROBLEM 315.

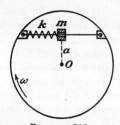

PROBLEM 316.

316. A horizontal turntable is rotated at uniform speed ω about its vertical axis O. It has attached to it a thin shaft or stretched wire at perpendicular distance a from the center. On this wire, a bead of mass m can slide without friction. The bead is attached to the table with a spring of stiffness k, so adjusted that when the spring is without force the bead is in the mid-point, at distance a from O.

a. Set up the equation of relative motion of the bead and find its natural frequency of vibration.

b. How does Coriolis enter the picture?

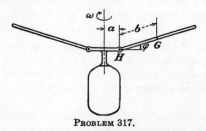

PROBLEM 317.

317. Helicopter rotor blades are attached to the central pylon through ball hinges H, which allow the blades (within limits) to flap freely up and down as well as in the plane of rotation. For this problem we consider only the up-and-down motion, and consider the blades to rotate at uniform angular speed ω in the horizontal plane, deviating from that plane only by small angles φ. Let the dimensions a and b be as shown in the sketch. Let w and I_G be the weight and moment of inertia of one blade, and let W be the weight of the aircraft per rotor blade. This weight W is sustained by one blade through an aerodynamic lift force W acting vertically upward on the blade, and we assume for simplicity that this lift force also acts through G.

a. Calculate the steady-state value of the angle φ for which dynamic equilibrium exists, with constant ω and the aircraft hovering in the air.

b. Write the equation of up-and-down flapping motion of a blade about the above equilibrium position, and from it deduce the natural frequency of flapping motion of a blade for small flapping angles.

318. In internal-combustion engines, particularly in aircraft engines, loose masses have been installed in the counterweights of the crankshaft for the purpose of reducing torsional vibration. These masses are aptly called "centrifugal pendulums," and the sketch illustrates the principle. O is the center of rotation of the engine, represented by a disk of constant angular speed ω. At an eccentric point C the thread of a simple pendulum is attached. Call $OC = a$ and $Cm = l$. The pendulum can swing in the centrifugal field, which in practice is so large that poor g is completely drowned and can be neglected.

a. Set up the equation of the motion of the pendulum relative to the disk and do this for small angles only (for large φ it gets too complicated).

b. Compare this result to that for a simple pendulum in a gravity field g.

c. Look at the Coriolis force in this problem and state how it affects the motion.

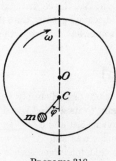

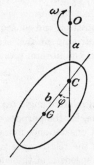

PROBLEM 318.   PROBLEM 319.

319. Replace the simple pendulum of the previous problem by a compound one of mass m and moment of inertia I_G, and use the notation $OC = a$, $CG = b$. Find the frequency of this compound pendulum in the centrifugal field.

320. Replace the pendulums of the two previous problems by a bifilar one (page 247) of mass m with the dimensions $C_1D_1 = C_2D_2 = l$. Introduce such other letters as you may need, and find the point for which the centrifugal acceleration gives the frequency of the pendulum correctly, when written instead of g in the formula for the simple gravity pendulum.

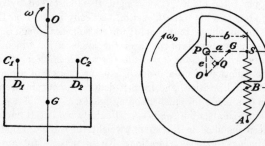

PROBLEM 320.   PROBLEMS 321, 322, and 323.

321. An engine governor is mounted in a flywheel, rotating about O at angular speed ω. It consists of an eccentric mass m, pivoted on the flywheel about a center P. The distance $OP = e$ is called the "eccentricity"; $PG = a$ is the distance of the center of gravity to the pivot. The centrifugal force tends to turn the weight about P in a counterclockwise direction relative to the flywheel. This is held in equilibrium by a stretched spring attached at S, of which the natural length is AB, so that $BS = l$ is the elastic elongation. If the stiffness of this spring is k, if the normal speed of the engine is ω_0, if PG is perpendicular to OP as shown, and if P,G, and S are collinear as shown, prove that $klb = m\omega_0^2 ae$.

322. If in the previous problem the engine speed $\omega = \omega_0 + \Delta\omega$ becomes greater than the normal speed ω_0 by a *small* amount $\Delta\omega$, the centrifugal force becomes greater, and consequently the angle OPG becomes greater than 90 deg by a *small* amount $\Delta\varphi$, stretching the spring somewhat. Sketch the new (dynamic) equilibrium position, and prove that

$$\frac{\Delta\omega}{\Delta\varphi} = 2\omega_0 \left(\frac{b}{l} - \frac{a}{e}\frac{e^2 - a^2}{e^2 + a^2} \right)$$

The small relative angle $\Delta\varphi$ is used to close the steam- or fuel-supply valve of the engine, for the purpose of slowing it down again to its desired speed ω_0.

323. Consider once more the governor of Problems 321 and 322. In those problems the relative position of the eccentric mass was studied for various constant speeds ω of the engine, and it was seen that for larger speeds ω the line PG is turned about the pivot P in a counterclockwise sense through a small angle, stretching the spring. This shuts off the steam or fuel supply, slowing down the motion.

Now imagine the speed to be $\omega_0 + \dot\omega t$, which means that the disk at the instant $t = 0$ has the normal speed ω_0 but has an angular acceleration $\dot\omega$. This will cause a relative counterclockwise acceleration of PG about point P, and thus will initiate a closing of the fuel valve even before the speed has increased, by the acceleration $\dot\omega$ alone.

a. Prove that this relative angular acceleration is

$$\dot{\omega} I_P = \dot{\omega} m(k_G^2 + a^2)$$

b. Describe the Coriolis forces. How do they influence the problem?

324. A uniform circular disk of radius R and weight W is rigidly keyed to a weightless shaft at an angle α. The shaft is supported in two bearings, distance $2a$ apart, and the assembly has an angular speed ω. The angular momentum vector $\mathfrak{M}$ will then include an angle β with the axis of rotation ω.

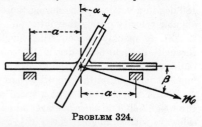

PROBLEM 324.

a. Derive a formula for $\tan \beta$ in terms of the angle α.

b. Calculate the rotating bearing reaction forces for the case that $W = 10$ lb, $R = 8$ in., $a = 10$ in., $\alpha = 30°$, and $N = 600$ rpm.

325. The disk of Problem 324 can be considered as the main element of a "wobble-plate" engine. Imagine eight pistons and cylinders arranged 90 deg apart around the shaft. Only one pair of these is shown in the sketch; two pairs lie in the plane of the drawing; one pair is situated in front of the paper, and another pair behind it, and of course the cylinders and pistons are stationary (do not rotate).

a. Prove that, when the shaft rotates at uniform speed ω, each piston executes a harmonic motion.

b. In the position shown, the two pairs of pistons in the plane of the paper have no velocity, while the two pairs above and below the paper have maximum velocity. If each piston weighs w_p lb, and is at distance r_p from the axis center line, calculate the magnitude and direction of the angular-momentum vector $\mathfrak{M}_p$ of these four pistons.

c. The angular-momentum vector of the entire wobble-plate engine is the vector sum of $\mathfrak{M}_{disk}$ and $\mathfrak{M}_p$. Find the relation for which this $\mathfrak{M}_{total}$ lies along the axis center line, so that the engine is completely balanced.

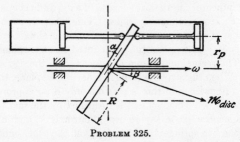

PROBLEM 325.

326. A man standing on a turntable has a spinning bicycle wheel in his hands. The man, turntable, and dead bicycle wheel have a total weight of

200 lb and a radius of gyration about a vertical axis of 8 in. The bicycle-wheel rim weighs 5 lb and is at 13 in. radius. At the start the man is not rotating; the wheel is held horizontally spinning at 200 rpm. The man then turns the wheel upside down at uniform speed in 2 sec, *i.e.*, the wheel axis turns through 180 deg in a vertical plane during those 2 sec. Calculate and plot the man's angular velocity during that interval.

327. A quickly maneuvering single-seat single-engined airplane turns through an angle of 90 deg in 10 sec. The propeller weighs 500 lb, has a radius of gyration of 3 ft, and rotates at 1,200 rpm. Calculate the gyroscopic moment imposed on the propeller-shaft bearings.

328. A bicycle is going at a speed of 10 mph. The front-wheel rim and tire weigh 5 lb, and their average radius is 13 in. If the bicycle is falling over sideways with an angular speed of $\frac{1}{4}$ radian/sec (which is barely noticeable), calculate the gyroscopic torque of the front wheel.

329. The rotor of an aircraft artificial horizon (Fig. 288, page 331) consists of a solid steel disk, 3 in. in diameter and $\frac{1}{2}$ in. thick, spinning at 3,000 rpm. The instrument is off the vertical, so that one of the small pendulums in the bottom blocks off an air passage and the opposing pendulum opens its passage wide. This passage has a moment arm of 4 in. with respect to the center of the instrument, and from it issues a jet of air, $\frac{1}{8}$ in. in diameter at a speed of 100 ft/sec, of atmospheric pressure ($\frac{1}{800}$ the density of water). Calculate the angular speed with which the instrument returns to its true position.

330. An ocean liner of 40,000 (short) tons weight has a radius of gyration of 20 ft about its longitudinal axis of rolling motion. It carries three Sperry anti-roll gyros (Fig. 287, page 329), each with a rotor of 2 tons weight and 5 ft radius of gyration. The gyro axes can be made to precess fore and aft through an angle from -30 deg to $+30$ deg with respect to the vertical. Without the gyros the ship can roll, performing a harmonic motion (page 159) with a period of $T = 20$ sec for a full swing.

Calculate the rpm of the three rotors necessary to impart to the ship sufficient angular rolling speed to ensure a roll angle of 1 deg (of the ship) for one precession from -30 deg to $+30$ deg of all three gyros.

331. A turn indicator for aircraft consists of a gyro wheel *a* mounted with a horizontal axis *bb* in a rectangular frame. This frame is supported in rigid bearings *c*, so that it can turn about another horizontal axis *cc* perpendicular to the gyro axis *bb*. The apparatus is mounted in an airplane with axis *cc* fore and aft. The rectangular frame is connected to the "ground" through two springs *d*. When the airplane turns in a horizontal plane, the gyro axis *bb* is forced to turn likewise, and this is possible only if an appropriate moment is exerted on the gyro. This moment is furnished by the fact that the frame turns through a small angle φ, say 10 deg, about axis *cc*, and in this deviated position one of the two springs *d* is in tension, and the other one is in compression. Thus the angle φ, which can be observed, is a measure for the rate of turn of the airplane.

Let the gyro disk weigh 1 lb, have a radius of gyration of 2 in., and rotate at 1,000 rpm. Let the horizontal distance between the two springs, $2a$, be 2 in. Calculate the stiffness k of the springs required if $\varphi = \frac{1}{5}$ radian is to represent a rate of turn of the airplane of one full turn per minute.

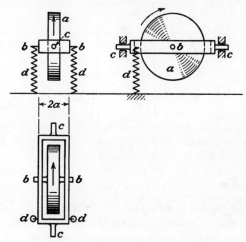

PROBLEM 331.

332. Refer to the primitive form of a ship's gyrocompass shown in Fig. 294 (page 335) and to its operational diagram (Fig. 295). Let the gyro rotor weigh 20 lb; let its radius of gyration be 3 in., and let it rotate at 4,000 rpm. Consider the pendulum to be a simple one (page 184) of 6 in. length. Calculate the necessary weight of the pendulum bob in order to cause in point C of Fig. 295 a westward precessional speed of 1 deg/min for a compass-needle elevation of 1 deg.

333. The problem of the hummingbird in its cage on the scale of a balance: A balance has on one of its scales an empty bird cage, and on the other side, sufficient weight to ensure neutral balance. A 1-oz live hummingbird is put in the cage. Everybody agrees that when the bird sits down in the cage, an additional 1-oz dead weight must be placed on the other scale for balance. What weight is necessary when the bird hovers in the air inside the cage? Discuss this for the following cases:

  a. A closed cage.

  b. A completely open cage, bottom and all.

  c. A cage with some side openings and a solid bottom, like the usual ones.

334. A perpetual-motion machine: An endless belt passes over two pulleys as shown. It carries a large number of identical and equidistant small cylinders, containing some air and a piston w. Only two of these pistons are shown. The entire apparatus is immersed in water and is assumed to be without friction and without air leakage. Obviously the pistons on the right side are farther down in their cylinders than those on the left side, so that the

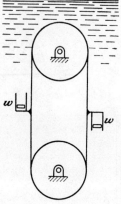

PROBLEM 334.

buoyancy on the right is greater than on the left, and the belt will rotate in a counterclockwise direction. Analyze the system in detail.

ANSWERS TO PROBLEMS

1. $\dfrac{W_3}{W_1} = \dfrac{2(h/l)}{\sqrt{1 + (h/l)^2}}$; plot W_3/W_1 versus h/l with an asymptote at $W_3/W_1 = 2$.

2. 90 deg.   **3.** $W_2 = 17.3$ lb, $W_3 = 20$ lb.

4. $\angle P_1AR = 22$ deg, $\angle P_2AR = 49$ deg.   **5.** (a) 14.4 lb, (b) 1.15 lb.

6. $200\sqrt{2}$ lb directed at 45 deg pointing to the right and downward; intersects the bottom edge of the square at 4 in. to the left of the bottom left-hand corner.

7. 381 tons = $\sqrt{75^2 + 373^2}$ at 46.7 ft behind bow.

8. Forward force only of 606 lb, located at 3.17 ft inboard from starboard (right) side.

9. $T_1 = 100$ lb, $T_2 = 141$ lb, $T_3 = 158$ lb, $T_4 = 115$ lb.

10. (a) 13.05, (b) 19.9.   **11.** (a) 2,588 ft-lb, (b) $7.07W$ ft-lb, (c) 365 lb.

12. (a) 200 lb-in. clockwise, (b) 600 lb-in. counterclockwise.

13. 2,030 ft-lb.   **14.** 500 lb.

15. 112 lb directed forward slightly to right with angle of about 30 deg with respect to the forward direction. The force intersects the center line of the boat at a point 11.6 in. forward of the stern.

16. 1,200 lb downward, through center of disk.   **17.** 2,400 lb.

18. 138 lb, (b) 120-lb tension.

19. (a) $x = aW_1/W_2$, (b) $W_{1\,max} = 12$ lb, linear scale.

20. (a) $W = \dfrac{b}{a+b}\, p\, \dfrac{\pi d^2}{4} - w\,\dfrac{b+c}{b+a}$, (b) $W = 125 - \tfrac{3}{15}w$ lb.

21. $P = 36$ lb, $F_w = 84$ lb.   **22.** $R = 12\frac{1}{2}$ lb, $G = 62\frac{1}{2}$ lb.

23. $P = 7.14$ lb.   **24.** $T_1a = T_2b$.

25. Rope pull 485 lb; hinge force 820 lb, of which 780 lb vertical and 242 lb horizontal.

26. Front 875 lb, rear, 1,625 lb.

27. (a) $F = W \tan \alpha$, (b) vertical balls F, horizontal balls W, inclined balls $\sqrt{F^2 + W^2}$. (c) 5°43′.

28. (a) $W_2 = W_1 \sin \alpha$, (b) 25 lb.   **29.** $L_1 = 14,000$ lb, $L_2 = 6,610$ lb.

30. 150 lb.   **31.** $Q = 2P \tan \alpha$.   **32.** $x = 4.21$ ft.

33. (a) 200 lb, (b) 200 lb.

34. Gripping pressure 5,920 lb, tension in crossbar 11,840 lb.

35. $T_1 = T_2 = T_3 = W/2$.   **36.** $A = -B = P\,\dfrac{a}{c}$; $C = P$.

37. $P/W = \frac{3}{2} \cot \alpha$.   **38.** 500 lb.

39. (a) 37.5 lb, (b) 37.5 lb horizontally.   **40.** $T_{AD} = P(4l/h)$.

41. (a) $\tan \alpha = \dfrac{W_1 - W_2}{w}\,\dfrac{b}{a}$, (b) make a/b very small.

42. (a) $T = W \dfrac{l}{\sqrt{l^2 - (b + d)^2}}$

Upper wheel: $\dfrac{U}{W} = -\dfrac{a - d}{2e} + \dfrac{(b + d)(c + e)}{2e\sqrt{l^2 - (b + d)^2}}$

Lower wheel: $\dfrac{L}{W} = \dfrac{a - d}{2e} - \dfrac{(b + d)(c - e)}{2e\sqrt{l^2 - (b + d)^2}}$

(b) $\dfrac{b + d}{\sqrt{l^2 - (b + d)^2}} = \dfrac{a - d}{c - e}.$

Interpret this by observing where the rope force intersects the weight force.
43. 154 lb. **44.** (a) 85 cu ft, (b) This answer is independent of the angle.
45. $T_2 = T_1 r_2/r_1$; tooth force $= T_1/r_1 = T_2/r_2$; each bearing force equals
the tooth force; $M_{\text{frame}} = T_1 \dfrac{r_1 + r_2}{r_1} = T_1 + T_2.$

46. $P = \dfrac{a}{b}\dfrac{c}{d} W.$ **47.** $l_1 = \dfrac{P}{2W} l_2.$

48. $H_L = H_R = 634$ lb, $V_L = 616$ lb, $V_R = 384$ lb.
49. Left support down, 386 lb, left support to right, 250 lb; right support
up, 34 lb, right support to left, 603 lb, hinge force, $\sqrt{250^2 + 386^2} = 460$ lb.
50. $X = 800$ lb. Left support force is 100 lb horizontal and 750 lb vertical.
Right support force is 500 lb horizontal and 750 lb vertical. Left hinge force
is 900 lb horizontal and 250 lb vertical. Right hinge force is 250 lb vertical,
while the horizontal component is 500 lb if the 400-lb load is applied to the
horizontal bar, and 900 lb if that load is applied to the 45-deg bar.
51. $X = -250$ lb (pulling to the left), $R_L = 500\sqrt{2}$ lb (compression in
45-deg bar), $R_R = 500$ lb up from ground on bar $+250$ lb to right from ground
on bar.
52. $R_B = 450$ lb horizontally to right $+50$ lb vertically down, $R_A = 50$ lb
horizontally to left $+ 150$ lb vertically up. Both directions are from the bars
onto the foundation.
53. The horizontal components are $11\tfrac{1}{4}$ lb to right on top hinge and
$11\tfrac{1}{4}$ lb to left on bottom hinge. The sum of the two vertical components is
30 lb, divided between the two hinges in a statically indeterminate manner.
56. $x/a = \tfrac{7}{9}$, $y/a = \tfrac{4}{9}.$
57. 0.95 in. to the right and 2.95 in. below the top left corner.
58. $\tfrac{3}{8}r$ above base.
59. $x_G = y_G = 0.41a$, with the origin in the upper left outside corner.
60. $\dfrac{2}{3}\dfrac{h^2 - 2r^2}{2h + \pi r}$ above center of semicircle.

61. $\dfrac{h^2 - 3r^2}{4h + 8r}$ above center of hemisphere.

62. (a) Drawing the straight line at 0.20 in. from the apex of the arcs,
the center of gravity is 1.28 in. from the center of the pin; (b) for lower arc:
0.35 per cent or 0.0045 in., for upper arc: 0.31 per cent or 0.0040 in.
63. (a) $\tfrac{3}{13}h$, (b) $x_G = \dfrac{h^4 - 4hh_1^3 + 3h_1^4}{4(h^3 - h_1^3)}$, independent of the shape of the
base, where h is the height of the complete cone and h_1 is the height of the
truncated top piece.

64. $w_1/w_2 = \frac{1}{2}\pi \cot \alpha$.   **65.** $W_1/W_2 = 2 \cot \alpha$.

68. Hinge force 14,700 lb, lip force 18,400 lb, both directed vertically to the door.

69. 8,000 lb.   **70.** $b = h \sqrt{2/15} = 0.37h$.   **71.** $b = h/\sqrt{3} = 0.58h$.

72. (a) $F = \gamma d \left(h - \dfrac{\pi}{8} d \right)$ per running ft, (b) 3,400 lb/ft.

73. 3.52 ft.

74. In top position, 139 lb/in.², in bottom position: 127 lb/in.²

75. Less deep by 7 in.   **76.** 46 per cent.

77.
$$\delta = \alpha \left[\frac{b^2}{12d} - \frac{d}{2} \right].$$

78. A round number.

79. (a) Ask the inventor. (b) Upward push equals the buoyancy of the cylinder in a liquid of $\gamma = \frac{1}{2}(\gamma_{water} + \gamma_{mercury})$; sidewise push to left is $dhl(\gamma_{mercury} - \gamma_{water})$, where h is the depth of liquid at the center of the cylinder.

80. (a) 216 lb; (b) $1\frac{1}{8}$ in.; of the 10 ft length, 8.9 ft is wet and 1.1 ft is dry.

81. Bar 1: $-P2\sqrt{2}$; bar 2 and 3: $+2P$; bar 4: $+P$.

82. Compression $2P$ in all three top members; tension $P\sqrt{5}$ in the two inclined bottom bars; tension $2P$ in the central bottom bar; compression P in vertical struts.

83. Top member, 9,000-lb compression; bottom member, 9,300-lb tension; diagonal, 1,760-lb compression.

84. See Fig. 63 or 64, page 58.

85. 12: $-P/2$; 23: zero; 34: $-5P/2$; 45: $-P\sqrt{5}/4$; 56: $-P$; 67: $-2P\sqrt{2}$.

87. Maximum force in top bar next to right support: $3.67P$ compression.

88. Bottom horizontal, $2P\sqrt{3}$ tension; diagonal, $P\sqrt{3}/4$ compression; upper girder, $15P/4$ compression.

89. 4 sq in.; worst bar is the upper horizontal above the right support.

90. (a) Six inside bars are stressless; bottom stringer, $3P$ compression; upper stringer, $P\sqrt{10}$ tension; upright at left, P compression. (c) Maximum compression in horizontal bottom bar next to left support is $13.5Q$; maximum tension in upper bar next to left support is $12.6Q$.

91. Bar 1: $-3P/4$; bar 2: $+15P/4$. It is necessary to weight down the shore support C. By how much?

92. nth upper bar, $-nP$; nth lower bar, $+(n - \frac{1}{2})P$; nth right diagonal, $-P/\sqrt{2}$; nth left diagonal, $+P/\sqrt{2}$.

93. Maximum deflection is 1.84 ft, under the 200-lb load.

94. 5,200 lb, 50 deg.   **95.** Sag is $l/2\sqrt{3}$, tensile force is P.

96. The lowest point is E, about $1.2l$ below A.

97. $H = 1,370$ lb, lowest point 165.7 ft from left tower.   **98.** $6\frac{1}{4}$ ft.

99. (a) Two different, but symmetrically located parabolas, whose tangents at the center include an angle 2α, determined by $2H \tan \alpha = W_{center}$, so that $\alpha = .025$ radians $= 1.43$ deg; (b) 2.05 ft.

100. 77.5 ft.

101. Maximum shear force is 10 tons at right support, maximum bending moment is 90 ft tons in the stretch between locomotive and tender.

102. Maximum bending moment under left wheel is 39.0 ft-tons.

103. Maximum shear is P, maximum bending moment is $Pa/2$ at the middle support.

104. Maximum bending moment is $15\frac{5}{8}$ ft from right support and is $(15\frac{5}{8})^2 = 244$ ft-tons.

105. Counting x from the right support to the left, the bending moment under the triangular load is $\frac{3}{16}wlx - \frac{2}{3}w(x^3/l)$.

106. The force in the nth upper stringer bar is $-\dfrac{NQ}{2}\dfrac{nl}{h} + Qn\dfrac{nl}{2h}\cdot$ The force in the nth lower stringer bar is $+\dfrac{NQ}{2}\cdot\dfrac{(n+\frac{1}{2})l}{h} - Qn\dfrac{(n-1)l}{2h}\cdot$ The vertical component of the force in the nth right slanting diagonal bar is $\dfrac{NQ}{2} - nQ$.

107. Cable tension: $3P/\sqrt{2}$; bar AE: $-P\sqrt{2}$, no bending; bar AC: maximum bending moment Pa at B, and the section AB is in tension $5P/2$; bar BD has a bending moment Pa in the middle and is in compression $2P\sqrt{2}$.

108. Zero bending in the good design. In Rube's horizontal bar the maximum bending occurs at the strut connection; it is $(Wl/3)(1 + 1/\sqrt{2})$. In Rube's strut the maximum bending moment is in the center: $Wl/6$.

109. In table top $M_{\max} = wl^2/8$ at the leg hinges. The maximum bending moment in the legs is $wl^2/2$ in the center. The upper halves of the legs have $3wl/\sqrt{2}$ compression; the lower halves only $wl/\sqrt{2}$.

110. $f = \tan(\alpha/2)$.

111. (a) $f = \tan\alpha$, (b) total force $100/\sqrt{3}$ lb consisting of 50 lb normal and $50/\sqrt{3}$ lb frictional component.

112. $\dfrac{w}{W} = \dfrac{2fr}{R - fr}\cdot$ **113.** $f_{\text{pulley}} = \dfrac{2}{\pi}\dfrac{r}{R}f_{\text{axle}}$.

114. (a) Tips at 100 lb; (b) 50 per cent; (c) slides at 120 lb, 100 per cent.

115. (a) Downward, (b) $f = \frac{1}{3}$. **116.** Torque $= \frac{2}{3}rfW$.

117. Torque $= \frac{2}{3}\dfrac{rfW}{\sin\alpha}\cdot$

118. (a) $x = l + fh - \sqrt{2flh + f^2h^2}$, $H = fw_1x$; (b) $V = w_1(l - x)$; (c) $l - x = 16$ ft, $V = 32$ lb, $H = 84$ lb, $F_{\text{total}} = 91$ lb.

119. $f = \tan^{-1}\left(\dfrac{\varphi}{2} - \dfrac{\pi}{8}\right)\cdot$ **121.** $\dfrac{1 + f^2}{f}\sin^2\alpha = 2\dfrac{r}{l}\cdot$

122. (a) $f = \dfrac{\cos^2\alpha + a/b}{\sin\alpha\cos\alpha}$, (b) $f = 2$; $\cdots$!, (c) add horizontal cross member close to ground.

123. $300e^{-\frac{\pi}{8}} < P < 300e^{+\frac{\pi}{8}}$, $F = \sqrt{W^2 + P^2}$.

124. 141 lb, the lower one pulling aft and the upper one forward.

125. (a) F up on black, F down on white; (b) F down on black, F up on white; (c) F down on extreme ends of both handles, F up on disk ends of both handles.

126. Force $F\sqrt{3}$ along diagonal of cube; moment $Fa\sqrt{3}$ in opposite direction, forming a left-handed screw.

127. Location of point of action is at $4a/5$ from origin, and at $a/5$ from force $2P$. The force of the screw is $F\sqrt{5}$ and its moment is $2Fa/\sqrt{5}$.

128. Same answer as 127, except that we write F_2 instead of F and $a/2$ instead of a.

129. $f \geqslant \tan\alpha$.

130. AC and BC have tension P; OC has compression $P\sqrt{2}$.

131. At A, downward 4,000 lb and to right 3,366 lb; at B, downward 10,500 lb and to right 3,950 lb.

132. $AE = +224$ lb, $CG = +378$ lb, $D_{\text{vert.}} = 1,000$ lb, $D_x = 444$ lb, $D_y = 134$ lb, maximum bending moment at $C = \sqrt{3,490^2 + 700^2} = 3,550$ ft-lb.

133. In direction of C_1, $x = 0.341$ in.; in direction of C_2, $y = 0.059$ in.; in the plane of C_2, $z = 0$.

134. (b) $\dfrac{2W}{P}\sin\alpha = \sqrt{\left(\dfrac{l}{r}\right)^2 - 4\sin^2\dfrac{\alpha}{2}}$; (c) for 90 deg $P = \dfrac{2W}{\sqrt{(l/r)^2 - 2}}$; for 180 deg $P = 0$; (d) the center of gravity is the center of rotation

$$\alpha = \frac{P}{2W}\frac{rl}{r^2 - a^2}.$$

135. (a) $x = \dfrac{Pl}{\sqrt{P^2 + W^2}}$; (b) same as (a); (c) same as 134(b), where $M = Pr/2$; (d) superposition of (a) and (c) with $M = Pr$.

136. (a) $\delta_{xy} = \dfrac{P}{kab}\left(\dfrac{1}{4} + \dfrac{3}{4}\dfrac{cx}{a^2} + \dfrac{3}{4}\dfrac{dy}{b^2}\right)$; (b) $\dfrac{c}{a} + \dfrac{d}{b} = \dfrac{1}{3}$.

137. BE and $DE = +P\sqrt{2}$, AE and $CE = -P\sqrt{2}$, AB, BC, CD, and $DA = -P/\sqrt{2}$, $AC = +2P$.

138. $AB, BC, AF, CF, AE, EF =$ zero force, AD and $CD = -P$, $BE = +P$, $AC = +P\sqrt{2}$, $AE = -P\sqrt{2}$, $DE = +P\sqrt{3}$.

139. Horizontal forces are -29.5 lb at A; $+43.1$ lb at C_1, zero at C_2, -29.2 lb at C_3, and $+15.6$ lb at B. The maximum horizontal bending moment of 44 in.-lb occurs at C_1.

140. (a) Horizontal bending moment: from $6Pa/5$ at left to $4Pa/5$ at right. Vertical bending moment: from $4Pa/5$ at left to $6Pa/5$ at right. Twist: Pa constant along section. (b) Maximum horizontal bending moment at left end of center section; maximum vertical bending moment at right end of center section; maximum twist moment all along center section.

141. Maximum bending moment in the apex of the semicircle, of value $P\sqrt{h^2 + (3a/2)^2}$. Maximum twisting moment all along the horizontal sections, of value Ph.

142. (a) 100 ft-lb, (b) -100 ft-lb, (c) zero. **143.** -400π in.-lb.

144. (a) 9,000 ft-lb, (b) zero, (c) 9,000 ft-lb.

145. (a) $a/c = b/d$; (b) $ew = aW$, where W is understood to include the weight of the platform.

146. $P = \dfrac{W}{2}\dfrac{r}{l}\dfrac{1}{\sin^2\alpha}$.

147. (a) $X = 4\frac{1}{2}P\sqrt{3}$; (b) $R_1 = 2P\sqrt{3}$, $R_2 = 2R_1$, $R_n = nR_1$, all R's being horizontal.

148. String force is $3W_b$, independent of the length of the string.

149. $M_B = M_C = \frac{1}{2}\frac{n_D}{n_A}M_A$, $M_{\text{housing}} = M_A\sqrt{1 + \left(\frac{n_D}{n_A}\right)^2}$.

150. $\dfrac{P}{Q} = \dfrac{r_1 R}{r_2 a}$.

151. (a) $P\delta\dfrac{R+r}{r} = W_{\text{drum}}\delta + 2W_{\text{bucket}}\epsilon$, where δ is the upward displacement of C, A standing still, and ϵ is the corresponding upward displacement of G; (b) $P = 10 + 40\epsilon/\delta$; (c) $P = 10$ lb; (d) $\alpha = \tan^{-1}\frac{1}{2} = 26.5$ deg, C being lower than B.

152. (a) $a/b = c/d$, (b) $\dfrac{x}{g}\cdot\dfrac{e+f}{e}\cdot\dfrac{a+b}{a} = T/H$, (c) in lever DF right under A.

153. (a) Frame torque $+3M_A$, propeller torque $-4M_A$; (b) frame torque $(1 + 2d_B/d_A)M_A$, propeller torque $-2(1 + d_B/d_A)M_A$.

154. Stable for $h/2 < r$.

155. Weights must be lower than 5 in. below knife edge.

156. Stable if $h > r_1^2/(r_1 + r_2)$, where h is the distance between the center of the upper half cylinder and its center of gravity, or $h = 4r_1/3\pi$.

157. (a) Same as 156 with $h = 3r_1/8$, (b) same formula with h for the upper object.

158. (b)(1) Call it zero; (2) the center of gravity rises 0.16 ft, A rises 1 ft, while B rises 0.11 ft; (3) the center of gravity rises 0.015 ft, A falls 1 ft, while B rises 0.68 ft.

159. Always stable.

160. Indifferent equilibrium in all angular positions.

161. (a) $\ddot{x} = -r\omega^2\sin\omega t$, (b) 308 ft/sec².

162. 11.3 drops/sec, $(\Delta s)_n = \frac{1}{2}g(\Delta t)^2(2n - 1)$; tap is $l/24$ above boards, distances are $3l/24$, $5l/24$, $7l/24$, and $9l/24$.

163. 105 sec.   **164.** $\frac{3}{4}$ mile.

165. $\dot{x} = -a\omega(\sin\omega t + \frac{1}{8}\sin 2\omega t)$, $\ddot{x} = -a\omega^2(\cos\omega t + \frac{1}{4}\cos 2\omega t)$, $\dot{x}_{\max} = 1.03a\omega$.

166. (a) $11\frac{2}{3}$ in., (b) $358\frac{1}{3}$ in.   **167.** 1,042 revolutions.

168. (a) 4,000 miles/hour, (b) 6 minutes.

169. (a) 29.8 ft/sec², of which the centripetal component is 25 ft/sec² and the tangential component is 16.1 ft/sec.² (b) Half the answer (a).

170. 11 ft/sec.   **171.** 2 ft/sec², upwards.

172. (a) ω_1/r_0, (b) 120 revolutions.   **173.** $n_D/2n_A$.

174. $\dot{y}/\dot{x} = -y/x = -y^2/(L^2 - l^2)$; $\dot{y} = -7.68$ in./sec.

175. $\ddot{z} = r_0\ddot{\varphi} = 0$, $\ddot{r} = \dfrac{v_0^2}{r_0[(h/2\pi r_0)^2 + 1]}$.

176. (a) $v_r = v_0$, $v_\varphi = v_0(r_0 + v_0 t)$, $v_x = \omega_0 r\cos\varphi + v_0\sin\varphi$,
$v_y = -\omega_0 r\sin\varphi + v_0\cos\varphi$, $v_{\text{total}} = \sqrt{(\omega_0 r)^2 + v_0^2}$,
(b) $\ddot{x} = -\omega_0^2 r\sin\varphi + 2\omega_0 v_0\cos\varphi$,
$\ddot{y} = -\omega_0^2 r\cos\varphi - 2\omega_0 v_0\sin\varphi$,
acceleration $= \sqrt{\ddot{x}^2 + \ddot{y}^2} = \omega_0 v_0\sqrt{3 + (v/v_0)^2}$; (c) $v_{\text{total}} = 65.2$ in./sec, $\dot{v}_{\text{total}} = 410$ in./sec.²

177. $\omega = \omega_0 + \dfrac{\omega_0(a/l)\sin\varphi}{\sqrt{1 - (a\cos\varphi/l)^2}}$, $\dot{\omega} = \omega_0^2\dfrac{a}{l}\dfrac{(1 - a^2/l^2)\cos\varphi}{(1 - a^2\cos^2\varphi/l^2)^{3/2}}$.

178. (a) Circle of radius $l/2$ about O as center; (b) $\dot{y}_C = \dfrac{-v_0^2 t}{\sqrt{l^2 - v_0^2 t^2}}$,

$\dot{y}_B = \frac{1}{2}\dot{y}_C$, $\dot{x}_B = \frac{1}{2}v_0$; (c) $\ddot{y}_C = \dfrac{-v_0^2 l^2}{(l^2 - v_0^2 t^2)^{3/2}}$, $\ddot{y}_B = \frac{1}{2}\ddot{y}_C$, $\ddot{x}_B = 0$.

179. (a) $x = r\cos\omega t + l\sqrt{1 - \dfrac{r^2}{l^2}(1 - \sin\omega t)^2}$;

$x \approx r\cos\omega t + \dfrac{r^2}{l}\sin\omega t + \dfrac{r^2}{4l}\cos 2\omega t + \text{const.}$, where $\omega = V/R$. With respect to rail $x_{\text{rail}} = x + Vt$.

(b) $\dot{x} = -\omega r\left(\sin\omega t - \dfrac{r}{l}\cos\omega t + \dfrac{r}{2l}\sin 2\omega t\right)$,

$\ddot{x} = -\omega^2 r\left(\cos\omega t + \dfrac{r}{l}\sin\omega t + \dfrac{r}{l}\cos 2\omega t\right)$;

(c) $\ddot{x}_{\max} = 2{,}340$ ft/sec^2.

180. $\dfrac{\dot{\psi}}{\dot{\varphi}} = \dfrac{2\cos\varphi - 1}{5 - 4\cos\varphi}$.

181. (a) $\sin(\varphi + \psi) = \sqrt{2}\sin\psi$, (b) $\dfrac{\dot{\psi}}{\dot{\varphi}} = \dfrac{\cos(\varphi + \psi)}{\sqrt{2}\cos\psi - \cos(\varphi + \psi)}$,

(c) $\dfrac{r_1}{d} = \sin\dfrac{\pi}{n}$.

182. (a) $v_B = \dfrac{\pi}{4}\dfrac{a\omega}{\cos\varphi}$, (b) 171 in./sec. **183.** 22 miles/hour.

184. Horizontal line. **185.** $v = 7.01$ ft/sec, $t = 0.855$ sec.

186. (a) $t = 4.01$ sec, (b) 45.3 ft. **187.** 66 ft.

188. (a) $t = \dfrac{v_0}{g\sin\alpha}\left(1 + \dfrac{1}{\sqrt{2}}\right)$, $v = v_0/\sqrt{2}$, (b) $t = 2.12$ sec, $v = 14.1$ ft./sec.

189. (a) $k = 20$ lb/in., (b) $f = 6.25$ cycles/sec, (c) $f = \sqrt{g/\delta}/2\pi$.

190. $P = \dfrac{W}{2}\dfrac{b_2}{h_2}$; the forward wheel lifts off. **191.** 6.35 sec.

192. (a) lb ft^{-1} sec, (b) $v_\infty = \dfrac{mg}{c}\dfrac{\gamma_{\text{body}} - \gamma_{\text{water}}}{\gamma_{\text{body}}}$, (c) $v = v_\infty(1 - e^{-\frac{c}{m}t})$, (d) 10.5 sec.

193. 1.18 ft. **194.** $t = 0.67$ sec, $T = 0.732$ lb.

195. String tension $= 0.866 + 0.85N^2 l 10^{-4}$ lb;

pressure on cone $= 0.500 - 1.47N^2 l 10^{-4}$ lb,

in which $N = $ rpm.

196. $g/\cos\alpha = \omega^2 r$, 91 rpm.

197. $v_{\text{top}} = 24.1$ ft/sec, $v_{\text{bottom}} = 33.1$ ft/sec, $F_{\text{top}} = 0.40W$,

$F_{\text{bottom}} = 2.14W$.

198. $\alpha = 41.8$ deg. **199.** 17.5 deg or 72.5 deg.

200. (a) $\tan\alpha = v^2/Rg$, (b) 960 ft.

201. (a) $\dfrac{r}{r_1} = \dfrac{1}{1 - m\omega^2/k}$, (b) $P = m\omega^2 r_2 - k(r_2 - r_1)$, (c) 815 rpm.

202. (a) Torque $= (w/2g)\,\omega^2 r^2\sin 2\omega t$, (b) 148 ft-lb, at 45 and 135 deg.

203. $v_{\min} = \sqrt{gl}$, water pressure force $= m_1\left(\dfrac{v^2}{l} - g\right)$, string tension

force $= (m_1 + m_2)\left(\dfrac{v^2}{l} - g\right)$.

204. (a) $\ddot{s} = g\left[\sin\alpha - \dfrac{f_1 m_1 + f_2 m_2}{m_1 + m_2}\cos\alpha\right]$; $T = \dfrac{m_1 m_2}{m_1 + m_2} g\cos\alpha(f_2 - f_1)$,

(b) $T = 0.115$ lb.

205. $T = \dfrac{m_1 g(\sin\alpha + f_1\cos\alpha) + m_2 g(\sin\alpha + f_2\cos\alpha) + (m_1 + m_2)\ddot{s}}{\cos\beta - f_2\sin\beta}$.

206. $\ddot{x}_1 = g\,\dfrac{\sin\alpha\cos\alpha}{1 + m_1\sin^2\alpha/m_2}$ to left

$\ddot{x}_2 = m_1\ddot{x}_1/m_2$ to right

$\ddot{y}_1 = \ddot{x}_1(1 + m_1/m_2)\tan\alpha$.

207. $\ddot{x}_1 = \left(g + \dfrac{T}{m_2}\tan\alpha\right)\dfrac{\sin\alpha\cos\alpha}{1 + m_1\sin^2\alpha/m_2}$.

$\ddot{x}_2 = -\dfrac{T}{m_2} + \dfrac{m_1}{m_2}\ddot{x}_1$

$T = (m_1 + m_2)g\tan\alpha$.

208. $T = \dfrac{b}{h}(W_b + W_c)$. **209.** 64 deg. **210.** $t = 0.725$ sec.

211. (a) $0.634l$ and $0.366l$, so that the two ends of the chain have the

same height, (b) $s = a\cosh\left(\sqrt{\dfrac{(1 + \sqrt{3})g}{2l}} \cdot t\right)$, (c) $t = 0.815$ sec.

212. (a) The instantaneous axis of rotation is horizontal, passing through the two points of contact of the lower race. (b) The upper contact point closest to the vertical center line should be lower, so that the ratio of the distances to the instantaneous axis of rotation of the ball and of the upper race is the same for both upper contact points. (c) Any location of the bottom contact point is permissible. The instantaneous axis is the line connecting the bottom contact point with the point of intersection of the vertical center line with the line connecting the two upper contact points.

213. (a) The velocity pole is the mirrored image of O with respect to AC, (b) circle of radius l about O as center, (c) circle with AC as diameter.

215. (a) $V_R = 1.87V_P$, $V_Q = 2.24V_P$, (b) midway between O and Q.

216. $V_Q = 2.1V_A$. **217.** Nearly zero. **218.** $V_A = 2V_C$.

219. (a) Vertical straight line, (c) the point of tangency d, e.

220. $CE = 2DE$.

222. $V_D = 0.707$ ft/sec, $\omega_{BD} = 0.25$ radians/sec; (b) $\dot{V}_C = 1$ ft/sec^2; (c) tangential $= 3\sqrt{2}/4$ ft/sec^2, normal $= \sqrt{2}/8$ ft/sec.^{2}

224. (a) The pole is the origin O; accelerations are $4v_0^2 r_0/l^2$, directed toward O. (b) The pole O' is the mirrored image of O with respect to AC; the accelerations are $2vr_{0'}/l$, directed tangentially, i.e., perpendicular to the radii $r_{0'}$, (c) A circle passing through O, A, O' and C with B as center, and the location of the pole on the circle is determined by $r_0/r_{0'} = v_0 l/2v_0^2$.

226. In line with BC, 2 ft to the right of C.

227. Acceleration of Q is zero; acceleration of S is $0.84\ v_p^2$.

228. The velocity pole is at the contact point between the planet gear and the outer gear. The acceleration pole is between the centers of the planet and sun gears at distance $r_2^2/(r_1+r_2)$ from the center of the planet gear.

229. (a) $\dfrac{m\omega^2 d}{2}\dfrac{b}{a+b+c}$ on each bearing, (b) 63.9 lb.

230. $F_{max} = 1,140$ lb.

231. (a) 4,000 lb radially outward $+75$ lb downward, (b) parabola with apex in center at 3,200 ft-lb..

232. (a) Stress $= \dfrac{m\omega^2 a}{2\pi A}$, where m = total mass; (b) 2,560 lb/in.²

233. (a) $\dfrac{dy}{dx} = \dfrac{\omega^2 x}{g}$, (b) 118 rpm.

234. (a) Inertia forces are radially directed from O; intensity at B is $2V_B^2 dm/l$; the end points of the vectors are on a line parallel to the rod. (b) Bending moment diagram is parabolic with a peak at B of $mV_B^2/4$. (c) Forces at A and C are $V_B^2 m\sqrt{2}/l$.

235. $I_{AA} = 1.67$ lb-ft-sec², $I_{BB} = 11.6$ lb-ft-sec.²

236. $I_{AA} = 22.1$ lb-ft-sec², $I_{BB} = 147$ lb-ft-sec.²

237. $I = \dfrac{35\pi}{8,192}\dfrac{D_2^5 - D_1^5}{D_2 - D_1}\rho l.$

238. (a) $I_x = 0.0302$, $I_y = 0.0075$, $I_{xy} = 0.0109$, all expressed in lb-in.-sec², (b) $\alpha = 22$ deg, $I_{max} = 0.0346$ lb-in.-sec.²

239. (a) $I_x = \dfrac{3}{8}\dfrac{Wa^2}{g}$, $I_y = \dfrac{3}{2}\dfrac{Wa^2}{g}$, $I_{xy} = \dfrac{7}{16}\dfrac{Wa^2}{g}$; (b) $x_G = 5a/6$, $y_G = 5a/12$, $I_{xG} = \dfrac{11}{96}\dfrac{Wa^2}{g}$, $I_{yG} = \dfrac{11}{24}\dfrac{Wa^2}{g}$, $I_{xyG} = -\dfrac{1}{12}\dfrac{Wa^2}{g}$, (c) $\alpha = 77$ deg.

240. $WR^2 = 32,800$ lb-ft². $I = 1,020$ lb-ft. sec².

241. $\dfrac{\pi}{10}\rho r_0^4 h.$ **242.** $\dfrac{\pi}{6}\rho r_0^4 h.$

243. (d) $I_1 = 0.315$ lb-in.-sec², $I_2 = 0.865$ lb-in.-sec², $x_G = 2.83$ in, $I_G = 0.19$ lb-in.-sec².

244. Vertical force is 138.5 lb steady downward due to dead weight plus a revolving force of 55.6 lb. The horizontal forces are just ± 55.6 lb.

245. (a) $\ddot{x} = g\dfrac{w}{w+2W+W_p(k/r)^2} = g\dfrac{w}{\text{Den}}$, $T_1 = W\left(1+\dfrac{w}{\text{Den}}\right); T_2 = (W+w)\left(1-\dfrac{w}{\text{Den}}\right).$ (b) $\ddot{x} = 0.018g = 6.9$ in./sec², $T_1 = 2.036$ lb; $T_2 = 2.063$ lb.

246. 16.5 minutes.

247. (a) $\ddot{\varphi} = \dfrac{M_t}{I_3 + I_2\left(\dfrac{d_3}{d_2}\right)^2 + (I_1+I_f)\left(\dfrac{d_3}{d_1}\right)^2}$, (b) $\ddot{\varphi} = 0.24$ radians/sec, (c) $I_{equiv} = 21.2$ lb-in.-sec².

248. $\dfrac{M}{\ddot{\varphi}} = I_3 + I_4 + \left(\dfrac{r_3}{r_2}\right)^2 (I_1 + I_2) + \left(\dfrac{r_1 r_3}{r_2}\right)^2 \dfrac{W_c}{g} + \left(\dfrac{r_1 r_3}{r_2 a}\right)^2 I_r.$

249. (a) $R_v = \left(1 - \dfrac{a}{2l}\right) mg, \ R_h = \dfrac{a}{2l} mg, \ F = \dfrac{a}{2l} mg \sqrt{2}$; (b) $R_h = \dfrac{m^2 a^2 g}{2I_0}$,

$R_v = mg - \dfrac{m^2 a^2 g}{2I_0}.$

251. Torque $= I\dot{\varphi}^2$ and $F = \dfrac{I\dot{\varphi}^2}{r_1}$, occurring at the instants that the pin either enters or leaves contact.

252. (a) Not possible, (b) $x = l/2$, added at the bottom end.

253. $x = x_0 \cosh kt$, where $\dfrac{1}{k^2} = \dfrac{I_0 + \pi r^3 \mu}{2g\mu r^2} + \dfrac{a}{g}.$

254. $R_x = \dfrac{2\mu r k^2}{g} x_0 \cosh kt, \ R_y = (\pi r + 2a)\mu - \dfrac{2\mu k^2}{g} x_0^2 \cosh kt.$

255. $\ddot{x} = 0.045g = 17.4$ in./sec.^{2}

256. $\ddot{x} = \tfrac{4}{15}g, \ T_1 = 1\tfrac{1}{15}W, \ T_2 = \tfrac{9}{15}W, \ T_3 = \tfrac{8}{15}W.$

257. (a) $f = \dfrac{1}{2\pi} \sqrt{\dfrac{g}{l}}$, (b) $f = \dfrac{1}{4\pi} \sqrt{\dfrac{3g}{l}}.$

258. $\alpha < \cos^{-1}(d_1/d_2)$, or in words: The thread force T must intersect the floor behind the point where the spool rests on the floor.

260. $x = l/3.$　**261.** $7r/5$ above the table.

262. (a) a_x, (b) $a_x/\sqrt{2}$, (c) $\dot{\omega} = \dfrac{a_x \sqrt{2}}{l}$, (d) $H_A = w\left(\dfrac{1}{2} - \dfrac{2}{3}\dfrac{a_x}{g}\right)$,

(e) $a_x = g\dfrac{w}{3W + 4w/3}.$

263. (a) $\ddot{\varphi}(I_G + mr^2) + \dfrac{r^2}{R - r} mg \sin \varphi = 0$, (b) $f = \dfrac{1}{2\pi} \sqrt{\dfrac{r^2}{R - r} \dfrac{g}{k_C^2}}$,

(c) $f = \dfrac{1}{2\pi} \sqrt{\dfrac{g}{R - r}}.$

264. $I = 506$ lb-ft-sec.^{2}　**265.** $\ddot{x} = \dfrac{3\sqrt{3}}{8} g - \dfrac{l}{2} \dot{\varphi}^2.$

266. $\ddot{x}_C\left(\dfrac{W_2}{g} + \dfrac{3}{2}\dfrac{W_1}{g}\right) + \dfrac{W_2 a}{g}(\ddot{\theta} \cos \theta - \dot{\theta}^2 \sin \theta) = 0,$

$\ddot{x}_C \dfrac{W_2}{g} a \cos \theta + \left(I_G + \dfrac{W_2 a^2}{g}\right)\ddot{\theta} + W_2 a \sin \theta = 0.$

267. $v_1^2 = v_0^2 + 2gs(\sin \alpha - f \cos \alpha).$　**268.** 11.3 ft/sec.

269. (a) $4\tfrac{1}{2}$ in., (b) 59 in./sec.　**270.** (a) 52 lb, (b) 62 lb.

271. 11.5 ft.　**272.** 400,000 hp.　**273.** 1,010 kw.　**274.** 116 ihp.

275. High-speed torques: 10,500 ft-lb, intermediate torques: 52,500 ft-lb, propeller-shaft torque: 1,050,000 ft-lb.

(b) 42,000 lb on the high-speed gears; 105,000 lb on the low-speed ones.

276. (a) 1,520 ft, 206 sec; (b) drawbar pull $= 242$ tons, driver load $= 968$ tons.

277. (a) hp $= \dfrac{2\pi N}{33,000} \dfrac{Plr}{a}\left(1 - e^{\frac{-3\pi f}{2}}\right)$, (b) hp $= \dfrac{2\pi N}{33,000} \dfrac{Plr}{a} \sqrt{2}.$

278. (b) $\dot{\varphi}^2 = \dfrac{4Pr}{I_C + (m_p + \frac{1}{3}m_r)r^2}$. **279.** 78 ft.

280. (a) $x = \dfrac{\pi}{3600}\dfrac{N^2I}{rfP}\dfrac{b}{c}\dfrac{d}{a}$, (b) 91 revolutions. **281.** 94 per cent.

282. (a) 69.5 hp, (b) empty cab 19.5 per cent, counterweight 26 per cent and motor 54.5 per cent.

283. (a) $22.3\ m_1v_0^2$, (b) $44.6\ m_1$.

284. (a) $V^2 = \dfrac{2m_Cgr_c(1 - \cos\varphi)}{m_A + m_B + m_C(1 - r_c/r_w)^2 + I_A/r_w^2}$,

(b) $f = \dfrac{1}{2\pi}\sqrt{\dfrac{m_cgr_c/r_w^2}{m_A + m_B + m_C(1 - r_c/r_w)^2 - I_A/r_w^2}}$.

285. (a) $\dfrac{a}{r} \leq \sin\alpha$; (b) G is to the right of C and on the same level; the equilibrium is unstable. (c) $1.09\sqrt{gr}$ at 90 deg, $1.86\sqrt{gr}$ at 180 deg, $2.12\sqrt{gr}$ at 270 deg, $2.84\sqrt{gr}$ at 360 deg.

286. $\dfrac{v_x^2}{gl} = \dfrac{3}{4}\left(\dfrac{3}{\sqrt{2}} - 1\right)$; the top point goes down with speed v_x; the bottom point has horizontal and vertical components each equal to v_x.

287. Position II: $v_A^2 = 6lg\sqrt{2}\,\dfrac{w_b + w_j}{7w_b + 9w_j}$

Position III: $v_A^2 = 1.28lg\,\dfrac{w_b + w_j}{2w_b + 3w_j}$

288. $v_C = \dfrac{41.2}{1 + 4(k/a)^2}$ in./sec, where k is the radius of gyration of one bucket about point B and $a = BC = 36$ in.

289. The light carriage goes four times as far as the heavy one.

290. 47.5 lb at 45 deg. **291.** 2,000 ft/sec; 6×10^6 ft/sec^2.

292. (a) 46.7 ft/sec, (b) 1.76 per cent in wood, 6.25 per cent in bullet, 92 per cent dissipated.

293. (a) 180,000 lb or five times the weight of the airplane, (b) 2.06 ft/sec.

294. $t = \dfrac{v_0}{g}\dfrac{W_1 + W_2 + W_3k^2/r^2}{W_2 - W_1}$.

295. (a) 11.72 radians/sec, (b) 1.87 revolutions. **296.** 74 rpm.

297. The top deck recoils at 0.49 ft/sec; the center of gravity recoils at 0.15 ft/sec.

298. $D = 1.90$ ft, 565 ihp.

299. $V^2 = g_0r\,\dfrac{r}{r + h}$; $V = 25,800$ ft/sec, or approximately 23 times the speed of sound in atmospheric air.

300. $V_0^2 = 2gr_{\text{earth}}$ for infinite distance, $V_0^2 = {}^{79}\!\!\diagup_{80}2gr_{\text{earth}}$ for the moon. In both cases substantially 37,000 ft/sec

301. $m_{\text{full}}/m_{\text{empty}} = e^{\frac{7.4}{\sqrt{2}}} = 187$. **302.** (a) 8.05 ft, (b) 6.65 ft.

303. (a) The dead weight goes up with the man, remaining always at the same height as the man. (b) The dead weight remains at its original location if the man climbs gently, taking care to keep the moment of his own inertia force about the pulley center less than the friction torque. (c) No difference whatever. (d) Same answer as (a).

304. $\frac{2}{7}$ of the original energy is dissipated; $\frac{5}{7}$ is preserved, of which $\frac{25}{49}$ is translational and $\frac{10}{49}$ is rotational.

305. 116 (short) tons.

306. $v_{1x} = \dfrac{m_1 - m_2}{m_1 + m_2} V \cos \alpha; \; v_{1y} = V \sin \alpha,$

$\quad\;\; v_{2x} = \dfrac{2m_1}{m_1 + m_2} V \cos \alpha; \; v_{2y} = 0.$

307. $\alpha = \ddot{x}/g$ for small α only.

308. (a) $\ddot{\varphi} + \dfrac{g}{l}\varphi = \dfrac{a}{l}$, (b) $\varphi = \dfrac{a}{g}\left(1 - \cos t \sqrt{\dfrac{g}{l}}\right).$

309. (a) $x/h = \ddot{s}/g$, (b) $\dot{x} = \dfrac{h}{g}\ddot{s}$, (c) 10 ft/sec.

310. (b) $C_b/C_w = \beta(\alpha - \beta)$, (c) $\beta = \alpha/2 = 10$ deg; $V_{\text{boat max}} = 2.73 V_{\text{wind}}$.

311. $y = \dfrac{8\sqrt{2}}{3}\,\Omega\sqrt{\dfrac{h^3}{g}} = 3$ miles or 16,000 ft.

312. (a) East deviation of 375 ft. (b) West deviation of 750 ft; both answers proportional to v_0^2, so that for $v_0 = 3,000$ ft/sec the deviation (b) is about half a mile.

313. 4 in. higher on west bank.

314. (a) $\ddot{\phi} + \dfrac{g}{r}\sin\phi = w^2 \sin\phi\cos\phi$. (b) $2mr\dot{\phi}\omega\cos\varphi$, directed perpendicular to the plane of the circular wire. This force does not affect the motion; it only causes a sidewise pressure against the wire frame.

315. (a) $g/\omega^2 r = \tan\alpha$, for steel as well as for air. (b) Unstable for both. (c) First stage (low speed): steel ball down, air bubble up; second stage (medium speed): both up; third stage (high speed): steel up, air down; fourth stage (medium speed): both down; last stage (slow speed): like first stage.

316. (a) $f = \dfrac{1}{2\pi}\sqrt{\dfrac{k - m\omega^2}{m}}$ for $k > m\omega^2$; when $k < m\omega^2$, the bead sits with pressure against the end stop. (b) Does not affect the motion; causes only a force against the wire.

317. (a) $\varphi \approx \tan\varphi = \dfrac{W - w}{w}\dfrac{gb}{\omega^2(ab + b^2 + k_G^2)}$,

$\quad\;\;$ (b) $f = \dfrac{\omega}{2\pi}\sqrt{\dfrac{ab + b^2 + k_G^2}{b^2 + k_G^2}}.$

318. (a) $\ddot{\varphi} + \dfrac{\omega^2 a}{l}\varphi = 0.$ (b) The pendulum acts as a gravity pendulum where the value of g is replaced by that of the centrifugal field at C: the point of suspension (and *not* the point of the mass!). (c) The Coriolis force does not affect the motion; it increases the string tension when the mass swings forward with the disk ω, and diminishes it when the mass swings against ω.

319. A compound pendulum as if in a field $g = \omega^2 a$.

320. A point at distance l above G.

323. (b) Directed toward P radially; they do not influence the motion, merely increase the pull on the pivot pin P.

324. (a) $\tan \beta = \dfrac{\tan \alpha}{2 + \tan^2 \alpha}$, (b) 71 lb.

325. (a) $x_p = r_p \tan \alpha \sin \omega t$, (b) $\mathfrak{M}_{4\,\mathrm{pis}} = 4 m_p \omega r_p^2 \tan \alpha$ vertically upward in plane of drawing, (c) $I_{\mathrm{all\ pistons}}/I_{\mathrm{disk}} = \cos^2 \alpha$.

326. Increases from 0 to 28.2 rpm in 2 sec along a curve which is 180 deg of cosine wave vertically displaced with horizontal tangents at the beginning and at the end.

327. 2800 ft-lb.   **328.** 7.4 in.-lb.   **329.** 0.009 radians/sec.

330. 5,550 rpm.   **331.** $k = 0.28$ lb/in.   **332.** 0.54 lb.

333. (a) 1 oz, (b) zero, (c) between 0 and 1 oz.

334. Ask your friends tonight at dinner.

LIST OF EQUATIONS

$$\ddot{x} = -\frac{k}{m}x$$

$$x = x_0 \cos\left(\sqrt{\frac{k}{m}}\,t\right) + v_0\sqrt{\frac{m}{k}}\sin\left(\sqrt{\frac{k}{m}}\,t\right)$$

9. $$\frac{1}{T} = f = \frac{1}{2\pi}\sqrt{\frac{k}{m}}$$ 182–185; and 235

$$f = \frac{1}{2\pi}\sqrt{\frac{g}{l}}$$

$$f = \frac{1}{2\pi}\sqrt{\frac{Wa}{I_O}}$$

10. $$\text{Moment} = \dot{\omega}\int r^2\,dm = \dot{\omega}I_O$$
$$\text{Moment} = \int \omega^2 xy\,dm = \omega^2\int xy\,dm = \omega^2 I_{xy}$$ 215, 216

11. $$I_{z'} = I_{z_G} + a^2 m$$
$$I_{x'y'} = I_{xyG} + abm$$ 220, 221

12. $$I = \int r^2\,dm = mk^2$$ 224

13. $$I_G = \frac{ml^2}{12}, \qquad I_{\text{end}} = \frac{ml^2}{3}$$ 224

14. $$I_{\text{polar}} = \frac{mR^2}{2}, \qquad I_{\text{diametral}} = \frac{mR^2}{4}$$ 226

15. $$\text{Moment of external forces about } O = I_O\dot{\omega}$$ 233

16. $$l_{\text{equiv}} = \frac{I_O}{ma}$$ 235

17. $$\Sigma F_x = m\ddot{x}_G$$
$$\Sigma F_y = m\ddot{y}_G$$
$$\Sigma M_G = I_G\ddot{\varphi}$$ 241

18. $$ab = k_G^2$$ 246

19. $$F\,dx = d(\tfrac{1}{2}mv^2) = dT$$ 251

20. $$W = \int_0^s P(ds\sin\alpha) = \int_0^h P\,dy = \int_0^h W\,dy = Wh = V$$ 256

21. $$W = \int_0^x P\,dx = \int_0^x kx\,dx = k\left.\frac{x^2}{2}\right|_0^x = \frac{kx^2}{2} = V$$ 256

22. $$\eta = \frac{\text{work output}}{\text{work input}} = \frac{\text{work input} - \text{work dissipated}}{\text{work input}}$$ 260

23. $$1 \text{ hp} = 33{,}000 \text{ ft-lb/minute} = 550 \text{ ft-lb/sec}$$
$$1 \text{ kw} = 1.34 \text{ hp}$$ 260

24. $$T = \int \frac{1}{2} r^2 \omega^2 \, dm = \frac{\omega^2}{2} \int r^2 \, dm = \frac{1}{2} I_O \omega^2$$ 261

25. $$T = \tfrac{1}{2} m v_G^2 + \tfrac{1}{2} I_G \omega^2$$ 262

26. $$F = \frac{d}{dt}(mv)$$ 271

$$M = I_O \ddot{\varphi} = I_O \frac{d\omega}{dt} = \frac{d}{dt}(I_O \omega) = \frac{d\mathfrak{M}}{dt}$$

27. $$\left. \begin{aligned} F_x &= \frac{d}{dt}(m\dot{x}) \\ F_y &= \frac{d}{dt}(m\dot{y}) \\ M_G &= \frac{d}{dt}(I_G \omega) \end{aligned} \right\}$$ 277

28. $$m_1 V_1 + m_2 V_2 = m_1 v_1 + m_2 v_2$$ 289

29. $$\left. \begin{aligned} v_2 &= \frac{m_2 V_2 - m_1 e V_2 + m_1 V_1 (1 + e)}{m_1 + m_2} \\ v_1 &= \frac{m_1 V_1 - m_2 e V_1 + m_2 V_2 (1 + e)}{m_1 + m_2} \end{aligned} \right\}$$ 290

30. $$\Delta T = \frac{1 - e^2}{2} \frac{m_1 m_2}{m_1 + m_2} (V_1 - V_2)^2$$ 290

31. $$\begin{aligned} \dot{\mathbf{v}}_{\text{abs}} &= \dot{\mathbf{v}}_{\text{rel}} + \dot{\mathbf{v}}_{\text{veh}} \\ \mathbf{F} &= m\dot{\mathbf{v}}_{\text{abs}} = m(\dot{\mathbf{v}}_{\text{rel}} + \dot{\mathbf{v}}_{\text{veh}}) \\ \mathbf{F} - m\dot{\mathbf{v}}_{\text{veh}} &= m\dot{\mathbf{v}}_{\text{rel}} \end{aligned}$$ 298

32. $$\begin{aligned} \mathbf{F} &= m\dot{\mathbf{v}}_a = m(\dot{\mathbf{v}}_{\text{rel}} + \dot{\mathbf{v}}_{\text{veh}} + \dot{\mathbf{v}}_{\text{Cor}}) \\ \mathbf{F} - m\dot{\mathbf{v}}_{\text{veh}} - m\dot{\mathbf{v}}_{\text{Cor}} &= m\dot{\mathbf{v}}_{\text{rel}} \end{aligned}$$ 306

33. $$\mathbf{M} = \frac{d}{dt}(\mathfrak{M})$$ 315

INDEX

A CATALOGUE OF
SELECTED DOVER BOOKS
IN ALL FIELDS OF INTEREST

A CATALOGUE OF SELECTED DOVER
BOOKS IN ALL FIELDS OF INTEREST

CELESTIAL OBJECTS FOR COMMON TELESCOPES, T. W. Webb. The most used book in amateur astronomy: inestimable aid for locating and identifying nearly 4,000 celestial objects. Edited, updated by Margaret W. Mayall. 77 illustrations. Total of 645pp. 5⅜ x 8½.
20917-2, 20918-0 Pa., Two-vol. set $8.00

HISTORICAL STUDIES IN THE LANGUAGE OF CHEMISTRY, M. P. Crosland. The important part language has played in the development of chemistry from the symbolism of alchemy to the adoption of systematic nomenclature in 1892. ". . . wholeheartedly recommended,"—Science. 15 illustrations. 416pp. of text. 5⅝ x 8¼.
63702-6 Pa. $6.00

BURNHAM'S CELESTIAL HANDBOOK, Robert Burnham, Jr. Thorough, readable guide to the stars beyond our solar system. Exhaustive treatment, fully illustrated. Breakdown is alphabetical by constellation: Andromeda to Cetus in Vol. 1; Chamaeleon to Orion in Vol. 2; and Pavo to Vulpecula in Vol. 3. Hundreds of illustrations. Total of about 2000pp. 6⅛ x 9¼.
23567-X, 23568-8, 23673-0 Pa., Three-vol. set $26.85

THEORY OF WING SECTIONS: INCLUDING A SUMMARY OF AIR-FOIL DATA, Ira H. Abbott and A. E. von Doenhoff. Concise compilation of subatomic aerodynamic characteristics of modern NASA wing sections, plus description of theory. 350pp. of tables. 693pp. 5⅜ x 8½.
60586-8 Pa. $6.50

DE RE METALLICA, Georgius Agricola. Translated by Herbert C. Hoover and Lou H. Hoover. The famous Hoover translation of greatest treatise on technological chemistry, engineering, geology, mining of early modern times (1556). All 289 original woodcuts. 638pp. 6¾ x 11.
60006-8 Clothbd. $17.50

THE ORIGIN OF CONTINENTS AND OCEANS, Alfred Wegener. One of the most influential, most controversial books in science, the classic statement for continental drift. Full 1966 translation of Wegener's final (1929) version. 64 illustrations. 246pp. 5⅜ x 8½.   61708-4 Pa. $3.00

THE PRINCIPLES OF PSYCHOLOGY, William James. Famous long course complete, unabridged. Stream of thought, time perception, memory, experimental methods; great work decades ahead of its time. Still valid, useful; read in many classes. 94 figures. Total of 1391pp. 5⅜ x 8½.
20381-6, 20382-4 Pa., Two-vol. set $13.00

HISTORY OF BACTERIOLOGY, William Bulloch. The only comprehensive history of bacteriology from the beginnings through the 19th century. Special emphasis is given to biography-Leeuwenhoek, etc. Brief accounts of 350 bacteriologists form a separate section. No clearer, fuller study, suitable to scientists and general readers, has yet been written. 52 illustrations. 448pp. 5⅝ x 8¼.   23761-3 Pa. $6.50

THE COMPLETE NONSENSE OF EDWARD LEAR, Edward Lear. All nonsense limericks, zany alphabets, Owl and Pussycat, songs, nonsense botany, etc., illustrated by Lear. Total of 321pp. 5⅜ x 8½. (Available in U.S. only)   20167-8 Pa. $3.00

INGENIOUS MATHEMATICAL PROBLEMS AND METHODS, Louis A. Graham. Sophisticated material from Graham Dial, applied and pure; stresses solution methods. Logic, number theory, networks, inversions, etc. 237pp. 5⅜ x 8½.   20545-2 Pa. $3.50

BEST MATHEMATICAL PUZZLES OF SAM LOYD, edited by Martin Gardner. Bizarre, original, whimsical puzzles by America's greatest puzzler. From fabulously rare Cyclopedia, including famous 14-15 puzzles, the Horse of a Different Color, 115 more. Elementary math. 150 illustrations. 167pp. 5⅜ x 8½.   20498-7 Pa. $2.50

THE BASIS OF COMBINATION IN CHESS, J. du Mont. Easy-to-follow, instructive book on elements of combination play, with chapters on each piece and every powerful combination team—two knights, bishop and knight, rook and bishop, etc. 250 diagrams. 218pp. 5⅜ x 8½. (Available in U.S. only)   23644-7 Pa. $3.50

MODERN CHESS STRATEGY, Ludek Pachman. The use of the queen, the active king, exchanges, pawn play, the center, weak squares, etc. Section on rook alone worth price of the book. Stress on the moderns. Often considered the most important book on strategy. 314pp. 5⅜ x 8½.   20290-9 Pa. $3.50

LASKER'S MANUAL OF CHESS, Dr. Emanuel Lasker. Great world champion offers very thorough coverage of all aspects of chess. Combinations, position play, openings, end game, aesthetics of chess, philosophy of struggle, much more. Filled with analyzed games. 390pp. 5⅜ x 8½.   20640-8 Pa. $4.00

500 MASTER GAMES OF CHESS, S. Tartakower, J. du Mont. Vast collection of great chess games from 1798-1938, with much material nowhere else readily available. Fully annotated, arranged by opening for easier study. 664pp. 5⅜ x 8½.   23208-5 Pa. $6.00

A GUIDE TO CHESS ENDINGS, Dr. Max Euwe, David Hooper. One of the finest modern works on chess endings. Thorough analysis of the most frequently encountered endings by former world champion. 331 examples, each with diagram. 248pp. 5⅜ x 8½.   23332-4 Pa. $3.50

THE DEPRESSION YEARS AS PHOTOGRAPHED BY ARTHUR ROTH-
STEIN, Arthur Rothstein. First collection devoted entirely to the work of
outstanding 1930s photographer: famous dust storm photo, ragged children,
unemployed, etc. 120 photographs. Captions. 119pp. 9¼ x 10¾.
23590-4 Pa. $5.00

CAMERA WORK: A PICTORIAL GUIDE, Alfred Stieglitz. All 559 illustrations and plates from the most important periodical in the history of
art photography, Camera Work (1903-17). Presented four to a page, reduced in size but still clear, in strict chronological order, with complete
captions. Three indexes. Glossary. Bibliography. 176pp. 8⅜ x 11¼.
23591-2 Pa. $6.95

ALVIN LANGDON COBURN, PHOTOGRAPHER, Alvin L. Coburn. Revealing autobiography by one of greatest photographers of 20th century
gives insider's version of Photo-Secession, plus comments on his own work.
77 photographs by Coburn. Edited by Helmut and Alison Gernsheim.
160pp. 8⅛ x 11. 23685-4 Pa. $6.00

NEW YORK IN THE FORTIES, Andreas Feininger. 162 brilliant photographs by the well-known photographer, formerly with Life magazine, show
commuters, shoppers, Times Square at night, Harlem nightclub, Lower
East Side, etc. Introduction and full captions by John von Hartz. 181pp.
9¼ x 10¾. 23585-8 Pa. $6.00

GREAT NEWS PHOTOS AND THE STORIES BEHIND THEM, John
Faber. Dramatic volume of 140 great news photos, 1855 through 1976,
and revealing stories behind them, with both historical and technical information. Hindenburg disaster, shooting of Oswald, nomination of Jimmy
Carter, etc. 160pp. 8¼ x 11. 23667-6 Pa. $5.00

THE ART OF THE CINEMATOGRAPHER, Leonard Maltin. Survey of
American cinematography history and anecdotal interviews with 5 masters—
Arthur Miller, Hal Mohr, Hal Rosson, Lucien Ballard, and Conrad Hall.
Very large selection of behind-the-scenes production photos. 105 photo-
graphs. Filmographies. Index. Originally Behind the Camera. 144pp.
8¼ x 11. 23686-2 Pa. $5.00

DESIGNS FOR THE THREE-CORNERED HAT (LE TRICORNE),
Pablo Picasso. 32 fabulously rare drawings—including 31 color illustrations
of costumes and accessories—for 1919 production of famous ballet. Edited
by Parmenia Migel, who has written new introduction. 48pp. 9⅜ x 12¼.
(Available in U.S. only) 23709-5 Pa. $5.00

NOTES OF A FILM DIRECTOR, Sergei Eisenstein. Greatest Russian
filmmaker explains montage, making of Alexander Nevsky, aesthetics; comments on self, associates, great rivals (Chaplin), similar material. 78 illustrations. 240pp. 5⅜ x 8½. 22392-2 Pa. $4.50

CATALOGUE OF DOVER BOOKS

GEOMETRY, RELATIVITY AND THE FOURTH DIMENSION, Rudolf Rucker. Exposition of fourth dimension, means of visualization, concepts of relativity as Flatland characters continue adventures. Popular, easily followed yet accurate, profound. 141 illustrations. 133pp. 5⅜ x 8½.
23400-2 Pa. $2.75

THE ORIGIN OF LIFE, A. I. Oparin. Modern classic in biochemistry, the first rigorous examination of possible evolution of life from nitrocarbon compounds. Non-technical, easily followed. Total of 295pp. 5⅜ x 8½.
60213-3 Pa. $4.00

THE CURVES OF LIFE, Theodore A. Cook. Examination of shells, leaves, horns, human body, art, etc., in "the classic reference on how the golden ratio applies to spirals and helices in nature . . . ."—Martin Gardner. 426 illustrations. Total of 512pp. 5⅜ x 8½.   23701-X Pa. $5.95

PLANETS, STARS AND GALAXIES, A. E. Fanning. Comprehensive introductory survey: the sun, solar system, stars, galaxies, universe, cosmology; quasars, radio stars, etc. 24pp. of photographs. 189pp. 5⅜ x 8½. (Available in U.S. only)   21680-2 Pa. $3.00

THE THIRTEEN BOOKS OF EUCLID'S ELEMENTS, translated with introduction and commentary by Sir Thomas L. Heath. Definitive edition. Textual and linguistic notes, mathematical analysis, 2500 years of critical commentary. Do not confuse with abridged school editions. Total of 1414pp. 5⅜ x 8½.   60088-2, 60089-0, 60090-4 Pa., Three-vol. set $18.00

DIALOGUES CONCERNING TWO NEW SCIENCES, Galileo Galilei. Encompassing 30 years of experiment and thought, these dialogues deal with geometric demonstrations of fracture of solid bodies, cohesion, leverage, speed of light and sound, pendulums, falling bodies, accelerated motion, etc. 300pp. 5⅜ x 8½.   60099-8 Pa. $4.00